Silicon Nanodevices

Silicon Nanodevices

Editors

Henry H. Radamson
Guilei Wang

MDPI • Basel • Beijing • Wuhan • Barcelona • Belgrade • Manchester • Tokyo • Cluj • Tianjin

Editors
Henry H. Radamson
Optoelectronics
GIICS
Guangzhou
China

Guilei Wang
Beijing Superstring Academy
of Memory Technology
Beijing
China

Editorial Office
MDPI
St. Alban-Anlage 66
4052 Basel, Switzerland

This is a reprint of articles from the Special Issue published online in the open access journal *Nanomaterials* (ISSN 2079-4991) (available at: www.mdpi.com/journal/nanomaterials/special_issues/nano_sili_devi).

For citation purposes, cite each article independently as indicated on the article page online and as indicated below:

LastName, A.A.; LastName, B.B.; LastName, C.C. Article Title. *Journal Name* **Year**, *Volume Number*, Page Range.

ISBN 978-3-0365-4678-0 (Hbk)
ISBN 978-3-0365-4677-3 (PDF)

© 2022 by the authors. Articles in this book are Open Access and distributed under the Creative Commons Attribution (CC BY) license, which allows users to download, copy and build upon published articles, as long as the author and publisher are properly credited, which ensures maximum dissemination and a wider impact of our publications.

The book as a whole is distributed by MDPI under the terms and conditions of the Creative Commons license CC BY-NC-ND.

Contents

About the Editors vii

Preface to "Silicon Nanodevices" ix

Henry H. Radamson and Guilei Wang
Special Issue: Silicon Nanodevices
Reprinted from: *Nanomaterials* **2022**, *12*, 1980, doi:10.3390/nano12121980 1

Yongliang Li, Fei Zhao, Xiaohong Cheng, Haoyan Liu, Ying Zan and Junjie Li et al.
Four-Period Vertically Stacked SiGe/Si Channel FinFET Fabrication and Its Electrical Characteristics
Reprinted from: *Nanomaterials* **2021**, *11*, 1689, doi:10.3390/nano11071689 3

Rui-Zi Hu, Rong-Long Ma, Ming Ni, Xin Zhang, Yuan Zhou and Ke Wang et al.
An Operation Guide of Si-MOS Quantum Dots for Spin Qubits
Reprinted from: *Nanomaterials* **2021**, *11*, 2486, doi:10.3390/nano11102486 13

Qide Yao, Xueli Ma, Hanxiang Wang, Yanrong Wang, Guilei Wang and Jing Zhang et al.
Investigate on the Mechanism of $HfO_2/Si_{0.7}Ge_{0.3}$ Interface Passivation Based on Low-Temperature Ozone Oxidation and Si-Cap Methods
Reprinted from: *Nanomaterials* **2021**, *11*, 955, doi:10.3390/nano11040955 29

Jian-Huan Wang, Ting Wang and Jian-Jun Zhang
Epitaxial Growth of Ordered In-Plane Si and Ge Nanowires on Si (001)
Reprinted from: *Nanomaterials* **2021**, *11*, 788, doi:10.3390/nano11030788 39

Lu Xie, Huilong Zhu, Yongkui Zhang, Xuezheng Ai, Junjie Li and Guilei Wang et al.
Investigation on $Ge_{0.8}Si_{0.2}$-Selective Atomic Layer Wet-Etching of Ge for Vertical Gate-All-Around Nanodevice
Reprinted from: *Nanomaterials* **2021**, *11*, 1408, doi:10.3390/nano11061408 49

Yangyang Li, Huilong Zhu, Zhenzhen Kong, Yongkui Zhang, Xuezheng Ai and Guilei Wang et al.
The Effect of Doping on the Digital Etching of Silicon-Selective Silicon–Germanium Using Nitric Acids
Reprinted from: *Nanomaterials* **2021**, *11*, 1209, doi:10.3390/nano11051209 65

Yuanhao Miao, Guilei Wang, Zhenzhen Kong, Buqing Xu, Xuewei Zhao and Xue Luo et al.
Review of Si-Based GeSn CVD Growth and Optoelectronic Applications
Reprinted from: *Nanomaterials* **2021**, *11*, 2556, doi:10.3390/nano11102556 79

Xuewei Zhao, Guilei Wang, Hongxiao Lin, Yong Du, Xue Luo and Zhenzhen Kong et al.
High Performance p-i-n Photodetectors on Ge-on-Insulator Platform
Reprinted from: *Nanomaterials* **2021**, *11*, 1125, doi:10.3390/nano11051125 123

Yong Du, Buqing Xu, Guilei Wang, Yuanhao Miao, Ben Li and Zhenzhen Kong et al.
Review of Highly Mismatched III-V Heteroepitaxy Growth on (001) Silicon
Reprinted from: *Nanomaterials* **2022**, *12*, 741, doi:10.3390/nano12050741 135

Md. Hasan Raza Ansari, Udaya Mohanan Kannan and Seongjae Cho
Core-Shell Dual-Gate Nanowire Charge-Trap Memory for Synaptic Operations for Neuromorphic Applications
Reprinted from: *Nanomaterials* **2021**, *11*, 1773, doi:10.3390/nano11071773 181

Prem. C. Pandey, Shubhangi Shukla and Roger J. Narayan
Organotrialkoxysilane-Functionalized Prussian Blue Nanoparticles-Mediated Fluorescence Sensing of Arsenic(III)
Reprinted from: *Nanomaterials* **2021**, *11*, 1145, doi:10.3390/nano11051145 . . . . **195**

Na Han, Jianjiang Li, Xuechen Wang, Chuanlong Zhang, Gang Liu and Xiaohua Li et al.
Flexible Carbon Nanotubes Confined Yolk-Shelled Silicon-Based Anode with Superior Conductivity for Lithium Storage
Reprinted from: *Nanomaterials* **2021**, *11*, 699, doi:10.3390/nano11030699 . . . . **215**

About the Editors

Henry H. Radamson

Henry H. Radamson received Ph.D. degree in semiconductor materials from Linköping University in Sweden in 1996. He joined KTH Royal Institute of Technology in Stockholm in 1997 as senior scientist. In 2016, he became full professor in Chinese Academy of Science in Beijing and since 2017, he is foreign expert in China. In 2019, he became manager of Optoelectronics Innovation Center in Guangzhou. Henry Radamson is a member of European Academy of Sciences since 2020. His research field is nanomaterials, and nanodevices towards integration of electronics and photonics.

He is author of two books: Monolithic Nanoscale Photonics-Electronics Integration in Silicon and Other Group IV Elements, Elsevier 2014 & CMOS past, present, and future, Elsevier, 2018.

Henry Radamson is a member of Executive Committee in European Material Research Society (EMRS) where has also organized several symposiums.

Henry Radamson is in Editorial board of Springer-Nature and he has been Guest Editor for the *Nanomaterials* journal.

He has been also awarded several times for his teaching efforts by Chinese Academy of Sciences and European academy.

Guilei Wang

Guilei Wang, a professor at the Beijing Superstring Academy of Memory Technology. He used to work as a professor at the Institute of Microelectronics, Chinese Academy of Sciences. So far, more than 120 research papers have been published in international journals. He has completed more than 100 patents and patent applications. He has published 1 English book (Guilei Wang: Investigation on SiGe Selective Epitaxy for Source and Drain Engineering in 22 nm CMOS Technology Node and Beyond) and 1 SiGe epitaxy chapter. He has delivered invited talks and short courses in several prestigious international conferences and academic institutions. He served as a member of the Technical Programme Committee (TPC) of the 2018 European Materials Research Society (E-MRS) Spring Conference. He is a guest editor for Nanomaterials and reviewers for the journals. He has won the "E-MRS Young Scientist", "Springer Excellent Doctorate Thesis", and "Excellent Member of the Youth Innovation Promotion Association of the Chinese Academy of Sciences" awards. His research interests focus on nanomaterials for microelectronic and semiconductor quantum computing technologies.

Preface to "Silicon Nanodevices"

Si nanodevices is a hot topic in the current technology. The start of Moore's law and the following of technology roadmap to scale down the transistors is one of the most important objectives in the semiconductor industry. As our research is proceeding towards to the unknown frontiers, the boundary in different parts of science is emerging together. Therefore, we have chosen a series of research articles in electronics and photonics to highlight the material synthesis and device manufacturing.

The editors would like to acknowledge valuable discussions with and material from all the authors and the valuable supports and help from Guangdong Greater Bay Area Institute of Integrated Circuit and System, Beijing Superstring Academy of Memory Technology, Institute of Microelectronics, Chinese Academy of Sciences.

Henry H. Radamson and Guilei Wang
Editors

Editorial

Special Issue: Silicon Nanodevices

Henry H. Radamson [1,*] and Guilei Wang [1,2,*]

1. Research and Development Center of Optoelectronic Hybrid IC, Guangdong Greater Bay Area Institute of Integrated Circuit and System, Guangzhou 510535, China
2. Beijing Superstring Academy of Memory Technology, Beijing 100176, China
* Correspondence: rad@kth.se (H.H.R.); guilei.wang@bjsamt.org.cn (G.W.)

In recent years, nanodevices have attracted a large amount of attention due to their low power consumption and fast operation in electronics and photonics, as well as their high sensitivity in sensor applications. For example, following Moore's law and the technology roadmap, the structure of transistors has been scaled down by a constant factor in order to obtain lower power consumption. Today, the scaling down of CMOS technology is focusing more on low voltages and cost-effective processes to match the requirements of mobile phone chips [1].

As a result of CMOS evolution, a lot of integration difficulties have been overcome, enabling the architecture of CMOS technology to change from planar or 2D to 3D. During this technology development, many issues, e.g., contact resistance, defects and reliability, have arisen and have been solved, which could affect device performance [2,3]. As an approach at the end of the technology roadmap (3 nm node), Si channel material is being changed to SiGe or Ge material, and even III-V materials could be integrated in the future. This is due to their material properties, as they have significantly higher carrier mobilities of ~40,000 cm^2 V^{-1} s^{-1} for InGaAs (for electrons) and 1900 cm^2 V^{-1} s^{-1} for Ge (for holes) compared to silicon, which has 1400 cm^2 V^{-1} s^{-1} for electrons and 450 cm^2 V^{-1} s^{-1} for holes [1]. In many cases, these are processed to be fin-like or nanowire transistors on insulator substrates. Special attention is paid to All Gate Around (GAA) transistors in vertical or horizontal directions. In such cases, GeSi/Ge multilayers are grown, and later, GeSi can be selectively etched to reach a nano-scale channel [4]. What is interesting is the choice of our transistor design beyond Moore's era. Meanwhile, as a future type of computation that could be used when Moore's Law ends, quantum computing is gaining considerable attention from academic and industrial communities. Group IV material devices will be an important aspect of quantum computing.

Currently, photonic and sensor devices have also attracted a large amount of interest. For example, materials are required for emissions in the infrared and terahertz range with high responsivity and low dark currents [5-7]. Therefore, many studies are seeking methods to decrease the defect density, in particular in device processing. The main goal is to find a monolithic solution where a material can be used for both photonic and electronic components on the chip. Ge and GeSn are materials that are recognized as excellent candidates as optoelectronic materials [8-12].

In this field, nanodevices are manufactured in 3D, and the emergence of electronics and photonics is inevitable. It is obvious that the complexity of our society requires technology to become more complicated in its design. In the future, electronic–photonic designs will be our ultimate goal in nanotechnology.

This Special Issue presents the fabrication and characterization of group IV nanostructures, nanodevices and nanosensors and their integration with photonics. The issue also covers optoelectronic materials and defect engineering as well as characterization.

Conflicts of Interest: The author declares no conflict of interest.

Citation: Radamson, H.H.; Wang, G. Special Issue: Silicon Nanodevices. *Nanomaterials* **2022**, *12*, 1980. https://doi.org/10.3390/nano12121980

Received: 26 May 2022
Accepted: 30 May 2022
Published: 9 June 2022

Publisher's Note: MDPI stays neutral with regard to jurisdictional claims in published maps and institutional affiliations.

Copyright: © 2022 by the authors. Licensee MDPI, Basel, Switzerland. This article is an open access article distributed under the terms and conditions of the Creative Commons Attribution (CC BY) license (https://creativecommons.org/licenses/by/4.0/).

References

1. Radamson, H.H.; Simoen, E.; Luo, J.; Zhao, C. *Past, Present and Future of CMOS*; Elsevier: Amsterdam, The Netherlands, 2018; pp. 95–114, ISBN 978-008-102-13.
2. Zhang, Q.; Yin, H.; Luo, J.; Yang, H.; Meng, L.; Li, Y.; Wu, Z.; Zhang, Y.; Zhang, Y.; Qin, C.; et al. FOI FinFET with ultra-low parasitic resistance enabled by fully metallic source and drain formation on isolated bulk-fin. In Proceedings of the International Electron Devices Meeting, San Francisco, CA, USA, 3–7 December 2016; pp. 17.3.1–17.3.4.
3. Ma, X.L.; Yin, H.X.; Hong, P.Z.; Xu, W.J. Self-Aligned Fin-On-Oxide (FOO) FinFETs for Improved SCE Immunity and Multi-V-TH Operation on Si Substrate. *ECS Solid State Lett.* 2015, *4*, Q13–Q16. [CrossRef]
4. Zhang, Y.; Ai, X.; Yin, X.; Zhu, H.; Yang, H.; Wang, G.L.; Li, J.J.; Du, A.Y.; Li, C.; Radamson, H.H. Vertical Sandwich GAA FETs with Self-Aligned High-k Metal Gate Made by Quasi Atomic Layer Etching Process. *IEEE Trans. Electron Devices.* 2021, *68*, 2604–2610. [CrossRef]
5. Zhao, X.; Wang, G.; Lin, H.; Du, Y.; Luo, X.; Kong, Z.; Su, J.; Li, J.; Xiong, W.; Radamson, H.H. High Performance pin Photodetectors on Ge-on-Insulator Platform. *Nanomaterials* 2021, *11*, 1125. [CrossRef] [PubMed]
6. Michel, J.; Liu, J.; Kimerling, L.C. High-performance Ge-on-Si photodetectors. *Nat. Photonics* 2010, *4*, 527–534. [CrossRef]
7. Zhao, X.; Moeen, M.; Toprak, M.S.; Wang, G.; Luo, J.; Ke, X.; Li, Z.; Liu, D.; Wang, W.; Zhao, C.; et al. Design impact on the performance of Ge PIN photodetectors. *J. Mater. Sci. Mater. Electron.* 2019, *31*, 18–25. [CrossRef]
8. Radamson, H.H.; Noroozi, M.; Jamshidi, A.; Thompson, P.E.; Ostling, M. Strain engineering in GeSnSi materials. *ECS Trans.* 2013, *50*, 527. [CrossRef]
9. Hu, T. *Synthesis and Properties of Sn-Based Group IV Alloys*; Arizona State University: Tempe, AZ, USA, 2019.
10. Vincent, B.; Gencarelli, F.; Bender, H.; Merckling, C.; Douhard, B.; Petersen, D.H.; Hansen, O.; Henrichsen, H.H.; Meersschaut, J.; Vandervorst, W.; et al. Undoped and in-situ B doped GeSn epitaxial growth on Ge by atmospheric pressure-chemical vapor deposition. *Appl. Phys. Lett.* 2011, *99*, 152103. [CrossRef]
11. Margetis, J.; Mosleh, A.; Ghetmiri, S.A.; Al-Kabi, S.; Dou, W.; Du, W.; Bhargava, N.; Yu, S.-Q.; Profijt, H.; Kohen, D.; et al. Fundamentals of Ge1−xSnx and SiyGe1−x-ySnx RPCVD epitaxy. *Mater. Sci. Semicond. Processing* 2017, *70*, 38–43. [CrossRef]
12. Aubin, J.; Hartmann, J.M.; Gassenq, A.; Rouviere, J.L.; Robin, E.; Delaye, V.; Cooper, D.; Mollard, N.; Reboud, V.; Calvo, V. Growth and structural properties of step-graded, high Sn content GeSn layers on Ge. *Semicond. Sci. Technol.* 2017, *32*, 094006. [CrossRef]

Article

Four-Period Vertically Stacked SiGe/Si Channel FinFET Fabrication and Its Electrical Characteristics

Yongliang Li, Fei Zhao, Xiaohong Cheng *, Haoyan Liu, Ying Zan, Junjie Li, Qingzhu Zhang, Zhenhua Wu, Jun Luo and Wenwu Wang *

Integrated Circuit Advanced Process Center, Institute of Microelectronics, Chinese Academy of Sciences, Beijing 100029, China; liyongliang@ime.ac.cn (Y.L.); zhaofei@ime.ac.cn (F.Z.); liuhaoyan@ime.ac.cn (H.L.); zanying@ime.ac.cn (Y.Z.); lijunjie@ime.ac.cn (J.L.); zhangqingzhu@ime.ac.cn (Q.Z.); wuzhenhua@ime.ac.cn (Z.W.); luojun@ime.ac.cn (J.L.)
* Correspondence: chengxiaohong@ime.ac.cn (X.C.); wangwenwu@ime.ac.cn (W.W.)

Abstract: In this paper, to solve the epitaxial thickness limit and the high interface trap density of SiGe channel Fin field effect transistor (FinFET), a four-period vertically stacked SiGe/Si channel FinFET is presented. A high crystal quality of four-period stacked SiGe/Si multilayer epitaxial grown with the thickness of each SiGe layer less than 10 nm is realized on a Si substrate without any structural defect impact by optimizing its epitaxial grown process. Meanwhile, the Ge atomic fraction of the SiGe layers is very uniform and its SiGe/Si interfaces are sharp. Then, a vertical profile of the stacked SiGe/Si Fin is achieved with HBr/O_2/He plasma by optimizing its bias voltage and O_2 flow. After the four-period vertically stacked SiGe/Si Fin structure is introduced, its FinFET device is successfully fabricated under the same fabrication process as the conventional SiGe FinFET. And it attains better drive current I_{on}, subthreshold slope (SS) and I_{on}/I_{off} ratio electrical performance compared with the conventional SiGe channel FinFET, whose Fin height of SiGe channel is almost equal to total thickness of SiGe in the four-period stacked SiGe/Si channel FinFET. This may be attributed to that the four-period stacked SiGe/Si Fin structure has larger effective channel width (W_{eff}) and may maintain a better quality and surface interfacial performance during the whole fabrication process. Moreover, Si channel of the stacked SiGe/Si channel turning on first also may have contribution to its better electrical properties. This four-period vertically stacked SiGe/Si channel FinFET device has been demonstrated to be a practical candidate for the future technology nodes.

Keywords: stacked SiGe/Si; epitaxial grown; Fin etching; FinFET

Citation: Li, Y.; Zhao, F.; Cheng, X.; Liu, H.; Zan, Y.; Li, J.; Zhang, Q.; Wu, Z.; Luo, J.; Wang, W. Four-Period Vertically Stacked SiGe/Si Channel FinFET Fabrication and Its Electrical Characteristics. *Nanomaterials* **2021**, *11*, 1689. https://doi.org/10.3390/nano11071689

Academic Editors: Antonio Di Bartolomeo, Henry Radamson and Guilei Wang

Received: 4 June 2021
Accepted: 26 June 2021
Published: 28 June 2021

Publisher's Note: MDPI stays neutral with regard to jurisdictional claims in published maps and institutional affiliations.

Copyright: © 2021 by the authors. Licensee MDPI, Basel, Switzerland. This article is an open access article distributed under the terms and conditions of the Creative Commons Attribution (CC BY) license (https://creativecommons.org/licenses/by/4.0/).

1. Introduction

High-mobility SiGe or Ge channel p-type Fin field effect transistor (FinFET) or gate-all-around (GAA) devices have been demonstrated to be a valid option as performance booster for future technology nodes [1–3]. The low-Ge-content SiGe channel will be implemented firstly on the FinFET owing to its advantages of higher hole mobility, better negative bias temperature instability (NBTI) reliability than silicon and more compatible with present silicon platform [4,5]. So far, a SiGe channel can be integrated in FinFET architectures in multiple ways, e.g., by shallow trench isolation (STI) last scheme [6] or by STI first strategy [7] or epitaxial cladding of Si Fins [8]. However, a high quality of low-Ge-content SiGe epitaxial grown on Si substrate is still a challenge task to solve the epitaxial thickness limit of SiGe film and the threading dislocations (TD) defects. This is because its theoretical critical thickness value is only ~10 nm [9]. Compared with the stable low-Ge-content SiGe layer, the thickness of metastable SiGe layer on Si can reach ~100 nm, but its quality is more easily affected by the following high temperature, implantation, and other processes. The other challenge about SiGe Fin channel is that it has relatively high interface trap charge (N_{it}) at the interfacial layer (IL)/SiGe channel due to the undesired GeO_x formation [10]. The passivation techniques of SiGe layer, such as Si-cap, O_3 low temperature oxidation, selective GeO$_x$-Scavenging, and fluorine/nitrogen plasma treatment, have been studied and the experimental demonstration on low-N$_{it}$ SiGe gate stack have been reported [11–15]. However, these passivation techniques may have compatibility problems with a state-of-the-art FinFET. Therefore, a high quality of low-Ge-content SiGe channel FinFET fabrication is still a challenge task and there are limited reports disclosing process details.

In this report, to solve the epitaxial thickness limit and the high interface trap density of SiGe channel FinFET, a four-period vertically stacked SiGe/Si multilayer with the thickness of each SiGe layer less than 10 nm grown on Si substrate is demonstrate by optimizing the epitaxial process. Then, an optimized stacked SiGe/Si Fin etching process with HBr/O$_2$/He plasma is also introduced to attain a perfect profile. Finally, the four-period vertically stacked SiGe/Si channel FinFET is successfully fabricated and it achieves better drive current I$_{on}$, subthreshold slope (SS) and I$_{on}$/I$_{off}$ ratio performance compared with the conventional SiGe channel FinFET under the similar fabrication process.

2. Materials and Methods

P-type FinFET device with a four-period vertically stacked SiGe/Si channel was fabricated on 8-inch p-type (100) wafers. Its fabrication flow is shown in Figure 1, where the fundamental differences with the conventional SiGe channel FinFET process are the stacked SiGe/Si Fin introduction (indicated with red color).

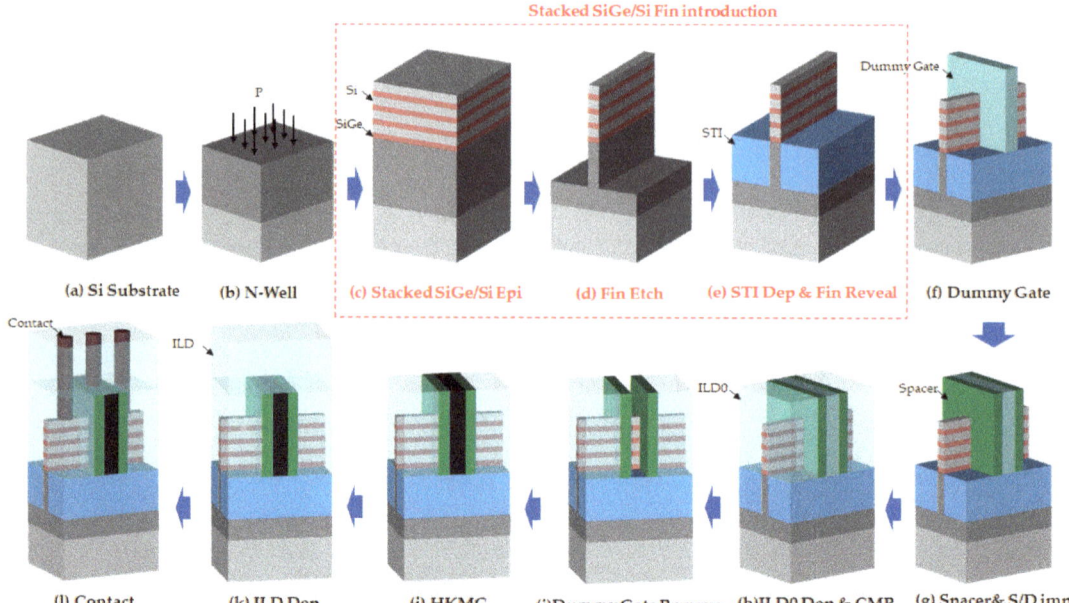

Figure 1. Fabrication flow of four-period stacked SiGe/Si Channel FinFET device.

After a standard nWell formation, four-period stacked SiGe/Si multilayer were epitaxially grown by reduced-pressure chemical vapor epitaxial deposition. Then, the vertical Fin pattern with stacked SiGe/Si multilayer on the top of Si substrate were formed by a spacer image transfer (SIT) technique under an optimal HBr/O$_2$/He plasma. After STI filling and planarization, a low temperature of 850 °C for 30 s STI densification anneal and 1:100 diluted HF solution Fin reveal was implemented to attain a stacked SiGe/Si Fin formation [16]. Then, a low temperature SiO$_2$ deposition and dummy gate patterning were performed. After spacer 1 and spacer 2 definition, lightly doped drain (LDD) and source and drain (S/D) implantation was implemented with B and BF$_2$ dopant respectively. A low temperature dopant activation of 850 °C for 30 s was performed to keep the stacked SiGe/Si Fin stability [16]. Inter layer dielectric (ILD) deposition and CMP was employed to exposure the dummy gate. A standard tetramethylammonium hydroxide (TMAH) solution at 70 °C was used to remove the dummy poly gate with high selectivity to the underlying oxide on the stacked SiGe/Si Fin. After removal of this oxide, an in-situ low temperature O_3 passivation treatment at 300 °C for 1 min was employed and the Al_2O_3/HfO_2 high-k (HK) dielectric and TiN-based/W metal gate (MG) stack were deposited by the atomic-layer-deposition (ALD) tool. Finally, the standard FinFET following processing was employed to complete the stacked SiGe/Si channel FinFET device fabrication.

3. Result and Discussion

3.1. Epitaxial Growth of Stacked SiGe/Si Multilayer

After a standard HF-last clean, to maintain the well doping profile and attain an excellent surface of the Si substrate, a H_2 pre-bake treatment of 825 °C for 5 min is performed. The epitaxial growth of stacked SiGe/Si multilayer is prepared using dichlorosilane (DCS) and GeH_4 as SiGe layer precursors and SiH_4 as Si layer precursor at 650 °C in H_2 ambient, respectively. And a four-period stacked SiGe/Si is fabricated successfully on the Si substrate by appropriately exchanging the sequences of introduced gases.

The crystalline microstructure of the four-period stacked SiGe/Si multilayer epitaxial grown on Si substrate is detected by high resolution X-ray diffraction (HRXRD, Bruker, Tel Aviv, Israel) in the vicinity of the (004) Bragg peak with a Cu peak radiation. Its $\omega - 2\theta$ HRXRD scan result is shown in Figure 2. It is worth to note that a series of obvious high-intensity satellite peaks are found, indicating that the epitaxial layers of the four-period stacked SiGe/Si multilayer are under strained due to the lattice constant mismatch of Si and SiGe. Moreover, the presence of small-intensity thickness fringes is the characteristic of high quality of the stacked SiGe/Si multilayer.

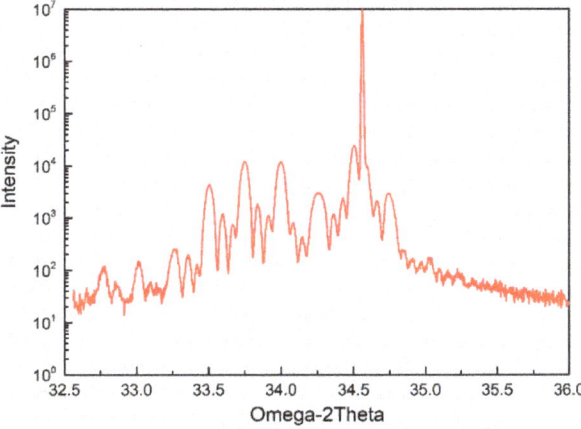

Figure 2. HRXRD spectra result on the four-period stacked SiGe/Si multilayer epitaxial grown on Si substrate.

In addition, its high-angle annular dark field scanning transmission electron microscopy (HAADF-STEM, FEI Talos, Brno, Czech Republic) analysis results are shown in Figure 3. It is found that there are no misfits at the SiGe/Si interfaces, nor threading dislocations crossing the stacked SiGe/Si epitaxial film. Therefore, a high crystal quality four-period stacked SiGe/Si multilayer with thin and distinct interfaces between SiGe and Si is successfully prepared. Meanwhile, the thickness of SiGe from top to bottom is 8.3, 8.2, 8.1 and 10.1 nm under the same time of epitaxial grown. Namely, the thickness of bottom SiGe is ~2 nm thicker than that of others. It is known that the epitaxial rate is strongly dependent on the crystallization of under-layer, that is, the epitaxial rate might be decreased if multi-crystallization occurs in the under-layer.

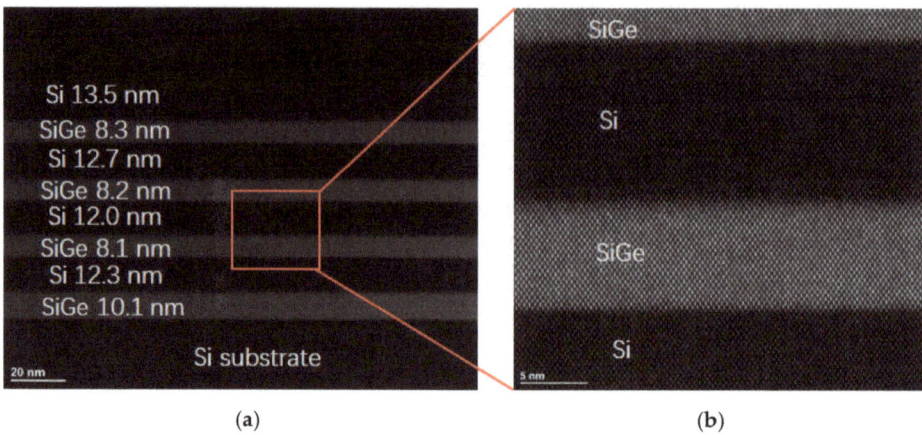

Figure 3. (a) Cross-section HAADF-STEM images of four-period stacked SiGe/Si multilayer; (b) its magnified images at the SiGe/Si interfaces.

Subsequently, EDX (FEI Talos, Brno, Czech Republic) line scan analysis of Ge and Si elements is also performed to determine the interfacial morphologies, and atomic fraction of the SiGe/Si layers for the four-period stacked SiGe/Si multilayer. The result is shown in Figure 4. It can be observed that the Ge atomic fraction of the SiGe layers from top to bottom are 23.2%, 23.5%, 23.8%, and 24.5%, respectively. Moreover, the width of transition layer of Si and SiGe are within 1.5 nm. Therefore, very uniform Ge atomic fraction of the SiGe layers with sharp SiGe/Si interfaces are achieved.

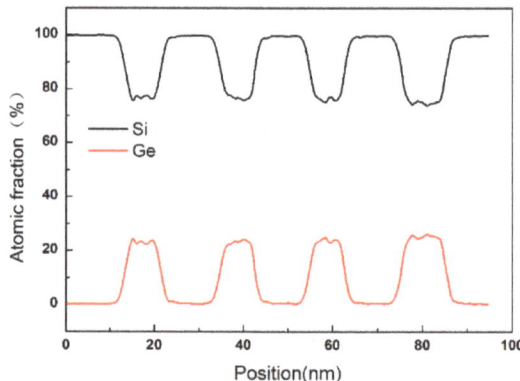

Figure 4. EDX line scan analysis of Ge and Si elements across the four-period stacked SiGe/Si multilayer.

3.2. Stacked SiGe/Si Fin Etching

Based on previous SiGe Fin etching result, HBr-based plasma is chosen as the etching gas for the four-period stacked SiGe/Si multilayer. Figure 5 presents the profiles of Si substrate/four-period stacked SiGe/Si multilayer Fin produced by $HBr/O_2/He$ plasma under different bias voltage and O_2 flow. The other etching process conditions are as follows: top power of 350 W, pressure of 6 mTorr, HBr flow of 110 sccm, He flow of 50 sccm. A more vertical profile of the Si substrate/four-period stacked SiGe/Si multilayer Fin structure can be achieved as its bias voltage increase from −70 V to −90 V and its O_2 flow increasing from 2.2 to 2.5 sccm. This is because that increasing the bias voltage can attain a larger ion bombardment effect and slightly increasing O_2 flow can help promote passivation films formation on the stacked SiGe/Si Fin sidewall and preserve its profile during etching.

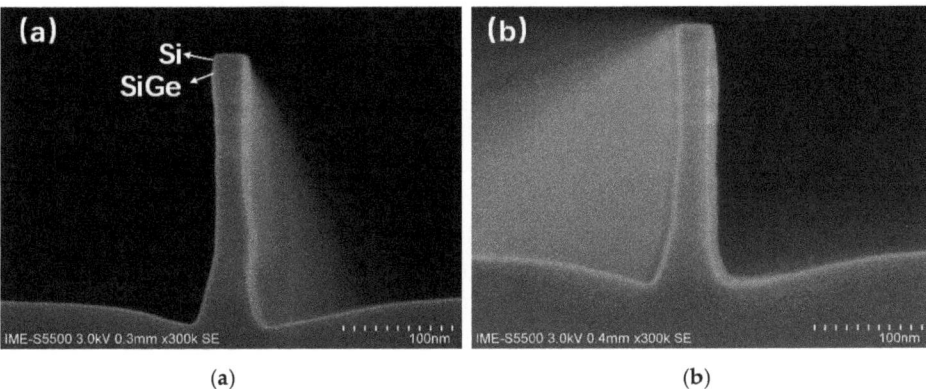

Figure 5. Scanning electron microscope (SEM) images of Fin profile under HBr/O_2/He plasma under (**a**) bias voltage of −70 V and O_2 flow of 2.2 sccm, (**b**) bias voltage of −90 V and O_2 flow of 2.5 sccm.

3.3. FinFET Device Fabrication

After these above newly developed epitaxial growth and etching technologies are implemented, the results of following major fabrication process of the four-period stacked SiGe/Si channel FinFET device are shown in Figure 6. Figure 6a presents the cross-sectional SEM image of the Fin reveal structure post STI recess by 1:100 diluted HF solution. It is found that the STI is well controlled and the stacked SiGe/Si Fin on the top of the Si has been revealed. Figure 6b shows the top view SEM image of stacked SiGe/Si channel FinFET device after gate formation with critical dimension (CD) of ~25 nm. And conformal spacer 1 and spacer 2 in Figure 6c are realized at two sides of dummy gate. Figure 6d,e present a top view of stacked SiGe/Si channel FinFET device after dummy gate CMP and dummy gate removal. As can be seen from the images, the surface of ILD is very smooth, and a dummy poly gate is successfully removed with a stacked SiGe/Si channel exposure in the open gate trench. After dummy gate removal, an in-situ low temperature O_3 passivation treatment at 300 °C for 1 min and the Al_2O_3/HfO_2 and TiN/TaN/TiN//W are implemented as HK and MG, respectively. Figure 7 shows cross-sectional transmission electron microscopy (TEM) image for the four-period stacked SiGe/Si channel under the HK/MG stack at the end of fabrication processing. And its HAADF-STEM and EDS mapping results are shown in Figure 8. It is found that a perfect four-period stacked SiGe/Si channel Fin structure with stable SiGe and Si layers is realized and the multilayer HK/MG are well wrapped around. At the same time, the Fin height of SiGe channel is 80.6 nm and the CD is 20 nm.

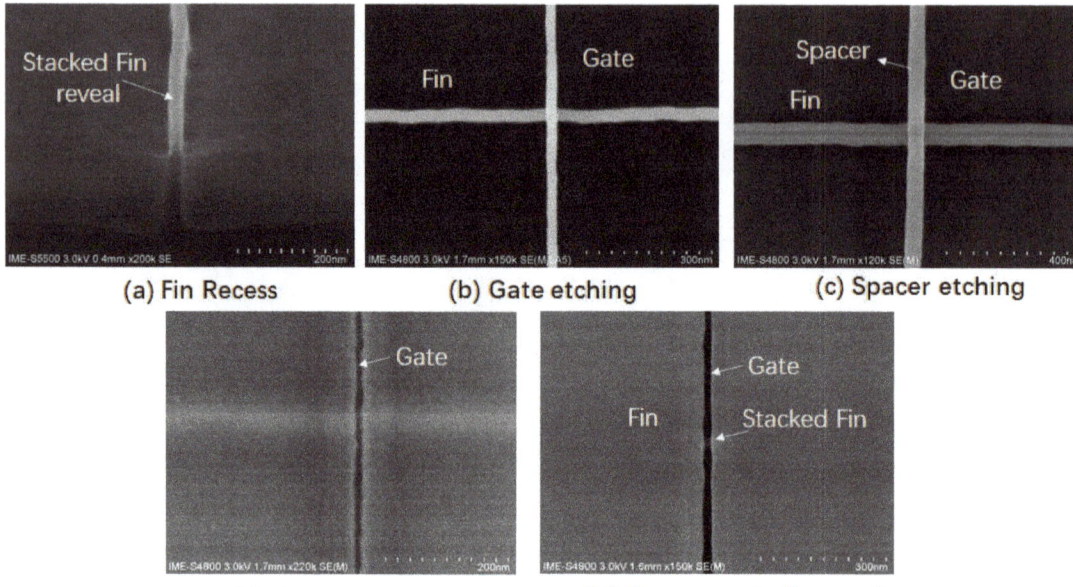

Figure 6. SEM images of stacked SiGe/Si channel FinFET device at different fabrication stages: (**a**) Fin reveal post STI recess, (**b**) dummy gate formation, (**c**) spacer formation, (**d**) poly gate open by CMP, (**e**) dummy gate removal.

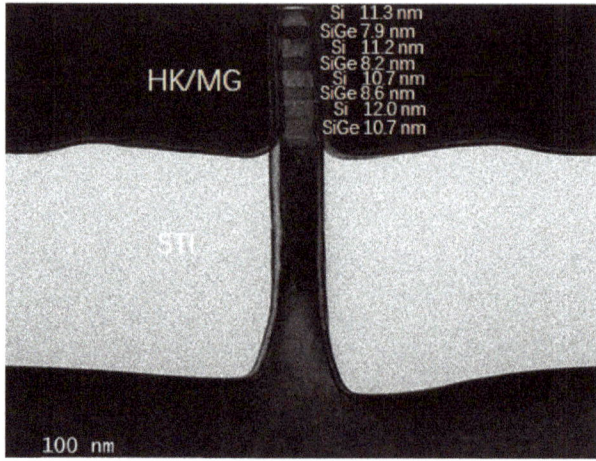

Figure 7. TEM result of four-period stacked SiGe/Si channel FinFET under the HK/MG stack at the end of fabrication processing.

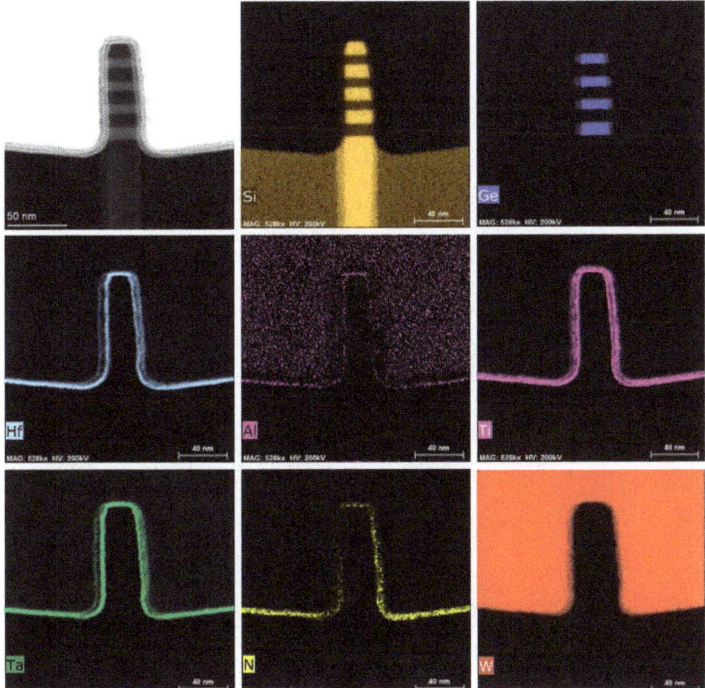

Figure 8. HAADF-STEM and EDS mapping results of the four-period stacked SiGe/Si channel FinFET under the HK/MG stack at the end of processing.

3.4. Electrical Performance

As a comparison, the conventional SiGe channel FinFET is also fabricated under the similar fabrication process.

As shown in Figure 9, the Fin height of its SiGe channel is 33 nm and the CD is 19.5 nm. The Fin height of SiGe channel is almost equal to total thickness of SiGe in the four-period stacked SiGe/Si channel FinFET. Moreover, the CD and profile of Fin of these two kinds of FinFET device are almost comparable. Therefore, these two kinds of FinFET device have almost the same footprint.

Figure 10 shows the I_{DS}-V_{GS} curves for the four-period stacked SiGe/Si channel FinFET and conventional SiGe channel FinFET. Compared with conventional SiGe channel FinFET, the I_{on}, SS and I_{on}/I_{off} ratio of the four-period stacked SiGe/Si channel FinFET under the same footprint show obvious benefit. And their electrical characteristic data are summarized in Table 1. Its I_{on} under $V_{DS} = V_{GS} = -0.8$ V increases 1.6 times, improved from 13.3 μA to 21.2 μA, and its I_{on}/I_{off} ratio can be improved from 1×10^5 to 1×10^6 due to the increase of I_{on} and the decrease of I_{off} at the same time. At the same time, the threshold voltage (V_{tsat}) extracting at I_{on} of 1×10^{-9} A can be improved from +0.38 V to +0.16 V. In addition, four-period stacked SiGe/Si channel FinFET is easier to obtain a better SS characteristic than the conventional SiGe FinFET under the same unoptimized O_3 passivation process. Its SS can be improved from 149 mV/dec to 90 mV/dec. The better SS characteristic should be related to the four-period stacked SiGe/Si Fin structure and it also can help to increase the I_{on} and decrease the I_{off}. The better SS can be attributed to the following two reasons: first, the stacked SiGe/Si with each layer SiGe of 8–10 nm in the stable stage may maintain a better quality and surface interfacial performance during the whole fabrication process compared with the the conventional SiGe of 33 nm in the metastable stage; second, the Si channel of the stacked SiGe/Si channel may turn on first due to its lower D_{it}. Moreover, the larger Ion can be attributed to the increasing of effective channel width (W_{eff}) because the W_{eff} of four-period stacked SiGe/Si Fin is 181 nm and the W_{eff} of conventional SiGe Fin is only 85.5 nm.

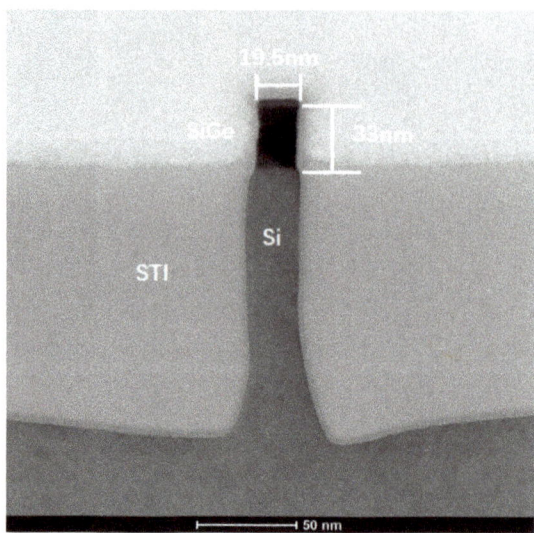

Figure 9. TEM result of the conventional SiGe channel Fin profile.

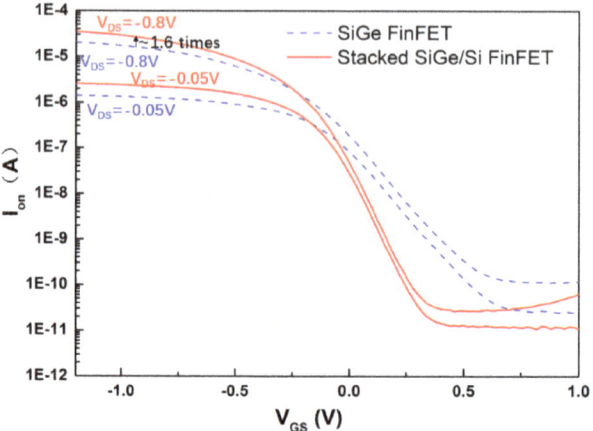

Figure 10. I_{DS}-V_{GS} characteristics of the four-period stacked SiGe/Si channel FinFET and conventional SiGe channel FinFET under the similar fabrication process.

Table 1. Electrical characteristic comparison of the four-period stacked SiGe/Si and conventional SiGe channel FinFET.

Category	Ion (μA)	SS (mV/dec)	V_{tsat} (V)	I_{on}/I_{off}
Conventional SiGe channel FinFET	13.3	149	0.38	1×10^5
Four-period stacked SiGe/Si channel FinFET	21.2	90	0.16	1×10^6

This four-period stacked SiGe/Si channel FinFET has been demonstrated to be a practical candidate for the future technology nodes. However, it is important to emphasize that these results are preliminary for the four-period stacked SiGe/Si channel FinFET, and there is still much room to improve its electrical characteristic, such as its SS and V_{tsat}. Therefore, we will employ the interfacial passivation, and gate stack engineering to further optimize its electrical performance in the future.

4. Conclusions

In a summary, a four-period vertically stacked SiGe/Si FinFET device was successfully fabricated by optimizing its epitaxial grown and Fin etching process. Compared with the conventional SiGe channel FinFET under the same footprint, its I_{on} increases 1.6 times, I_{on}/I_{off} ratio increases 1 order and SS can be improved from 149 to 90 mV/dec because the four-period stacked SiGe/Si Fin structure has larger W_{eff} and may maintain a better quality and surface interfacial performance during the whole fabrication process. Therefore, this device has been demonstrated to be a practical candidate for future technology nodes. Moreover, considering the compatibility of fabrication process, it also can be use as the I/O device for the vertically stacked Gate-All-Around horizontal nanowire/sheet technology if a thicker gate stack is employed and the channel release step is skipped.

Author Contributions: Methodology, Y.L., F.Z. and X.C.; investigation, Y.L., F.Z., H.L. and Y.Z.; data curation, Y.L., X.C., J.L. (Junjie Li) and Q.Z.; writing—original draft preparation, Y.L., F.Z., X.C., Z.W. and J.L. (Jun Luo); writing—review and editing, Y.L., F.Z. and X.C.; project administration, Y.L.; funding acquisition, Y.L. and W.W. All authors have read and agreed to the published version of the manuscript.

Funding: This work is supported in part by National Natural Science Foundation of China (Grant no. 62074160), in part by the Science and Technology Program of Beijing Municipal Science and Technology Commission (grant no. Z201100004220001), in part by the CAS Pioneer Hundred Talents Program, in part by Beijing Municipal Natural Science Foundation (Grant no. 4202078), and in part by the National Key Project of Science and Technology of China (Grant no. 2017ZX02315001-002).

Institutional Review Board Statement: Not applicable.

Informed Consent Statement: Not applicable.

Data Availability Statement: Not applicable.

Acknowledgments: We thank the Integrated Circuit Advanced Process Center (ICAC) at the Institute of Microelectronics of the Chinese Academy of Sciences for the devices fabricated on their advanced 200 mm CMOS platform.

Conflicts of Interest: The authors declare no conflict of interest.

References

1. Yeap, G.; Lin, S.S.; Chen, Y.M.; Shang, H.L.; Wang, P.W.; Lin, H.C.; Peng, Y.C.; Sheu, J.Y.; Wang, M.; Chen, X.; et al. 5 nm CMOS Production Technology Platform featuring full-fledged EUV, and High Mobility Channel FinFETs with densest 0.021 µm2 SRAM cells for Mobile SoC and High Performance Computing Applications. In Proceedings of the 2019 IEEE International Electron Devices Meeting (IEDM), San Francisco, CA, USA, 7–11 December 2019; pp. 36.7.1–36.7.4.
2. Mertens, H.; Ritzenthaler, R.; Arimura, H.; Franco, J.; Sebaai, F.; Hikavyy, A.; Pawlak, B.; Machkaoutsan, V.; Devriendt, K.; Tsvetanova, D. Si-cap-free SiGe p-channel FinFETs and gate-all-around transistors in a replacement metal gate process: Interface trap density reduction and performance improvement by high-pressure deuterium anneal. In Proceedings of the 2015 Symposium on VLSI Technology (VLSI Technology), Kyoto, Japan, 16–18 June 2015; pp. T142–T143.
3. Witters, L.; Arimura, H.; Sebaai, F.; Hikavyy, A.; Milenin, A.P.; Loo, R.; De Keersgieter, A.; Eneman, G.; Schram, T.; Wostyn, K.; et al. Strained Germanium Gate-All-Around pMOS Device Demonstration Using Selective Wire Release Etch Prior to Replacement Metal Gate Deposition. *IEEE Trans. Electron Devices* **2017**, *64*, 4587–4593. [CrossRef]
4. Franco, J.; Kaczer, B.; Roussel, P.J.; Mitard, J.; Cho, M.; Witters, L.; Grasser, T.; Groeseneken, G. SiGe Channel Technology: Superior Reliability Toward Ultrathin EOT Devices-Part I: NBTI. *IEEE Trans. Electron Devices* **2013**, *60*, 396–404. [CrossRef]
5. Lee, C.H.; Southwick, R.G.; Mochizuki, S.; Li, J.; Miao, X.; Wang, M.; Bao, R.; Ok, I.; Ando, T.; Hashemi, P.; et al. Toward High Performance SiGe Channel CMOS: Design of High Electron Mobility in SiGe nFinFETs Outperforming Si. In Proceedings of the 2018 IEEE International Electron Devices Meeting (IEDM), San Francisco, CA, USA, 1–5 December 2018; pp. 35.1.1–35.1.4.
6. Zhao, Z.Q.; Li, Y.; Zan, Y.; Li, Y.L.; Li, J.J.; Cheng, X.H.; Wang, G.L.; Liu, H.Y.; Wang, H.X.; Zhang, Q.Z.; et al. Fabrication technique of the $Si_{0.5}Ge_{0.5}$ Fin for the high mobility channel FinFET device. *Semicond. Sci. Technol.* **2020**, *35*, 045015. [CrossRef]

7. Zhao, Z.Q.; Li, Y.L.; Wang, G.L.; Du, A.Y.; Li, Y.; Zhang, Q.Z.; Xu, G.B.; Zhang, Y.K.; Luo, J.; Li, J.F.; et al. Process optimization of the $Si_{0.7}Ge_{0.3}$ Fin Formation for the STI first scheme. *Semicond. Sci. Technol.* **2019**, *34*, 125008. [CrossRef]
8. Mochizuki, S.; Colombeau, B.; Zhang, J.; Kung, S.C.; Stolfi, M.; Zhou, H.; Breton, M.; Watanabe, K.; Li, J.; Jagannathan, H.; et al. Structural and Electrical Demonstration of SiGe Cladded Channel for PMOS Stacked Nanosheet Gate-All-Around Devices. In Proceedings of the 2020 IEEE Symposium on VLSI Technology, Honolulu, HI, USA, 16–19 June 2020; pp. 1–2.
9. Paul, D.J. Si/SiGe heterostructures: From material and physics to devices and circuits. *Semicond. Sci. Technol.* **2004**, *19*, R75–R108. [CrossRef]
10. Arimura, H.; Wostyn, K.; Ragnarsson, L.Å.; Capogreco, E.; Chasin, A.; Conard, T.; Brus, S.; Favia, P.; Franco, J.; Mitard, J.; et al. Ge oxide scavenging and gate stack nitridation for strained $Si_{0.7}Ge_{0.3}$ pFinFETs enabling 35% higher mobility than Si. In Proceedings of the 2019 IEEE International Electron Devices Meeting (IEDM), San Francisco, CA, USA, 7–11 December 2019; pp. 29.2.1–29.2.4.
11. Jungwoo, O.; Majhi, P.; Jammy, R.; Joe, R.; Dip, A.; Takuya, S.; Yasushi, A.; Takanobu, K.; Tsunetoshi, A.; Masayuki, T. Additive mobility enhancement and off-state current reduction in SiGe channel pMOSFETs with optimized Si Cap and high-k metal gate stacks. In Proceedings of the 2009 International Symposium on VLSI Technology, Systems, and Applications, Hsinchu, Taiwan, 27–29 April 2009; pp. 22–23.
12. Zhang, B.; Yu, W.; Zhao, Q.T.; Hartmann, J.; Lupták, R.; Buca, D.; Bourdelle, K.; Wang, X.; Mantl, S. High mobility $Si/Si_{0.5}Ge_{0.5}$/strained SOI p-MOSFET with HfO_2/TiN gate stack. In Proceedings of the 2010 10th IEEE International Conference on Solid-State and Integrated Circuit Technology, Shanghai, China, 1–4 November 2010; pp. 911–913.
13. Ando, T.; Hashemi, P.; Bruley, J.; Rozen, J.; Ogawa, Y.; Koswatta, S.; Chan, K.K.; Cartier, E.A.; Mo, R.; Narayanan, V. High Mobility High-Ge-Content SiGe PMOSFETs Using Al_2O_3/HfO_2 Stacks With In-Situ O3 Treatment. *IEEE Electron Device Lett.* **2017**, *38*, 303–305. [CrossRef]
14. Lee, C.H.; Kim, H.; Jamison, P.; Southwick, R.G.; Mochizuki, S.; Watanabe, K.; Bao, R.; Galatage, R.; Guillaumet, S.; Ando, T.; et al. Selective GeOx-scavenging from interfacial layer on $Si_{1-x}Ge_x$ channel for high mobility $Si/Si_{1-x}Ge_x$ CMOS application. In Proceedings of the 2016 IEEE Symposium on VLSI Technology, Honolulu, HI, USA, 14–16 June 2016; pp. 1–2.
15. Chang, K.M.; Chen, C.F.; Lai, C.H.; Kuo, P.S.; Chen, Y.M.; Chang, T.Y. Electrical properties of SiGe nanowire following fluorine/nitrogen plasma treatment. *Appl. Surf. Sci.* **2014**, *289*, 581–585. [CrossRef]
16. Li, Y.; Cheng, X.; Zhong, Z.; Zhang, Q.; Wang, G.; Li, Y.; Li, J.; Ma, X.; Wang, X.; Yang, H.; et al. Key Process Technologies for Stacked Double $Si_{0.7}Ge_{0.3}$ Channel Nanowires Fabrication. *ECS J. Solid State Sci. Technol.* **2020**, *9*, 064009. [CrossRef]

Article

An Operation Guide of Si-MOS Quantum Dots for Spin Qubits

Rui-Zi Hu [1,2,†], Rong-Long Ma [1,2,†], Ming Ni [1,2], Xin Zhang [1,2], Yuan Zhou [1,2], Ke Wang [1,2], Gang Luo [1,2], Gang Cao [1,2], Zhen-Zhen Kong [3], Gui-Lei Wang [3,*], Hai-Ou Li [1,2,*] and Guo-Ping Guo [1,2,4]

1   CAS Key Laboratory of Quantum Information, University of Science and Technology of China, Hefei 230026, China; hrz@mail.ustc.edu.cn (R.-Z.H.); rlma@mail.ustc.edu.cn (R.-L.M.); mingni@mail.ustc.edu.cn (M.N.); xzhang16@mail.ustc.edu.cn (X.Z.); zy1995@mail.ustc.edu.cn (Y.Z.); wk0910@ustc.edu.cn (K.W.); rogone@ustc.edu.cn (G.L.); gcao@ustc.edu.cn (G.C.); gpguo@ustc.edu.cn (G.-P.G.)
2   CAS Center for Excellence and Synergetic Innovation Center in Quantum Information and Quantum Physics, University of Science and Technology of China, Hefei 230026, China
3   Key Laboratory of Microelectronics Devices & Integrated Technology, Institute of Microelectronics, Chinese Academy of Sciences, Beijing 100029, China; kongzhenzhen@ime.ac.cn
4   Origin Quantum Computing Company Limited, Hefei 230026, China
*   Correspondence: wangguilei@ime.ac.cn (G.-L.W.); haiouli@ustc.edu.cn (H.-O.L.)
†   These authors contributed equally to this work.

Abstract: In the last 20 years, silicon quantum dots have received considerable attention from academic and industrial communities for research on readout, manipulation, storage, near-neighbor and long-range coupling of spin qubits. In this paper, we introduce how to realize a single spin qubit from Si-MOS quantum dots. First, we introduce the structure of a typical Si-MOS quantum dot and the experimental setup. Then, we show the basic properties of the quantum dot, including charge stability diagram, orbital state, valley state, lever arm, electron temperature, tunneling rate and spin lifetime. After that, we introduce the two most commonly used methods for spin-to-charge conversion, i.e., Elzerman readout and Pauli spin blockade readout. Finally, we discuss the details of how to find the resonance frequency of spin qubits and show the result of coherent manipulation, i.e., Rabi oscillation. The above processes constitute an operation guide for helping the followers enter the field of spin qubits in Si-MOS quantum dots.

Keywords: Si-MOS; quantum dot; spin qubits; quantum computing

1. Introduction

As early as 1982, the famous physicist Feynman proposed that quantum computers can simulate problems that cannot be solved by classical computers [1]. Then, in 1994, Shor proposed the well-known quantum prime factor decomposition algorithm that can be used to crack classic RSA encrypted communications [2], and in 1996, Grover devised the quantum search algorithm which uses only $O(\sqrt{N})$ evaluations of the function [3]. After that, Loss and DiVincenzo proposed the Loss–DiVincenzo quantum computer in 1998 [4] and then in 2000, DiVincenzo presented the DiVincenzo Criteria for physical implementation of quantum computing [5]. These findings set off a wave of quantum computing research.

In this wave, researchers tried to build quantum computers in various systems. Trapped ions [6,7], nuclear magnetic resonance (NMR) [8,9], superconducting loops [10,11], nitrogen vacancy center [12,13], semiconductor quantum dots [14,15], and other systems have enabled the manipulation of single and two qubits and have demonstrated simple quantum algorithms. Among them, silicon quantum dots (QDs) have emerged as promising hosts for qubits to build a quantum processor due to their long coherence time [16,17], small footprint [18], potential scalability [19,20], and compatibility with advanced semiconductor manufacturing technology [21].

In recent decades, silicon QDs have engaged research participants all around the world and have developed fast. In 2012, a long-time singlet–triplet oscillation was realized in silicon double quantum dots (DQD) [22]. Then, high-quality single-spin control was developed in silicon QDs [16,23]. After that, a two-qubit controlled gate in silicon QDs was experimentally implemented [24–27]. Nowadays, the single-qubit operations of spin qubits achieve fidelities of 99.9% [28,29], the two-qubit operation fidelities are above 99% as reported [30], the spin–photon coupling rates are more than 10 megahertz [31–33], and the qubit operation temperature is higher than 1 kelvin [34,35]. In the meantime, experiments on other properties of silicon QDs, including valley states [36–38], orbital states [39], and noise spectra [40,41], have been carried out. Furthermore, experimental approaches and techniques for characterizing features of QDs from other systems, e.g., charge stability diagrams [42–44], random telegraph signals (RTS) [45–48], Elzerman readout [49,50], Pauli spin blockade (PSB) readout [51,52], electron spin resonance (ESR) [53,54], and electron dipole spin resonance (EDSR) [55,56], have been applied in silicon QDs as well. In addition, several reviews [57–59] and guides on fabrication [60] have been reported. However, the process from silicon QD to qubit manipulation is still challenging.

In this article, we give a brief introduction of how to realize a single spin qubit from QDs in a Si-MOS structure. First, we introduce the gate-defined DQD in an isotopically enriched ^{28}Si-MOS structure and the low-temperature measurement circuits. Second, by applying these circuits, we investigate the basic properties of silicon QD devices, i.e., charge states, excited orbital states, valley splitting, lever arms, electron temperature, tunneling rate, and noise spectrum. Then, we introduce two mainstream spin-state-readout methods named as the Elzerman readout and the PSB readout. Finally, we use the rapid adiabatic passage to find out the resonance frequency of the spin qubits and apply the Rabi pulsing schemes to coherently manipulate the spin qubit.

2. Materials and Methods
2.1. Spin Qubit Devices

Spin qubits are hosted in a pair of metal-oxide-semiconductor (MOS) dots with isotopically enriched silicon. By using the high vacuum activation annealing technique, we improve the mobility of Si-MOS devices by a factor of two, reaching 1.5 m^2/(V·s) [61]. In this work, we use a DQD that has a similar structure (Ref. [38]) and was fabricated in our lab's clean room. As shown in Figure 1a, the aluminum electrodes are vaporized on top of the silicon oxide by electron beam evaporation techniques. Between every two layers of the electrodes, an insulating layer of aluminum oxide is formed by thermal oxidation. Figure 1b shows that the electrons are confined in the potential wells and the DQD is formed by applying voltages to the electrodes [62]. In the quantum well, a single electron can tunnel between the two QDs by biasing the electrodes' voltages. The entire structure consists of a DQD and a single-electron transistor (SET) sensing the charge states in DQD.

2.2. Measurement Circuits

There are three main types of measurement circuits commonly used to characterize the properties of DQDs, as shown in Figure 2a,c,e:

- Figure 2a: Transport measurements based on a lock-in amplifier. The AC excitation is added to the SET source (S_1) by connecting the lock-in amplifier to an external 1000:1 voltage divider, and finally reaches S_1 at approximately 50 µV, with a lock-in frequency generally between 70 and 1000 Hz. In addition, the drain (D_1) is connected back to the lock-in amplifier to demodulate the signal and obtain the currents.
- Figure 2c: Charge detection based on the lock-in amplifier. The bias voltage at S_1 is connected to a Stanford Research Systems Isolated Voltage Source (SIM 928) through a 1000:1 voltage divider, reaching S_1 at around 500 µV, while the AC excitation of the lock-in amplifier (output at approximately 0.5 to 1.5 mV) is connected to LP through an analog summing amplifier (SIM 980, bandwidth of approximately 1 MHz).

- Figure 2e: Charge detection based on a current-voltage amplifier. The source-drain bias is the same as Figure 2c, except no excitation is applied to the LP and D_1 is connected to a current-voltage amplifier; here, we use a Femto DLPCA200, connected to a voltage amplifier (SIM 910), an analog filter (SIM 965), and finally to a voltmeter (Agilent 34410) for signal measurement or to a PCI-based waveform digitizer (ATS 460), oscilloscope, etc. for the real-time observation of electron tunneling.

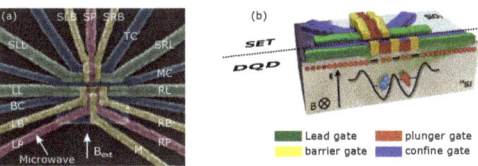

Figure 1. (a) Scanning electron micrograph image of a typical Si-MOS DQD. An SET, which is used as a charge sensor, is confined by the top confine gate (TC), middle confine gate (MC), left barrier gate (SLB), and right barrier gate (SRB) and is tuned by the plunger gate (SP). A DQD is composed of a left lead gate (LL), right lead gate (RL), left barrier gate (LB), middle gate (M), right barrier gate (RB), left plunger gate (LP), and right plunger gate (RP) and is confined by a bottom confine gate (BC) and middle confine gate (MC). We tune the left and right QD via the LP and RP, respectively. The tunneling rate of the QDs can be tuned by the LB and RB. The spin of electrons in the left QD is controlled by applying a microwave pulse to the LP. The right white arrow indicates the direction of an in-plane external magnetic field. (b) Cross-sectional view of a 3D model of the device. Electrodes for different functions are distinguished by different colors. The SET and DQD are on each side of the dotted line. The electrons in the DQD are located under the plunger gates.

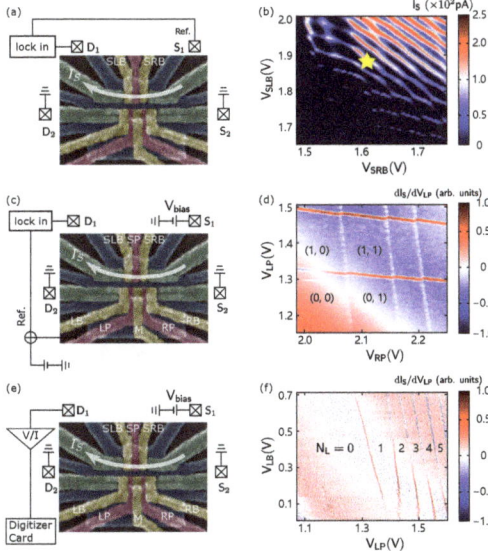

Figure 2. The three different measurement circuits. The white arrow above the SET indicates the direction of the current (I_S). (**a**,**b**) Measurement circuit diagram of the SET using the lock-in amplifier and the Coulomb peak diagram obtained by scanning the voltage of the SET barrier gates (SRB and SLB). The yellow star identifies a sensitive SET position. (**c**,**d**) Measurement circuit diagram of the DQD and the charge stability diagram of the DQD obtained by scanning the RP and LP. (**e**,**f**) Measurement circuit diagram for measuring the DQD using a current-voltage amplifier and the corresponding charge stability diagram of the left QD.

3. Results

3.1. Basic Properties of Silicon QDs

3.1.1. Charge Stability Diagram

Obtaining the QD charge stability diagram by the charge detection method is one of the most basic QD characterization methods [42–44]. As shown in Figure 2a–d, according to the method of measuring QDs using the modulation signal of the lock-in amplifier introduced in Section 2, the source (S_2) and drain (D_2) of the DQD are grounded. We set the voltages of the SLB and SRB near a Coulomb peak so that the SET works sensitively at this position, which is identified by the yellow star in Figure 2b. Then, a voltage of 2.60 V is applied to the LL and RL to ensure that the channel of DQD is turned on. After that, we measured the charge stability diagrams with different gate voltages to obtain the DQD electron occupation numbers and tunneling properties of the left QD. Figure 2d shows the charge stability diagram of the last few electrons in the DQD. The numbers in this figure indicate the electron occupation on the left and right QD. The slope is relatively symmetric with respect to the two electrodes. This indicates that two QDs are formed under electrodes LP and RP. When scanning the voltage of LB and LP, there are continuous electron tunneling lines observed, which correspond to the left QD. As shown in Figure 2f, the tunneling line of the last few electrons in the left QD becomes more invisible when the voltage of LB decreases. This is because a decrease in the LB voltage reduces the tunneling rate of the left QD to the reservoir of D_2.

3.1.2. Detection of Orbital Excited States in Silicon QDs

The orbital excited state in silicon QDs is several meV above the ground state, and it can be detected by the pulsed-voltage spectroscopy method [39,63]. Based on the measurement circuit in Figure 2c, we change the modulation signal output from the lock-in amplifier to a square waveform generated by an external arbitrary waveform generator that is synchronized with the lock-in amplifier. By zooming in and remapping, the single tunneling lines in Figure 3a split into pairs of lines in Figure 3b. As shown in Figure 3c, the principle of the pulsed-voltage spectroscopy method is illustrated. When the voltage of LP is set at the position of the blue square in Figure 3d, the electron can tunnel into the ground state of the QD. As the voltage increases, the energy level of the excited state gradually approaches the amplitude window of the square wave. When the excited state enters the window, the electron can tunnel into the excited state, so that another transport line appears parallel to the left line, identified by the green circle in Figure 3d. According to Figure 3d and the extracted lever arm of LP (α_{LP}, which will be discussed in Section 3.1.3), which is 0.33 meV/mV, the calculated energy of excited state is 1.3 meV.

3.1.3. Detection of Valley States in Silicon QDs

In solid-state physics, due to the six-fold degeneracy at the bottom of the conduction band of silicon, the energy levels at the bottom of the six conduction bands are named as the valley level. In the case of two-dimensional electron gas, the six-fold degeneracy is split into a four-fold degeneration and a double-fold degeneration. Due to the existence of the interfacial electric field, the quadruple degenerate and the double degenerate split further and form valley-level splits [57]. Unlike the orbital state, the splitting energy of the two lowest valley states (E_{VS}) in silicon QDs is similar to the Zeemen splitting energy (E_Z) under the applied magnetic field in our experiment [36–39,64,65]. Therefore, it is important to determine the splitting energy of the valley state. A commonly used method is to measure the electron tunneling line at different magnetic fields. Here, we tune the energy level of the first four electrons by changing the magnetic field of which the direction of is along the surface of the device and perpendicular to the one-dimensional channel formed by the QD, as shown in Figure 1a. Figure 4a shows the transition lines of the first four electrons in the device in Ref. [38]. The voltage of the first transition line of the QD decreases as the magnetic field increases, while the fourth line increases. Differently, the second transition line increases first and then decreases, and the third line is reversed to the second line.

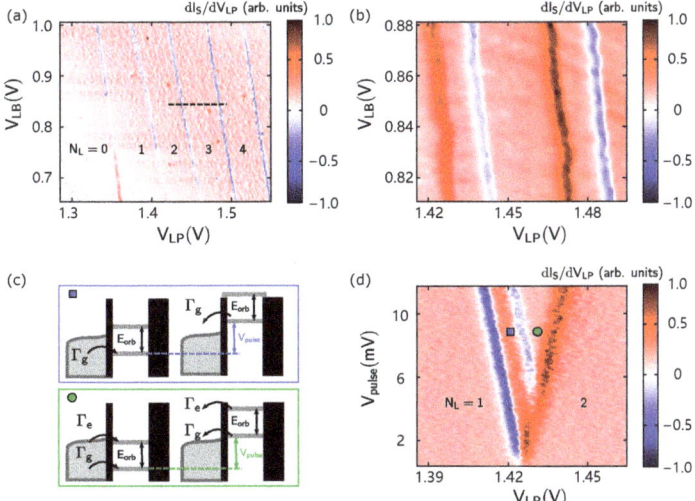

Figure 3. (**a**) Charge stability diagram of the QDs obtained by scanning the LB and LP voltages. (**b**) Zoom in on the QD charge stability diagram after applying a square pulse with a frequency of 687 Hz and an amplitude of 20 mV. (**c**) Schematic diagram of the square pulse spectrum measurement of the excited orbital state. When the LP voltage increases, the tunneling line of the electron in the excited state appears. (**d**) Diagram of the excited orbital state obtained by scanning the amplitude of the square pulse and the LP voltage.

We use the principle of minimum energy to simply explain this phenomenon, as shown in Figure 4b. When filling the first electron, the electron will be filled to the lowest energy level. As the magnetic field increases, E_Z increases, so the energy level of filling the first electron decreases. When filling the second electron, the first electron has been filled to the bottom level. In accordance with the principle of minimum energy, the second electron should be filled with the second-lowest level, but this second-lowest energy level depends on the magnetic field. When the magnetic field is small, the second-lowest energy state is the valley state v_- with spin state up, and vice versa. As shown in Figure 4b, the energy states of the third and fourth electron are mirror-symmetrical to the second and first electrons, respectively. It is obvious from Figure 4b, the position of the kink point is exactly where E_{VS} is equal to E_Z, so by using the position of the kink point and the Bohr magneton (μ_B), we can obtain:

$$E_{VS} = g\mu_B B_{kink} \tag{1}$$

According to Figure 4a, the E_{VS} of the second electron is 170 µeV, and the E_{VS} of the third one is 245 µeV. The difference between the E_{VS} of these two electrons is caused by the different LP voltages [38].

Additionally, we can estimate α_{LP} by:

$$\alpha_{LP} = \frac{g\mu_B \Delta B}{2\Delta V_{LP}} \tag{2}$$

Therefore, the lever arms of the first four electrons are shown in Table 1:

Table 1. Lever arm α_{LP} for the first four electrons.

Electron Number	Lever Arm α_{LP} (meV/mV)
1	0.33
2	0.32
3	0.31
4	0.34

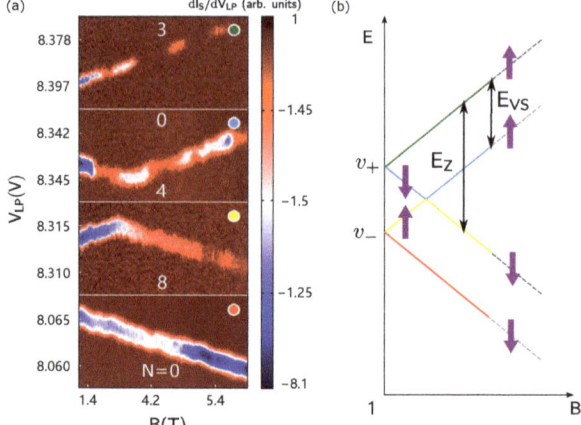

Figure 4. The magnetic spectrum and the corresponding diagram of the energy state for different electron numbers in the QDs. (**a**) The dependence of different electron tunneling lines on the magnetic field, where N = 0, 1, 2, 3, and 4 refer to the number of electrons in the QD. The slopes of the first four electron tunneling lines reveal the lever arms of LP for the first four electrons. (**b**) Energy state diagram of E_Z as a function of the magnetic field. The ordinate is the state energy, and the abscissa is the order of the magnetic field. The purple arrow indicates the direction of the spin. The arrow with E_{VS} represents the energy of the valley splitting.

3.2. Real-Time Observation of Electron Tunneling in Silicon QDs

The characterization of the orbital state, spin state, and valley state in QDs is based on steady-state measurement. However, to detect more properties of electrons, such as tunneling rate, electron temperature, noise spectrum, spin state, etc., we also need the ability to observe the tunneling process of electrons in QDs in real time [45–48]. The measurement circuit of real-time detection has been introduced in Section 2, as shown in Figure 2e. Next, we introduce the measurement results of tunneling rate, electron temperature, noise spectrum, and spin state, respectively.

3.2.1. RTS and the Measurement of Electron Temperature

When we align the electrochemical potential of the first electron in the QD with the Fermi surface of the electron reservoir, the electrons will continuously tunnel in and out of the QD (see the green circle in Figure 5a,b). At this time, on the oscilloscope or digitizer, we can see the signal as shown in the inset of Figure 5b. Since tunneling events happen randomly, we call the observed signal a RTS.

Ideally, electrons tunnel only when the electrochemical potential in the QD is aligned with the Fermi surface of the electron reservoir. However, in practice, due to the limited electron temperature, the Fermi surface of the electron reservoir will have a certain broadening. Therefore, changes in the electron tunneling events can be observed when the LP voltage is changed. The insets in Figure 5b show that when the electrode voltage increases, the electrochemical potential in the QD decreases, so the probability of the electrons occupying the energy state in the QDs gradually increases and vice versa. By fitting the Fermi distribution to the electron occupancy, we can extract the electron temperature. The specific form of the Fermi distribution function we used here is the following [48]:

$$N = \frac{1}{\exp[\alpha_{LP}(V_{LP0} - V_{LP})/(k_B T) + 1]} \tag{3}$$

where k_B represents the Boltzmann constant, α_{LP} has been calculated in Table 1, V_{LP0} and T are fitting parameters. By fitting this equation, the electron temperature of approximately 224 mK is obtained.

3.2.2. Measurement of the Tunneling Rate

For the RTS, we can mark the time of electron tunneling from the reservoir to the QDs as t_{on}, and the time of electron tunneling from the QDs to the reservoir as t_{off}. By counting the distribution of t_{on} and t_{off} over a long period of time, we can actually determine the time of electron tunneling in and out of the reservoir [45].

Here, we introduce another method based on RTS. As shown in Figure 5c, by applying a square waveform on the LP, the signal will also switch between low and high levels with an approximate square wave period. Figure 5d illustrates that the rising and falling edges of the signal are slower, unlike the square wave from the AWG. Excluding the bandwidth limitation of the SET, the width of the edges represents the electron tunneling times t_{on} and t_{off}. By fitting the rising and falling edges exponentially, we can obtain the exact tunneling time values: $t_{on} = 3.45$ ms and $t_{off} = 3.23$ ms.

3.2.3. Noise Spectrum

When observing the real-time electron tunneling signal, there will inevitably be noise interference. Analyzing the noise spectrum can help us analyze the source of the noise and then suppress the noise. Figure 6a shows a typical noise spectrum of a QD system but does not include the noise introduced by the measurement system. The QD system suffers from charge noise [29], random telegraph noise (RTN) [40] and nuclear noise [66] at low frequencies. Johnson Nyquist noise and phonon noise are relatively large at high frequencies and affect the spin relaxation time.

Figure 6b shows the noise spectrum under the different device conditions given in Ref. [38]. The red line is the noise spectrum when the QD is connected. This noise conforms to the law of $1/f$. In fact, this is typical charge noise from the QD. The green line is the noise spectrum when the QD is not connected and almost overlaps with the red line above 10 Hz, indicating that the noise above 10 Hz does not come from the QD. The blue line is the noise spectrum when the amplifier input is open, indicating that the noise above 10 Hz comes from the DC line. Compared with Ref. [41], the noise from the QD is low enough for the measurement of the qubit. On the other hand, the capacitance and resistance of the DC line is the main reason for this high-frequency noise. To further reduce the noise source, we can switch to a coaxial line with a smaller capacitance in the future.

3.3. Spin State Readout

After being able to detect the QD charge state and control and measure the tunneling of electrons through a simple square waveform, we now introduce two of the most commonly used methods for spin-to-charge conversion: the Elzerman readout [49,50] and PSB readout [14,51–53].

3.3.1. Elzerman Readout

The process of the Elzerman readout is shown in Figure 7a. We set the voltage to locate the Fermi surface of the electron reservoir between the energy state of the electrons with different spin states. Therefore, the spin-up electrons can tunnel to the electron reservoir (after a period of time to load spin down electrons from the electron reservoir), while the spin down electrons cannot. Since the signal of SET responds to the two events of electron tunneling in and out of the QD, a square wave is formed in the signal of the SET. By observing the change in the current, it can be determined whether electron tunneling occurs; then, it can be determined whether the spin state of the electron is up.

Figure 7b shows a series of the measured SET current signal while reading the spin state. The signal in the top panel has a square pulse, which corresponds to a spin up state. The signal in the bottom panel does not have such a pulse and indicates a spin down state. Based on the above process, we have achieved a single-shot readout of the electron spin state.

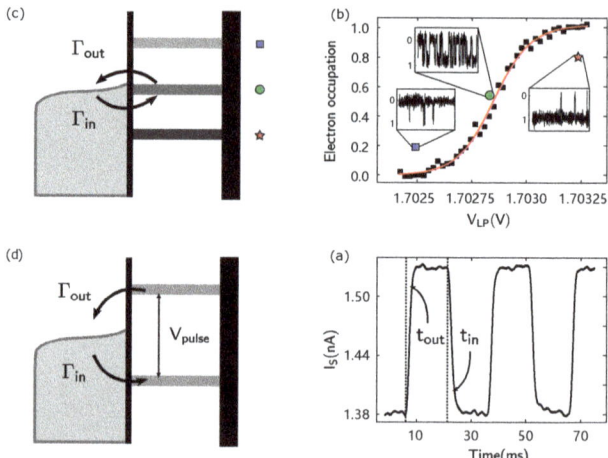

Figure 5. Measurement of the electron temperature and tunneling rate. (**a**) Schematic diagram of different positions of the QD energy state and the Fermi surface of the electron reservoir. (**b**) The probability of electron occupation probability. The electron temperature can be fitted as 223.8 ± 0.8 mK. The inset shows RTS for different electron occupation situations: the circle, square and star marks correspond to the alignment, negative bias and positive bias, respectively. (**c**) Electron tunnel in the QD when the voltage is high and vice versa. (**d**) The average current of the electron tunneling by applying a square wave with a 30 ms period. The average current decays exponentially with the tunnel time, and is characteristic of a Poisson process. A single exponential fitting can be used to obtain t_{on} and t_{off}.

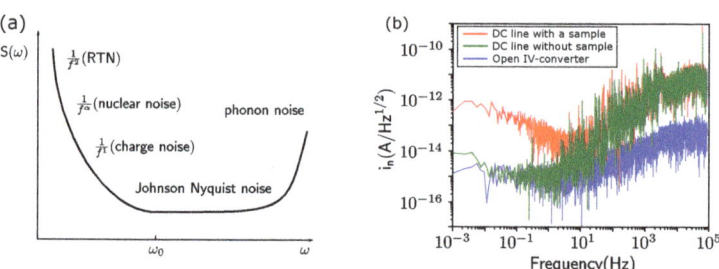

Figure 6. Measurement of the noise spectrum in silicon QD. (**a**) Typical noise spectrum of a silicon QD; the noise from the measurement system is not included. Here, ω_0 is the spin resonance frequency. (**b**) The noise spectrum is measured by a dynamic signal analyzer (SR785) in our system and the spectrum contains three conditions: DC line with a sample, DC line without sample and open I–V converter.

3.3.2. PSB

For the PSB readout, we first need to know the double spin eigenstates; the singlet (S) and triplet (T, include T_0, T_+ and T_-) states:

$$S = \frac{|\uparrow\downarrow\rangle - |\downarrow\uparrow\rangle}{\sqrt{2}}, T_0 = \frac{|\uparrow\downarrow\rangle + |\downarrow\uparrow\rangle}{\sqrt{2}}, T_+ = |\uparrow\uparrow\rangle, T_- = |\downarrow\downarrow\rangle \quad (4)$$

When there is no magnetic field, the singlet state is the ground state. The three-spin triplet state energies degenerate, which is referred to as the T state. This T state is an excited state. Now, we consider two charge states in a DQD: (1,1) and (0,2). For the (0,2) state, there are two electrons in one QD. According to the Pauli exclusion principle, the spin wave function of the electrons in the T state is symmetric, so two electrons must occupy different orbital states. Therefore, S(0,2) and T(0,2) are non-degenerate, as shown in Figure 7c,d. Δ_{ST} is the energy difference between S(0,2) and T(0,2). However, for the (1,1) state, two electrons are located in their respective QDs, thus avoiding the Pauli exclusion principle and two electrons can occupy one orbital state. Therefore, S(1,1) and T(1,1) are almost degenerate, as shown in Figure 7c,d.

Based on these energy states, we now introduce the PSB readout. As shown in Figure 7c, when a negative bias is applied (the Fermi surface of the source is higher than the drain), electrons in the source can first tunnel to the S(1,1) or T(1,1) state. When tunneling to the S(1,1) state, the electron can continue to tunnel to S(0,2) and then reach the drain to form current. When tunneling to the T(1,1) state, the electron cannot continue to tunnel to S(0,2) due to the PSB, and T(0,2) is higher than T(1,1), so the electron cannot enter any (0,2) charge states, and the current is suppressed. Figure 7d shows the positive bias condition. The Fermi surface of the source is lower than the drain. The electrons in the drain tunnel to the S(0,2) state, and then through the S(1,1) state to the source to form a current. No blockade occurs in the process, so there is current in the entire bias triangle region. In addition, for the PSB readout, we can use the SET to sense the charge states in DQD, and the operating temperature can be raised to higher than 1 kelvin [34,35].

3.3.3. Measurement of Spin Lifetime

After being able to perform a single-shot measurement to read the spin state, we can use the same waveform to measure the spin lifetime (T_1) [50]. The process of a typical single-shot readout is shown in Figure 8a. First, we reduce the voltage for electron evacuating from the QD; this is also referred to as "empty". Then, we raise the voltage so that electrons can tunnel from the electron reservoir to the QDs, which is also called "load". At this time the spin state of the electron in QD is random. Finally, we carefully reduce the voltage to locate the Fermi surface of the electron reservoir between the energy state of different spin electrons to "read" the spin state. We count the number of spin relaxation events for different load time periods. Figure 8b illustrates that the probability of the spin up state ($P_\uparrow$) decreases exponentially, so the T_1 can be obtained by fitting the exponential function.

3.4. Manipulation of the Spin Qubit

Now that we are able to read the spin state via the single-shot readout method, we introduce the manipulation of the spin qubit. There are two mainstream manipulation methods: ESR [16,53,54] and EDSR [23,52,55,56,67,68]. The ESR can be achieved by applying an alternating magnetic field B_1 (5–50 µT) perpendicular to the external magnetic field B_{ext} (typically 150–1500 mT) via an antenna structure. For EDSR, we apply an alternating electric field combined with spin-orbit coupling to flip the spin. However, the natural spin–orbit coupling in silicon is weak, so we need micromagnets to introduce a gradient magnetic field to construct synthetic spin-orbit coupling. The advantages of EDSR include a fast spin flip rate, low heating, ease of fabrication, etc. However, the additional magnetic field from the micromagnets makes it difficult to find the resonance frequency $\gamma_e(B_{ext} + B_1)$ of the qubit. Therefore, we introduce rapid adiabatic passage to solve this problem.

3.4.1. Rapid Adiabatic Passage

We use frequency chirped microwave bursts, and when the excitation frequency passes through the resonance frequency, the electron spin is inverted (see Figure 9b) [23,55,56]. Figure 9a shows the principle of the rapid adiabatic passage process. In the reference frame rotating at the resonance frequency, the Hamiltonian of the system is the following [56]:

$$H(t) = \frac{1}{2}\frac{\partial}{\partial t}(\Delta \nu) t \sigma_z + \nu_1 \sigma_x \tag{5}$$

Here, $\Delta \nu$ is the microwave frequency detuning from the resonance frequency, and ν_1 is the spin flip rate.

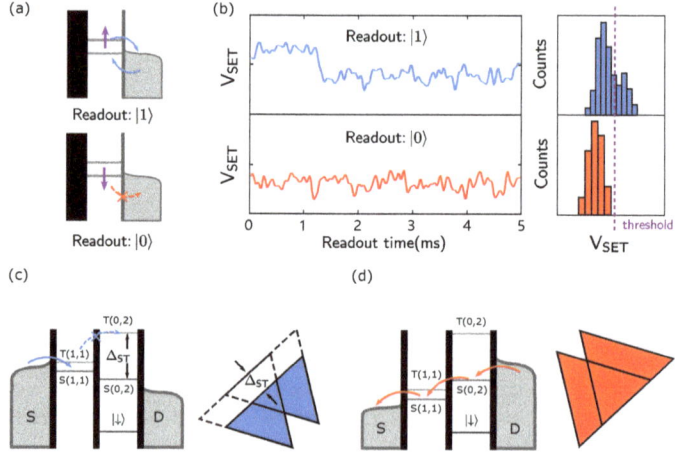

Figure 7. Spin-charge conversion. (**a**) Schematic diagram to read the spin state by the Elzerman readout. The spin-up electrons can tunnel out because the energy is higher than the Fermi surface of the electron reservoir and vice versa. (**b**) The measurement result of the spin state is read out by the Elzerman method in our experiment. When the electron in spin-up state tunnels out, there is a high level in the signal. The electron in spin-down state cannot tunnel out, so the signal remains at a low level. (**c**) Schematic diagram of the energy state and corresponding measurement results of the electron transition current with a negative bias. Δ_{ST} is the energy difference between the S and T states. When the energy detuning is less than Δ_{ST}, PSB occurs. (**d**) Schematic diagram of the energy state with a positive bias. Here, no PSB occurs.

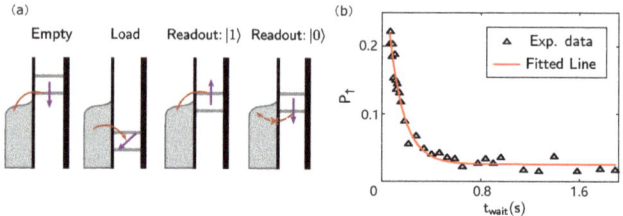

Figure 8. Schematic diagram and measurement result of T_1. (**a**) Schematic diagram of a single-shot readout for T_1 measurement. (**b**) Measured spin up probability ($P_\uparrow$) as a function of waiting time (t_{wait}). The fitting result of T_1 is 335 ± 5 ms for the left QD.

We use the Landau–Zener theory to solve this time evolution of a two-level system that is described by a linearly time-dependent Hamiltonian. The probability of adiabatic transition from one eigenstate to the other is given by [56]

$$P = 1 - \exp\left(-4\pi^2 \frac{\nu_1^2}{\left|\frac{\partial}{\partial t}(\Delta \nu)\right|}\right) \quad (6)$$

An electron spin in the $|\downarrow\rangle$ state will flip to the $|\uparrow\rangle$ state if the microwave frequency sweeps across the resonance frequency. To satisfy the adiabatic evolution condition, the sweep rate $\left|\frac{\partial}{\partial t}(\Delta \nu)\right|$ cannot be too fast compared with ν_1.

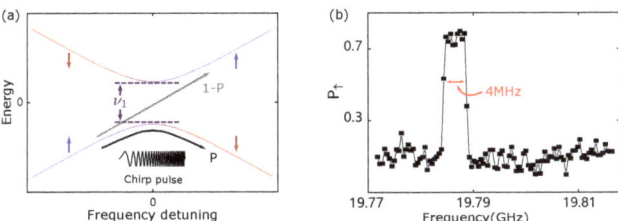

Figure 9. Schematic explanation and measurement result of rapid adiabatic passage. (**a**) Schematic explanation of rapid adiabatic passage in the rotating reference frame. (**b**) $P_\uparrow$ as a function of microwave frequency with a 0.5 ms burst time and a 4 MHz frequency modulation depth.

3.4.2. Rabi Oscillation

After calibrating the resonant frequency through the rapid adiabatic passage, we now use a single-frequency microwave combined with a single-shot readout to manipulate the qubit [54,56], as shown in the inset of Figure 10b. Figure 10a shows the Rabi pulsing scheme. First, we increase the voltage so that electrons in the $|\downarrow\rangle$ or $|\uparrow\rangle$ state cannot tunnel from the QDs to the electron reservoir. We apply the microwave pulse before the next stage to flip the electron spin. Then, as mentioned in Section 3.3.3, we carefully decrease the voltage to locate the Fermi surface of the electron reservoir between the energy states of different spin electrons to "read" the spin state. At the end of the "read" phase, the electron spin state will be $|\downarrow\rangle$ no matter the spin state at the beginning. Figure 10b shows the result of a Rabi oscillation. As the microwave duration time increases, the spin of the qubit continuously flips between $|\downarrow\rangle$ and $|\uparrow\rangle$ states. The amplitude of oscillation decreases with time due to noise. We fit the Rabi oscillation with the function $P(t) = A \cdot \exp(-t/T_2^{\text{Rabi}}) \cdot \sin(f_{\text{Rabi}} t)$. Here, $f_{\text{Rabi}} = 1.256 \pm 0.003$ MHz represents the spin flip rate, and $T_2^{\text{Rabi}} = 5.4 \pm 0.4$ μs represents the influence of the noise in Figure 10b.

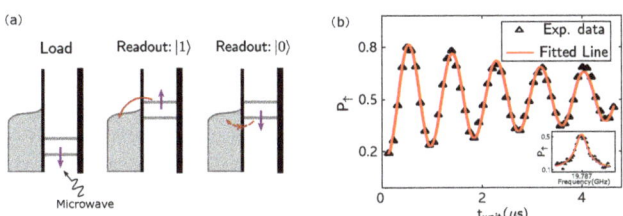

Figure 10. Schematic diagram and measurement result of Rabi oscillation. (**a**) Schematic diagram for Rabi oscillation. (**b**) $P_\uparrow$ as a function of t_{wait}. The inset shows $P_\uparrow$ as a function of microwave frequency around the resonance frequency $\nu = 19.787$ GHz.

4. Conclusions

In this paper, we provide an operation guide of Si-MOS QDs for spin qubits. First, we introduce the structure of the devices and the measurement circuit. Next, we show the charge stability diagram and detect the orbital and valley states. Then, we use a digitizer to detect the RTS and measure electron temperature and tunneling rate. Moreover, we introduce two commonly used methods, the Elzerman readout and the PSB readout, and use the single-shot readout method to measure the T_1. Finally, we give a brief introduction of ESR and EDSR, use rapid adiabatic passage to calibrate the resonance frequency of the spin qubit, and show the result of the Rabi oscillation. For future directions, researchers may be interested in hybrid qubits coupling [33], hot qubits [34,35], cryogenic control [69,70], foundry-fabrication [71,72], high fidelity readouts [73,74], and qubit number expansion [75].

Author Contributions: R.-Z.H., R.-L.M., M.N., X.Z. and H.-O.L. performed the experiments. R.-L.M. and M.N. fabricated the sample. G.C., K.W. and Y.Z. helped analyze the data. Z.-Z.K., G.L. and G.-L.W. prepared the silicon wafer. H.-O.L., R.-Z.H., X.Z., R.-L.M., M.N. and G.-P.G. designed the experiment, provided theoretical support, and analyzed the results. H.-O.L. and G.-P.G. supervised the project. R.-Z.H. and H.-O.L. wrote the manuscript, with input from all authors. All authors have read and agreed to the published version of the manuscript.

Funding: This work was supported by National Key Research and Development Program of China (Grant No. 2016YFA0301700), the National Natural Science Foundation of China (Grants No. 12074368, 61922074, 12034018, 11625419 and 62004185), the Anhui initiative in Quantum Information Technologies (Grant No. AHY080000). G.-L.W. acknowledges financial support by Youth Innovation Promotion Association of CAS (Grant No. 2020037). This work was partially carried out at the USTC Center for Micro and Nanoscale Research and Fabrication.

Data Availability Statement: The data presented in this study are available on request from the corresponding authors.

Conflicts of Interest: The authors declare no conflict of interest.

References

1. Feynman, R.P. Simulating physics with computers. *Int. J. Theor. Phys.* **1982**, *21*, 467–488. [CrossRef]
2. Shor, P.W. Algorithms for quantum computation: Discrete logarithms and factoring. In Proceedings of the 35th Annual Symposium on Foundations of Computer Science, Santa Fe, NM, USA, 20–22 November 1994; pp. 124–134.
3. Grover, L.K. A fast quantum mechanical algorithm for database search. In Proceedings of the Twenty-Eighth Annual ACM Symposium on Theory of Computing, Philadelphia, PA, USA, 22–24 May 1996; pp. 212–219.
4. Loss, D.; DiVincenzo, D.P. Quantum computation with quantum dots. *Phys. Rev.* **1998**, *57*, 120–126. [CrossRef]
5. Di Vincenzo, D.P. The physical implementation of quantum computation. *Fortschritte Phys.* **2000**, *48*, 771–783. [CrossRef]
6. Cirac, J.I.; Zoller, P. Quantum computations with cold trapped ions. *Phys. Rev. Lett.* **1995**, *74*, 4091–4094. [CrossRef]
7. Monroe, C.; Kim, J. Scaling the ion trap quantum processor. *Science* **2013**, *339*, 1164–1169. [CrossRef]
8. Vandersypen, L.M.K.; Steffen, M.; Breyta, G.; Yannoni, C.S.; Sherwood, M.H.; Chuang, I.L. Experimental realization of shor's quantum factoring algorithm using nuclear magnetic resonance. *Nature* **2001**, *414*, 883–887. [CrossRef]
9. Vandersypen, L.M.K.; Chuang, I.L. NMR techniques for quantum control and computation. *Rev. Mod. Phys.* **2005**, *76*, 1037–1069. [CrossRef]
10. Mooij, J.E.; Orlando, T.P.; Levitov, L.; Tian, L.; Wal, C.H.V.; Lloyd, S. Josephson persistent-current qubit. *Science* **1999**, *285*, 1036–1039. [CrossRef]
11. Krantz, P.; Kjaergaard, M.; Yan, F.; Orlando, T.P.; Gustavsson, S.; Oliver, W.D. A quantum engineer's guide to superconducting qubits. *Appl. Phys. Rev.* **2019**, *6*, 021318. [CrossRef]
12. Dutt, M.V.G.; Childress, L.; Jiang, L.; Togan, E.; Maze, J.; Jelezko, F.; Zibrov, A.S.; Hemmer, P.R.; Lukin, M.D. Quantum register based on individual electronic and nuclear spin qubits in diamond. *Science* **2007**, *316*, 1312–1316. [CrossRef]
13. Childress, L.; Hanson, R. Diamond NV centers for quantum computing and quantum networks. *MRS Bull.* **2013**, *38*, 134–138. [CrossRef]
14. Petta, J.R.; Johnson, A.C.; Taylor, J.M.; Laird, E.A.; Yacoby, A.; Lukin, M.D.; Marcus, C.M.; Hanson, M.P.; Gossard, A.C. Coherent manipulation of coupled electron spins in semiconductor quantum dots. *Science* **2005**, *309*, 2180–2184. [CrossRef]
15. Elzerman, J.M.; Hanson, R.; Beveren, L.H.W.V.; Tarucha, S.; Vandersypen, L.M.K.; Kouwenhoven, L.P. *Semiconductor Few-Electron Quantum Dots as Spin Qubits*; Springer: New York, NY, USA, 2005.

16. Veldhorst, M.; Hwang, J.C.C.; Yang, C.H.; Leenstra, A.W.; de Ronde, B.; Dehollain, J.P.; Muhonen, J.T.; Hudson, F.E.; Itoh, K.M.; Morello, A.; et al. An addressable quantum dot qubit with fault-tolerant control-fidelity. *Nat. Nanotechnol.* **2014**, *9*, 981–985. [CrossRef]
17. Veldhorst, M.; Yang, C.H.; Hwang, J.C.C.; Huang, W.; Dehollain, J.P.; Muhonen, J.T.; Simmons, S.; Laucht, A.; Hudson, F.E.; Itoh, K.M.; et al. A two-qubit logic gate in silicon. *Nature* **2015**, *526*, 410–414. [CrossRef]
18. Kandel, Y.P.; Qiao, H.; Fallahi, S.; Gardner, G.C.; Manfra, M.J.; Nichol, J.M. Adiabatic quantum state transfer in a semiconductor quantum-dot spin chain. *Nat. Commun.* **2021**, *12*, 2156. [CrossRef]
19. Vandersypen, L.M.K.; Bluhm, H.; Clarke, J.S.; Dzurak, A.S.; Ishihara, R.; Morello, A.; Reilly, D.J.; Schreiber, L.R.; Veldhorst, M. Interfacing spin qubits in quantum dots and donors—Hot, dense, and coherent. *Npj Quantum Inf.* **2017**, *3*, 34. [CrossRef]
20. Li, R.; Petit, L.; Franke, D.P.; Dehollain, J.P.; Helsen, J.; Steudtner, M.; Thomas, N.K.; Yoscovits, Z.R.; Singh, K.J.; Wehner, S.; et al. A crossbar network for silicon quantum dot qubits. *Sci. Adv.* **2018**, *4*, eaar3960. [CrossRef] [PubMed]
21. Zwerver, A.M.J.; Krähenmann, T.; Watson, T.F.; Lampert, L.; George, H.C.; Pillarisetty, R.; Bojarski, S.A.; Amin, P.; Amitonov, S.V.; Boter, J.M.; et al. Qubits made by advanced semiconductor manufacturing. *arXiv* **2021**, arXiv:2101.12650.
22. Maune, B.M.; Borselli, M.G.; Huang, B.; Ladd, T.D.; Deelman, P.W.; Holabird, K.S.; Kiselev, A.A.; Alvarado-Rodriguez, I.; Ross, R.S.; Schmitz, A.E.; et al. Coherent singlet-triplet oscillations in a silicon-based double quantum dot. *Nature* **2012**, *481*, 344–347. [CrossRef] [PubMed]
23. Kawakami, E.; Scarlino, P.; Ward, D.R.; Braakman, F.R.; Savage, D.E.; Lagally, M.G.; Friesen, M.; Coppersmith, S.N.; Eriksson, M.A.; Vandersypen, L.M.K. Electrical control of a long-lived spin qubit in a Si/SiGe quantum dot. *Nat. Nanotechnol.* **2014**, *9*, 666–670. [CrossRef]
24. Watson, T.F.; Philips, S.G.J.; Kawakami, E.; Ward, D.R.; Scarlino, P.; Veldhorst, M.; Savage, D.E.; Lagally, M.G.; Friesen, M.; Coppersmith, S.N.; et al. A programmable two-qubit quantum processor in silicon. *Nature* **2018**, *555*, 633–637. [CrossRef]
25. Zajac, D.M.; Sigillito, A.J.; Russ, M.; Borjans, F.; Taylor, J.M.; Burkard, G.; Petta, J.R. Resonantly driven cnot gate for electron spins. *Science* **2018**, *359*, 439–442. [CrossRef]
26. Huang, W.; Yang, C.H.; Chan, K.W.; Tanttu, T.; Hensen, B.; Leon, R.C.C.; Fogarty, M.A.; Hwang, J.C.C.; Hudson, F.E.; Itoh, K.M.; et al. Fidelity benchmarks for two-qubit gates in silicon. *Nature* **2019**, *569*, 532–536. [CrossRef] [PubMed]
27. Xue, X.; Watson, T.F.; Helsen, J.; Ward, D.R.; Savage, D.E.; Lagally, M.G.; Coppersmith, S.N.; Eriksson, M.A.; Wehner, S.; Vandersypen, L.M.K. Benchmarking gate fidelities in a Si/SiGe two-qubit device. *Phys. Rev.* **2019**, *9*, 021011. [CrossRef]
28. Chan, K.W.; Huang, W.; Yang, C.H.; Hwang, J.C.C.; Hensen, B.; Tanttu, T.; Hudson, F.E.; Itoh, K.M.; Laucht, A.; Morello, A.; et al. Assessment of a silicon quantum dot spin qubit environment via noise spectroscopy. *Phys. Rev. Appl.* **2018**, *10*, 044017. [CrossRef]
29. Yoneda, J.; Takeda, K.; Otsuka, T.; Nakajima, T.; Delbecq, M.R.; Allison, G.; Honda, T.; Kodera, T.; Oda, S.; Hoshi, Y.; et al. A quantum-dot spin qubit with coherence limited by charge noise and fidelity higher than 99.9%. *Nat. Nanotechnol.* **2018**, *13*, 102–106. [CrossRef] [PubMed]
30. Xue, X.; Russ, M.; Samkharadze, N.; Undseth, B.; Sammak, A.; Scappucci, G.; Vandersypen, L.M.K. Computing with spin qubits at the surface code error threshold. *arXiv* **2021**, arXiv:2107.00628.
31. Mi, X.; Benito, M.; Putz, S.; Zajac, D.M.; Taylor, J.M.; Burkard, G.; Petta, J.R. A coherent spin–photon interface in silicon. *Nature* **2018**, *555*, 599. [CrossRef] [PubMed]
32. Samkharadze, N.; Zheng, G.; Kalhor, N.; Brousse, D.; Sammak, A.; Mendes, U.C.; Blais, A.; Scappucci, G.; Vandersypen, L.M.K. Strong spin-photon coupling in silicon. *Science* **2018**, *359*, 1123–1127. [CrossRef]
33. Borjans, F.; Croot, X.G.; Mi, X.; Gullans, M.J.; Petta, J.R. Resonant microwave-mediated interactions between distant electron spins. *Nature* **2020**, *577*, 195–198. [CrossRef]
34. Petit, L.; Eenink, H.G.J.; Russ, M.; Lawrie, W.I.L.; Hendrickx, N.W.; Philips, S.G.J.; Clarke, J.S.; Vandersypen, L.M.K.; Veldhorst, M. Universal quantum logic in hot silicon qubits. *Nature* **2020**, *580*, 355–359. [CrossRef]
35. Yang, C.H.; Leon, R.C.C.; Hwang, J.C.C.; Saraiva, A.; Tanttu, T.; Huang, W.; Lemyre, J.C.; Chan, K.W.; Tan, K.Y.; Hudson, F.E.; et al. Operation of a silicon quantum processor unit cell above one kelvin. *Nature* **2020**, *580*, 350–354. [CrossRef]
36. Lim, W.H.; Yang, C.H.; Zwanenburg, F.A.; Dzurak, A.S. Spin filling of valley–orbit states in a silicon quantum dot. *Nanotechnology* **2011**, *22*, 335704. [CrossRef]
37. Zajac, D.M.; Hazard, T.M.; Mi, X.; Wang, K.; Petta, J.R. A reconfigurable gate architecture for Si/SiGe quantum dots. *Appl. Phys. Lett.* **2015**, *106*, 223507. [CrossRef]
38. Zhang, X.; Hu, R.; Li, H.; Jing, F.; Zhou, Y.; Ma, R.; Ni, M.; Luo, G.; Cao, G.; Wang, G.; et al. Giant anisotropy of spin relaxation and spin-valley mixing in a silicon quantum dot. *Phys. Rev. Lett.* **2020**, *124*, 57701. [CrossRef]
39. Yang, C.H.; Lim, W.H.; Lai, N.S.; Rossi, A.; Morello, A.; Dzurak, A.S. Orbital and valley state spectra of a few-electron silicon quantum dot. *Phys. Rev.* **2012**, *86*, 115319. [CrossRef]
40. Kamioka, J.; Kodera, T.; Takeda, K.; Obata, T.; Tarucha, S.; Oda, S. Charge noise analysis of metal oxide semiconductor dual-gate Si/SiGe quantum point contacts. *J. Appl. Phys.* **2014**, *115*, 203709. [CrossRef]
41. Freeman, B.M.; Schoenfield, J.S.; Jiang, H. Comparison of low frequency charge noise in identically patterned Si/SiO$_2$ and Si/SiGe quantum dots. *Appl. Phys. Lett.* **2016**, *108*, 253108. [CrossRef]
42. Elzerman, J.M.; Hanson, R.; Greidanus, J.S.; van Beveren, L.H.W.; Franceschi, S.D.; Vandersypen, L.M.K.; Tarucha, S.; Kouwenhoven, L.P. Few-electron quantum dot circuit with integrated charge read out. *Phys. Rev.* **2003**, *67*, 161308. [CrossRef]

43. Borselli, M.G.; Eng, K.; Ross, R.S.; Hazard, T.M.; Holabird, K.S.; Huang, B.; Kiselev, A.A.; Deelman, P.W.; Warren, L.D.; Milosavljevic, I.; et al. Undoped accumulation-mode Si/SiGe quantum dots. *Nanotechnology* **2015**, *26*, 375202. [CrossRef]
44. Lawrie, W.I.L.; Eenink, H.G.J.; Hendrickx, N.W.; Boter, J.M.; Petit, L.; Amitonov, S.V.; Lodari, M.; Wuetz, B.P.; Volk, C.; Philips, S.G.J.; et al. Quantum dot arrays in silicon and germanium. *Appl. Phys. Lett.* **2020**, *116*, 080501. [CrossRef]
45. Vandersypen, L.M.K.; Elzerman, J.M.; Schouten, R.N.; Beveren, L.H.W.V.; Hanson, R.; Kouwenhoven, L.P. Real-time detection of single-electron tunneling using a quantum point contact. *Appl. Phys. Lett.* **2004**, *85*, 4394–4396. [CrossRef]
46. Li, H.; Xiao, M.; Cao, G.; Zhou, C.; Shang, R.; Tu, T.; Guo, G.; Jiang, H.; Guo, G. Back-action-induced non-equilibrium effect in electron charge counting statistics. *Appl. Phys. Lett.* **2012**, *100*, 092112. [CrossRef]
47. House, M.G.; Xiao, M.; Guo, G.; Li, H.; Cao, G.; Rosenthal, M.M.; Jiang, H. Detection and measurement of spin-dependent dynamics in random telegraph signals. *Phys. Rev. Lett.* **2013**, *111*, 126803. [CrossRef]
48. Zajac, D.M.; Hazard, T.M.; Mi, X.; Nielsen, E.; Petta, J.R. Scalable gate architecture for a one-dimensional array of semiconductor spin qubits. *Phys. Rev. Appl.* **2016**, *6*, 054013. [CrossRef]
49. Elzerman, J.M.; Hanson, R.; van Beveren, L.H.W.; Witkamp, B.; Vandersypen, L.M.K.; Kouwenhoven, L.P. Single-shot read-out of an individual electron spin in a quantum dot. *Nature* **2004**, *430*, 431–435. [CrossRef]
50. Morello, A.; Pla, J.J.; Zwanenburg, F.A.; Chan, K.W.; Tan, K.Y.; Huebl, H.; Möttönen, M.; Nugroho, C.D.; Yang, C.; van Donkelaar, J.A.; et al. Single-shot readout of an electron spin in silicon. *Nature* **2010**, *467*, 687. [CrossRef]
51. Johnson, A.C.; Petta, J.R.; Marcus, C.M.; Hanson, M.P.; Gossard, A.C. Singlet-triplet spin blockade and charge sensing in a few-electron double quantum dot. *Phys. Rev.* **2005**, *72*, 165308. [CrossRef]
52. Zhang, X.; Zhou, Y.; Hu, R.; Ma, R.; Ni, M.; Wang, K.; Luo, G.; Cao, G.; Wang, G.; Huang, P.; et al. Controlling synthetic spin-orbit coupling in a silicon quantum dot with magnetic field. *Phys. Rev. Appl.* **2021**, *15*, 044042. [CrossRef]
53. Koppens, F.H.; Buizert, C.; Tielrooij, K.J.; Vink, I.T.; Nowack, K.C.; Meunier, T.; Kouwenhoven, L.P.; Vandersypen, L.M. Driven coherent oscillations of a single electron spin in a quantum dot. *Nature* **2006**, *442*, 766–771. [CrossRef]
54. Pla, J.J.; Tan, K.Y.; Dehollain, J.P.; Lim, W.H.; Morton, J.J.L.; Jamieson, D.N.; Dzurak, A.S.; Morello, A. A single-atom electron spin qubit in silicon. *Nature* **2012**, *489*, 541–545. [CrossRef] [PubMed]
55. Shafiei, M.; Nowack, K.C.; Reichl, C.; Wegscheider, W.; Vandersypen, L.M.K. Resolving spin-orbit- and hyperfine-mediated electric dipole spin resonance in a quantum dot. *Phys. Rev. Lett.* **2013**, *110*, 107601. [CrossRef]
56. Laucht, A.; Kalra, R.; Muhonen, J.T.; Dehollain, J.P.; Mohiyaddin, F.A.; Hudson, F.; McCallum, J.C.; Jamieson, D.N.; Dzurak, A.S.; Morello, A. High-fidelity adiabatic inversion of a ^{31}P electron spin qubit in natural silicon. *Appl. Phys. Lett.* **2014**, *104*, 092115. [CrossRef]
57. Zwanenburg, F.A.; Dzurak, A.S.; Morello, A.; Simmons, M.Y.; Hollenberg, L.C.L.; Klimeck, G.; Rogge, S.; Coppersmith, S.N.; Eriksson, M.A. Silicon quantum electronics. *Rev. Mod. Phys.* **2013**, *85*, 961–1019. [CrossRef]
58. Zhang, X.; Li, H.; Cao, G.; Xiao, M.; Guo, G. Semiconductor quantum computation. *Natl. Sci. Rev.* **2018**, *6*, 32–54. [CrossRef]
59. Zhang, X.; Li, H.; Wang, K.; Cao, G.; Xiao, M.; Guo, G. Qubits based on semiconductor quantum dots. *Chin. Phys.* **2018**, *27*, 020305. [CrossRef]
60. Spruijtenburg, P.C.; Amitonov, S.V.; Wiel, W.G.V.; Zwanenburg, F.A. A fabrication guide for planar silicon quantum dot heterostructures. *Nanotechnology* **2018**, *29*, 143001. [CrossRef]
61. Wang, K.; Li, H.; Luo, G.; Zhang, X.; Jing, F.; Hu, R.; Zhou, Y.; Liu, H.; Wang, G.; Cao, G.; et al. Improving mobility of silicon metal-oxide–semiconductor devices for quantum dots by high vacuum activation annealing. *EPL (Europhys. Lett.)* **2020**, *130*, 27001. [CrossRef]
62. Thorbeck, T.; Zimmerman, N.M. Formation of strain-induced quantum dots in gated semiconductor nanostructures. *AIP Adv.* **2015**, *5*, 087107. [CrossRef]
63. Elzerman, J.M.; Hanson, R.; van Beveren, L.H.W.; Vandersypen, L.M.K.; Kouwenhoven, L.P. Excited-state spectroscopy on a nearly closed quantum dot via charge detection. *Appl. Phys. Lett.* **2004**, *84*, 4617–4619. [CrossRef]
64. Tarucha, S.; Austing, D.G.; Honda, T.; Hage, R.J.V.; Kouwenhoven, L.P. Shell filling and spin effects in a few electron quantum dot. *Phys. Rev. Lett.* **1996**, *77*, 3613–3616. [CrossRef]
65. Jiang, L.; Yang, C.H.; Pan, Z.; Rossi, A.; Dzurak, A.S.; Culcer, D. Coulomb interaction and valley-orbit coupling in si quantum dots. *Phys. Rev.* **2013**, *88*, 085311. [CrossRef]
66. Nakajima, T.; Noiri, A.; Kawasaki, K.; Yoneda, J.; Stano, P.; Amaha, S.; Otsuka, T.; Takeda, K.; Delbecq, M.R.; Allison, G.; et al. Coherence of a driven electron spin qubit actively decoupled from quasistatic noise. *Phys. Rev.* **2020**, *10*, 011060. [CrossRef]
67. Nowack, K.C.; Koppens, F.H.; Nazarov, Y.V.; Vandersypen, L.M. Coherent control of a single electron spin with electric fields. *Science* **2007**, *318*, 1430–1433. [CrossRef] [PubMed]
68. Pioro-Ladrière, M.; Obata, T.; Tokura, Y.; Shin, Y.S.; Kubo, T.; Yoshida, K.; Taniyama, T.; Tarucha, S. Electrically driven single-electron spin resonance in a slanting zeeman field. *Nat. Phys.* **2008**, *4*, 776–779. [CrossRef]
69. Pauka, S.J.; Das, K.; Kalra, R.; Moini, A.; Yang, Y.; Trainer, M.; Bousquet, A.; Cantaloube, C.; Dick, N.; Gardner, G.C.; et al. A cryogenic CMOS chip for generating control signals for multiple qubits. *Nat. Electron.* **2021**, *4*, 64–70. [CrossRef]
70. Xue, X.; Patra, B.; van Dijk, J.P.G.; Samkharadze, N.; Subramanian, S.; Corna, A.; Wuetz, B.P.; Jeon, C.; Sheikh, F.; Juarez-Hernandez, E.; et al. CMOS-based cryogenic control of silicon quantum circuits. *Nature* **2021**, *593*, 205–210. [CrossRef]
71. Ansaloni, F.; Chatterjee, A.; Bohuslavskyi, H.; Bertrand, B.; Hutin, L.; Vinet, M.; Kuemmeth, F. Single-electron operations in a foundry-fabricated array of quantum dots. *Nat. Commun.* **2020**, *11*, 6399. [CrossRef]

72. Gilbert, W.; Saraiva, A.; Lim, W.H.; Yang, C.H.; Laucht, A.; Bertrand, B.; Rambal, N.; Hutin, L.; Escott, C.C.; Vinet, M.; et al. Single-electron operation of a silicon-CMOS 2 × 2 quantum dot array with integrated charge sensing. *Nano Lett.* **2020**, *20*, 7882–7888. [CrossRef] [PubMed]
73. Liu, Y.-Y.; Philips, S.G.J.; Orona, L.A.; Samkharadze, N.; McJunkin, T.; MacQuarrie, E.R.; Eriksson, M.A.; Vandersypen, L.M.K.; Yacoby, A. Radio-frequency reflectometry in silicon-based quantum dots. *Phys. Rev. Appl.* **2021**, *16*, 014057. [CrossRef]
74. Ruffino, A.; Yang, T.; Michniewicz, J.; Peng, Y.; Charbon, E.; Gonzalez-Zalba, M.F. Integrated multiplexed microwave readout of silicon quantum dots in a cryogenic cmos chip. *arXiv* **2021**, arXiv:2101.08295.
75. Takeda, K.; Noiri, A.; Nakajima, T.; Yoneda, J.; Kobayashi, T.; Tarucha, S. Quantum tomography of an entangled three-qubit state in silicon. *Nat. Nanotechnol.* **2021**. [CrossRef] [PubMed]

Article

Investigate on the Mechanism of HfO$_2$/Si$_{0.7}$Ge$_{0.3}$ Interface Passivation Based on Low-Temperature Ozone Oxidation and Si-Cap Methods

Qide Yao [1,2], Xueli Ma [2,*], Hanxiang Wang [1,2], Yanrong Wang [1,*], Guilei Wang [2], Jing Zhang [1], Wenkai Liu [1], Xiaolei Wang [2], Jiang Yan [1], Yongliang Li [2,*] and Wenwu Wang [2]

1. School of Information Science and Technology, North China University of Technology, Beijing 100144, China; yaoqide@ime.ac.cn (Q.Y.); wanghanxiang@ime.ac.cn (H.W.); zhangj@ncut.edu.cn (J.Z.); liuwk@ncut.edu.cn (W.L.); yanjiang@ncut.edu.cn (J.Y.)
2. Integrated Circuit Advanced Process Center, Institute of Microelectronics, Chinese Academy of Science, Beijing 100029, China; wangguilei@ime.ac.cn (G.W.); wangxiaolei@ime.ac.cn (X.W.); wangwenwu@ime.ac.cn (W.W.)
* Correspondence: maxueli@ime.ac.cn (X.M.); wangyanrong@ncut.edu.cn (Y.W.); liyongliang@ime.ac.cn (Y.L.)

Abstract: The interface passivation of the HfO$_2$/Si$_{0.7}$Ge$_{0.3}$ stack is systematically investigated based on low-temperature ozone oxidation and Si-cap methods. Compared with the Al$_2$O$_3$/Si$_{0.7}$Ge$_{0.3}$ stack, the dispersive feature and interface state density (D$_{it}$) of the HfO$_2$/Si$_{0.7}$Ge$_{0.3}$ stack MOS (Metal-Oxide-Semiconductor) capacitor under ozone direct oxidation (pre-O sample) increases obviously. This is because the tiny amounts of GeO$_x$ in the formed interlayer (IL) oxide layer are more likely to diffuse into HfO$_2$ and cause the HfO$_2$/Si$_{0.7}$Ge$_{0.3}$ interface to deteriorate. Moreover, a post-HfO$_2$-deposition (post-O) ozone indirect oxidation is proposed for the HfO$_2$/Si$_{0.7}$Ge$_{0.3}$ stack; it is found that compared with pre-O sample, the D$_{it}$ of the post-O sample decreases by about 50% due to less GeO$_x$ available in the IL layer. This is because the amount of oxygen atoms reaching the interface of HfO$_2$/Si$_{0.7}$Ge$_{0.3}$ decreases and the thickness of IL in the post-O sample also decreases. To further reduce the D$_{it}$ of the HfO$_2$/Si$_{0.7}$Ge$_{0.3}$ interface, a Si-cap passivation with the optimal thickness of 1 nm is developed and an excellent HfO$_2$/Si$_{0.7}$Ge$_{0.3}$ interface with D$_{it}$ of 1.53×10^{11} eV^{-1}cm^{-2} @ E−E$_v$ = 0.36 eV is attained. After detailed analysis of the chemical structure of the HfO$_2$/IL/Si-cap/Si$_{0.7}$Ge$_{0.3}$ using X-ray photoelectron spectroscopy (XPS), it is confirmed that the excellent HfO$_2$/Si$_{0.7}$Ge$_{0.3}$ interface is realized by preventing the formation of Hf-silicate/Hf-germanate and Si oxide originating from the reaction between HfO$_2$ and Si$_{0.7}$Ge$_{0.3}$ substrate.

Keywords: HfO$_2$/Si$_{0.7}$Ge$_{0.3}$ gate stack; ozone oxidation; Si-cap; interface state density; passivation

1. Introduction

High-mobility channel materials and novel device architectures, such as FinFETs (Fin Field-Effect Transistor) and nanowire FETs, are proposed to address the demand for scaling CMOS (Complementary Metal-Oxide-Semiconductor) technology [1,2]. In contrast to other potential materials, such as germanium (Ge) or III–V materials, silicon germanium (SiGe) is considered the most promising channel material for PMOS due to its tunability of band gaps and high hole mobility [3]. However, one of the main challenges in integrating SiGe into the novel devices is obtaining a high-quality interlayer (IL) between high-k gate oxide and SiGe substrate.

To control the interface quality, many methods have been extensively explored, such as plasma (N$_2$ or NH$_3$) nitridation passivation [4,5], sulfur passivation [6], thermal oxidation [7,8], low-temperature ozone passivation [9–12] and Si-cap passivation [13]. Among them, low-temperature ozone passivation with low thermal budge and Si-cap passivation with excellent properties of interface are considered the most promising passivation methods. For example, the interface state density (D$_{it}$) of 2.2×10^{12} eV^{-1}cm^{-2}

is attained by using a low-temperature ozone oxidation to passivate the interface of $Al_2O_3/Si_{0.7}Ge_{0.3}$ [11], and the D_{it} of 2×10^{11} eV^{-1}cm^{-2} for the interface of $HfO_2/Si_{0.8}/Ge_{0.2}$ is realized by using a Si-cap passivation method [14]. However, the technique and mechanism of interface passivation of the $HfO_2/SiGe$ via low-temperature ozone oxidation or Si-cap method still needs further investigation.

In this paper, we fabricated $HfO_2/IL/Si_{0.7}Ge_{0.3}$ gate stacks MOS capacitors by utilizing low-temperature ozone oxidation and Si cap passivation methods. We carefully compared their electrical properties, and the chemical structure of $HfO_2/IL/SiGe$ gate stacks. It is found that the post-HfO_2-depositon (post-O) ozone indirect oxidation is a better choice than a step-by-step procedure (pre-O) method in terms of D_{it} reduction. More importantly, the optimal Si cap method can realize a lower D_{it} of 1.53×10^{11} eV^{-1}cm^{-2} @ $E-E_v = 0.36$ eV by preventing the formation of Hf-silicate/Hf-germanate and Si oxide originating from the reaction between HfO_2 and $Si_{0.7}Ge_{0.3}$ substrate.

2. Materials and Methods

After standard HF-last cleaning, the 30 nm $Si_{0.7}Ge_{0.3}$ layer was epitaxially grown in a reduced pressure chemical vapor deposition system (ASM E2000 plus, Amsterdam, The Netherlands) on an 8-inch Si substrate. The low-temperature ozone passivation or Si-cap passivation was employed to passivate the interface of $HfO_2/Si_{0.7}Ge_{0.3}$. For low-temperature ozone passivation samples, the ozone oxidation can occur on the $Si_{0.7}Ge_{0.3}$ surface directly (step-by-step procedure (pre-O)) or post HfO_2 deposition (post-O). The ozone oxidation was carried out in 10% O_3/O_2 mixture ambience with the pressure of 3.1 Torr in an atomic-layer-deposition (ALD) chamber (Beneq TFS 200 system, Espoo, Finland). The temperature of the ozone oxidation was 300 °C. For Si-cap passivation, a Si-cap layer was in situ formed on the epitaxial $Si_{0.7}Ge_{0.3}$ layer in the same chamber. After the passivation treatment, the W/TiN or W/TiN/HfO_2 gate stack was deposited as the gate stack of MOS capacitors. Finally, W/TiN/HfO_2/IL/$Si_{0.7}Ge_{0.3}$ MOS capacitors were annealed in the forming gas (10% H_2, 90% N_2) at 350 °C for 30 min.

The chemical structures of the $HfO_2/IL/Si_{0.7}Ge_{0.3}$ stacks were studied by X-ray photoelectron spectroscopy (XPS), which was carried out in a Thermo Scientific ESCALAB 250xi (Waltham, MA, USA) system with a photon energy of 1486.7 eV (Al Kα source). The photoelectron emission take-off angle was 90° relative to the sample surface and the pass energy was 15 eV. Moreover, TEM (Transmission Electron Microscope) and EDX (Energy Dispersive X-Ray Spectroscopy) Mapping analysis were performed by using FEI Talos F200X (Hillsboro, MI, USA) to verify the gate stack lattice structure and element content. Multi-frequency capacitance-voltage (C-V) along with conductance-voltage (G-V) measurements were measured using a Keysight 4990 A (Santa Rosa, CA, USA), and leakage-voltage (I-V) was measured using an Agilent B-1500 semiconductor analyzer.

3. Results and Discussion

3.1. Low-Temperature Ozone Oxidation Passivation of $HfO_2/Si_{0.7}Ge_{0.3}$ Interface

In our previous work, the low-temperature ozone oxidation passivation method has been studied in detail based on $Al_2O_3/Si_{0.7}Ge_{0.3}$ gate stacks. It was found that oxidation time played an important role to obtain a high-quality interlayer (IL) and should be at least 5 minutes. Otherwise, the unoxidized Ge atoms would be trapped in the IL, causing the IL as well as the relevant electrical properties to deteriorate. Moreover, increasing oxidation time would result in an increase in the ratio of Si^{4+} to Si^{3+} of the oxide interlayer, which can help decrease the D_{it} [15]. Thus, we chose 30 min as the oxidation time, which has proven to be an optimal experimental condition, to passivate the $HfO_2/Si_{0.7}Ge_{0.3}$ interface in this work.

Figure 1a,b depicts the multi-frequency (1 kHz to 1 MHz) C-V characteristics of W/TiN/Al_2O_3/IL/$Si_{0.7}Ge_{0.3}$ (Al_2O_3 sample) and W/TiN/HfO_2/IL/$Si_{0.7}Ge_{0.3}$ (HfO_2-pre-O sample) MOS capacitors treated with 30 min ozone direct oxidation, respectively. The flat band voltages (V_{fb}) are also shown in the figures. The frequency dispersion features of the C–V curves observed at gate biases smaller than the V_{fb}, are caused by trapping and de-trapping of holes at traps with energies between approximately mid-gap and the $Si_{0.7}Ge_{0.3}$ valence band edge, corresponding to the depletion of the $Si_{0.7}Ge_{0.3}$ substrate. Comparing Figure 1b with Figure 1a, it is observed that the dispersion feature increases considerably. The energy distributions of the interface state density (D_{it}) were extracted using the conductance method [16], and given in their respective inset in Figure 1. We can see that both of the D_{it} of the two samples decreases along with SiGe band gap energy and the maximum D_{it} values appear near the valence band edge (E_v). However, the maximum value increases from 3.96×10^{12} eV^{-1}cm^{-2} for the Al_2O_3 sample to 2.67×10^{13} eV^{-1}cm^{-2} for the HfO_2-pre-O sample. According to our previous work [17], it is known that for 300 °C/30 min ozone oxidation, about 54% of the Ge atoms of the outermost atomic layer of $Si_{0.7}Ge_{0.3}$ can be oxidized in the initial stage of oxidation. No more Ge atoms would take part in the oxidation process as the oxidation time increases. The GeO_x and SiO_x thickness of the formed oxide layer are estimated to be 0.15 nm and 0.72 nm, respectively. Compared with Al_2O_3, GeO_x is more likely to diffuse into HfO_2 and cause the HfO_2/SiGe interface to deteriorate [18]. Therefore, the increased D_{it} of the HfO_2-pre-O sample can be attributed to tiny amounts of GeO_x in the formed oxide layer.

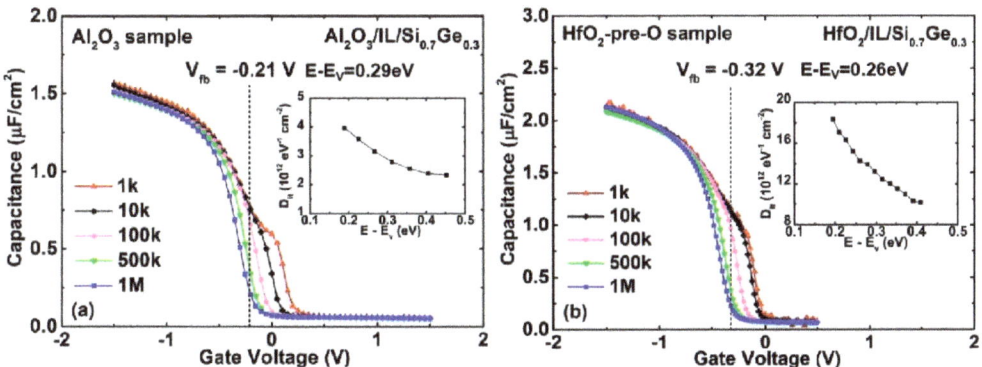

Figure 1. Multi-frequency C-V characteristics of (a) Al_2O_3 sample (b) HfO_2-pre-O sample with 30 min oxidation time (direct). The insets are their respective energy distributions of D_{it}.

Figure 2 depicts the multi-frequency (1 kHz to 1 MHz) C-V characteristics of W/TiN/HfO_2/IL/$Si_{0.7}Ge_{0.3}$ (HfO_2-post-O sample) MOS capacitor treated with 30 min ozone indirect oxidation, in which the ozone oxidation was carried out after the deposition of HfO_2. The corresponding energy distributions of D_{it} is also given in the inset. Compared with Figure 1b, an obvious improvement in the frequency dispersion feature is observed, and the D_{it} value decreases by about 50%. We infer that the improvement may arise from the following two factors. First, due to the barrier effect of the HfO_2 layer on the diffusion of the oxidizer, the amount of oxygen atoms reaching the interface becomes fewer. Because silicon oxidation is more favorable than germanium oxidation in view of thermodynamic considerations [19], germanium atoms are hardly oxidized in this case. Therefore, almost no GeO_x would diffuse into HfO_2 layer. In addition, the IL thickness of the HfO_2-post-O sample is smaller than that of the HfO_2-pre-O sample, which means the amounts of the Ge atoms accumulating at the IL/SiGe interface decrease accordingly. The experimental results prove that the post-O method is a promising technology to realize an HfO_2/IL/SiGe gate stack with small D_{it}.

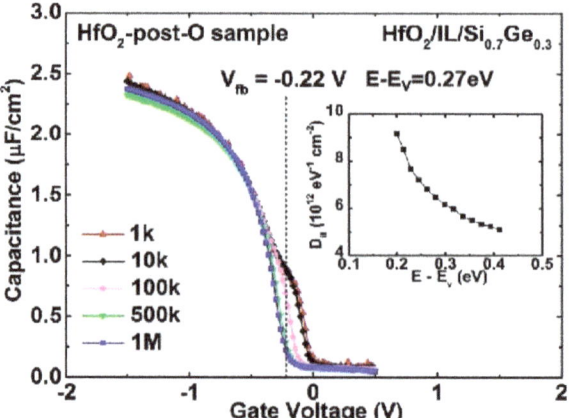

Figure 2. Multi-frequency C-V characteristics of HfO$_2$-post-O sample with 30 min oxidation time (indirect). The corresponding energy distributions of D_{it} is given in the inset.

The Al$_2$O$_3$ sample, HfO$_2$-pre-O sample and HfO$_2$-post-O sample were compared on capacitance equivalent oxide thickness (CET) at −1.5 V bias voltage in accumulation. The CETs of each are 2.28 nm, 1.5 nm and 1.37 nm respectively. Comparing the Al$_2$O$_3$ sample with the HfO$_2$, the CET of the Al$_2$O$_3$ sample is bigger. The HfO$_2$-post-O sample decreased the CET, compared to the HfO$_2$-pre-O sample. This is supposed to be related to the diffusion of GeO$_x$. In general, the diffusion of GeO$_x$ is less in Al$_2$O$_3$ and HfO$_2$-post-O. Using the post-O method can limit the diffusion of GeO$_x$ in HfO$_2$. The diffusion of GeO$_x$ affects not only the CET but also the leakage current.

Figure 3 shows the gate Leakage of the Al$_2$O$_3$ sample, HfO$_2$-pre-O sample and HfO$_2$-post-O sample. Because GeO$_x$ is not easily diffused in Al$_2$O$_3$, the leakage current is minimal for the Al$_2$O$_3$ sample. Comparing with the HfO$_2$-pre-O sample, the leakage current HfO$_2$-post-O sample can be reduced by an order of magnitude.

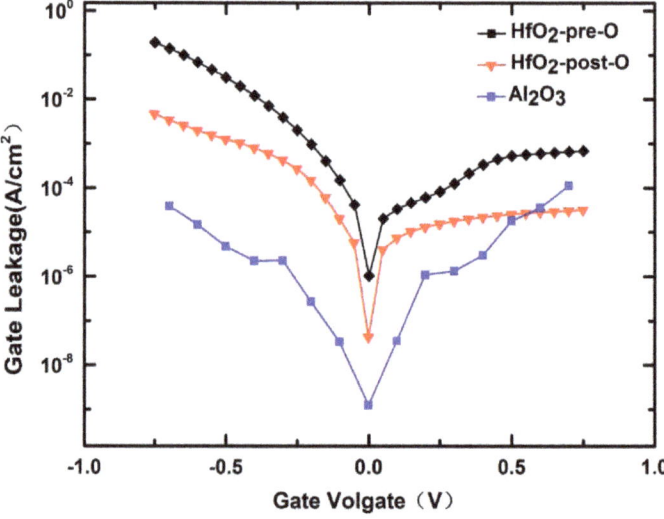

Figure 3. Gate leakage of Al$_2$O$_3$ sample, HfO$_2$-pre-O sample and HfO$_2$-post-O sample.

3.2. Si-Cap Passivation of $HfO_2/Si_{0.7}Ge_{0.3}$ Interface

To further reduce the D_{it} of the $HfO_2/Si_{0.7}Ge_{0.3}$ interface, Si-cap passivation is in situ performed on the $Si_{0.7}Ge_{0.3}$ layer with different thicknesses. It is found that if the Si cap thickness is larger than or equal to 2 nm, there is a step observed in its C-V curve because a second channel is formed in the Si cap layer. This can be avoided by further thinning of the Si cap layer to 1 nm. Moreover, multi-frequency C-V curves (1 kHz to 1 MHz) of the $W/TiN/HfO_2/IL/Si\text{-}cap/Si_{0.7}Ge_{0.3}$ MOS capacitor with 1 nm Si-cap are measured and shown in Figure 3. It is worthy to note that the frequency dispersive feature is obviously improved compared with the above ozone passivation. However, the CET of the Si-cap sample from Figure 4 may be inaccurate due to the large gate leakage in the accumulation region. In addition, it can be seen that the carriers are mainly confined in the $Si_{0.7}Ge_{0.3}$ layer under this optimal Si-cap thickness due to its large valance band offset. For quantitative analysis, the D_{it} of 1.53×10^{11} $eV^{-1}cm^{-2}$ @ $E-E_v = 0.36$ eV is attained by using the conductance method. Meanwhile, HRTEM, Si and Ge element EDX mapping of the $W/TiN/HfO_2/IL/Si\text{-}cap/Si_{0.7}Ge_{0.3}$ MOS capacitor with 1nm Si-cap is also implemented and shown in Figure 5. It is found that there is a ~0.6 nm Si capping on the $Si_{0.7}Ge_{0.3}$ with a smooth and high-quality interfacial layer. The reduction of Si cap thickness of 0.4 nm is due to the oxidation of Si cap layer in the process of MOS capacitor fabrication. Therefore, 1-nm Si-cap in situ epitaxial grown is chosen as the optimal Si-cap thickness.

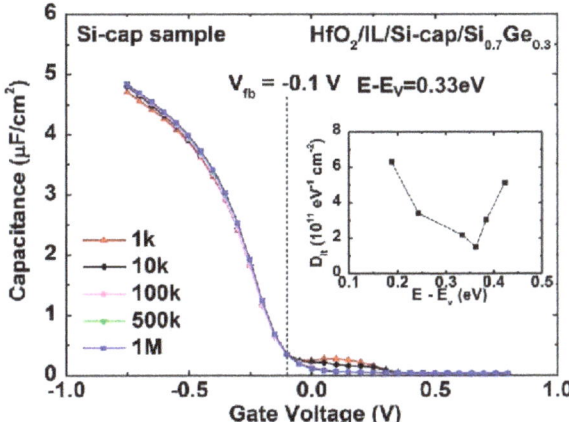

Figure 4. Multi-frequency C-V characteristic of $W/TiN/HfO_2/IL/Si\text{-}cap/Si_{0.7}Ge_{0.3}$ MOS capacitor with 1 nm Si-cap.

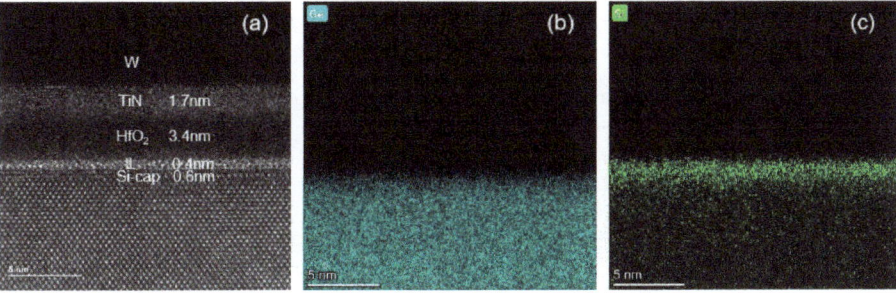

Figure 5. (a) HRTEM, (b) Ge, and (c) Si element EDX mapping of the $W/TiN/HfO_2/IL/Si\text{-}cap/Si_{0.7}Ge_{0.3}$ MOS capacitor with 1 nm Si-cap.

For the purpose of investigating the chemical structure of the HfO_2/IL/Si-cap/$Si_{0.7}Ge_{0.3}$ gate stack (Si-cap sample), X-ray photoelectron spectroscopy (XPS) technology is implemented. The chemical structure of the HfO_2/$Si_{0.7}Ge_{0.3}$ gate stack (SiGe sample), in which HfO_2 is deposited on $Si_{0.7}Ge_{0.3}$ directly, is also analyzed as a control sample. Gaussian-Lorentzian line shapes are used for deconvolution of all the spectra after standard Shirley background subtraction [20]. Figure 6a,b shows the Hf 4f core-level spectra of the Si-cap sample and SiGe sample, respectively. The spectra are both fitted with two component peaks. For the Si-cap sample (shown in Figure 6a), the Hf 4f spectrum consists of a main component at 16.8 eV related to the Hf-O bands in HfO_2, and a second component shifted by ~0.9 eV to higher binding energy, which is from the Hf-O-Si and/or Hf-O-Ge bonds. Because the electro-negativities of the Hf second neighbors (i.e., Si and Ge) are similar, it is difficult to distinguish the two contributions of Hf-O-Si and Hf-O-Ge bonds by XPS. It is worth noting that for the SiGe sample (shown in Figure 6b), the areal intensity of Hf-O-Si/Hf-O-Ge is much more than that of Hf-O. This suggests that a large portion of HfO_2 would react with SiGe to form Hf-silicate/Hf-germanate during the HfO_2 ALD deposition process. In addition, no feature of lower banding energy (14.3 eV–14.8 eV) is detected, indicating that no metallic Hf-Si and/or Hf-Ge are formed in the two samples.

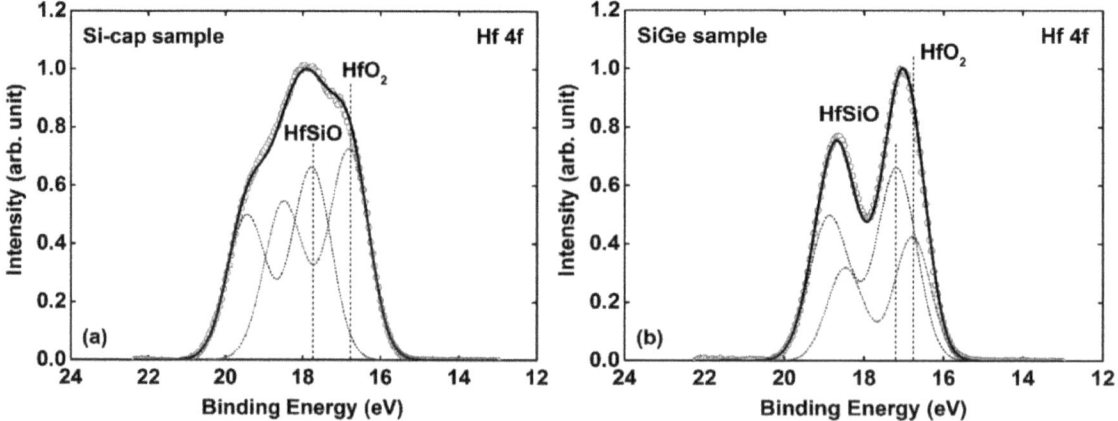

Figure 6. The fitted Hf 4f core-level spectra of (**a**) Si-cap sample and (**b**) SiGe sample. The blue and red dot lines denote the Hf 4f photoelectron from Hf-O-Si and/or Hf-O-Ge bonds and Hf-O bonds in HfO_2, respectively.

Figure 7a,b shows the Si 2p core-level spectra of the Si-cap sample and the SiGe sample, respectively. The spectra are decomposed into four component peaks i.e., Si 2p photoelectron from SiGe (~99.7 eV), SiO_x (~101.2 eV), HfSiO (~102.8 eV), and SiO_2 (~103.9 eV). For the Si-cap sample (shown in Figure 7a), the high-binding energy shoulder (101 eV~105 eV) contains few amounts of Si oxide (SiO_x and SiO_2) and Hf-silicate (HfSiO). When compared with the Si-cap sample, an obvious increase in the areal intensity of the high-binding energy shoulder (101 eV~105 eV) can be observed for the SiGe sample, and there is no peak corresponding SiO_x. Figure 8a,b shows the O 1s core-level spectra of the SiGe sample and Si-cap sample, respectively. The spectra are fitted by the O 1s of SiO_x (~532.8 eV), HfSiO (~532.08 eV) and HfO_2 (~531 eV). We can see that the O 1s photoelectron mainly originates from HfO_2 for the Si-cap sample, while that of the SiGe sample is mainly from SiO_x and HfSiO. This is consistent with the previous discussions about Hf 4f and Si 2p spectra. All of these results indicate that the interfacial region of the HfO_2/SiGe (SiGe sample) is a composite of large amounts of HfSiO (and/or HfGeO) and Si oxide (SiO_2). In other words, Si-cap can prevent the formation of Hf-silicate/Hf-germanate and Si oxide originating from the reaction between HfO_2 and SiGe substrate, and obtain an excellent HfO_2/SiGe interface.

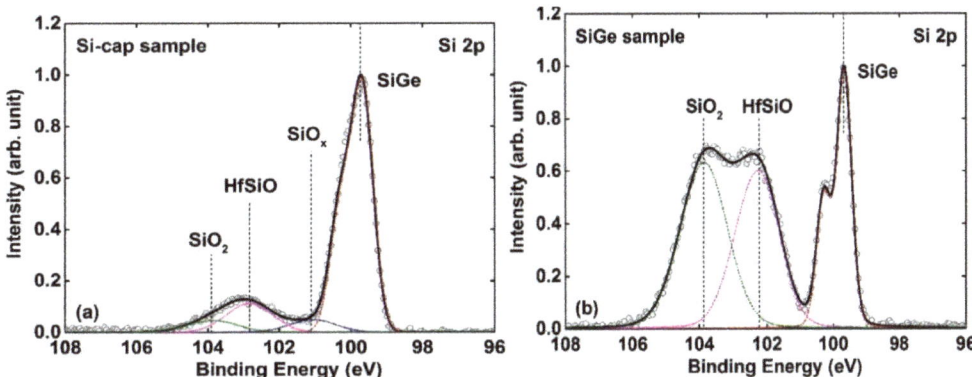

Figure 7. The fitted Si 2p core-level spectra of (**a**) Si-cap sample (**b**) SiGe sample. The red, blue, magenta, and green dot lines denote the Si 2p of SiGe, SiO$_x$, HfSiO, and SiO$_2$, respectively.

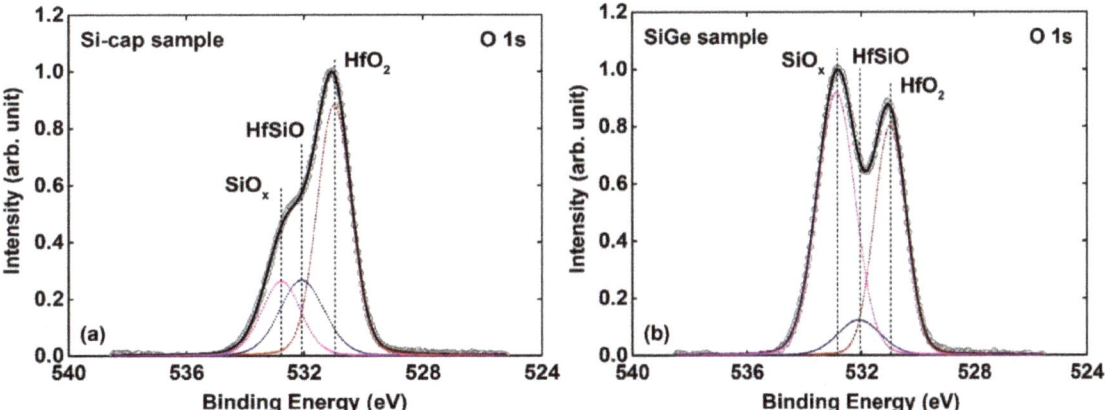

Figure 8. The fitted O 1s core-level spectra of (**a**) Si-cap sample (**b**) SiGe sample. The red, blue, and magenta dot lines denote the O1s photoelectron from HfO$_2$, HfSiO, and SiO$_x$, respectively.

4. Conclusions

In summary, the interface passivation of the HfO$_2$/Si$_{0.7}$Ge$_{0.3}$ stack is systematically investigated based on low-temperature ozone oxidation and Si-cap methods. Compared with pre-O method, the D$_{it}$ of the post-O sample decreases by about 50% due to less GeO$_x$ available in the IL layer. However, the D$_{it}$ of the HfO$_2$/IL/Si$_{0.7}$Ge$_{0.3}$ gate stack still has room to be further optimized. Finally, an excellent HfO$_2$/Si$_{0.7}$Ge$_{0.3}$ interface with a D$_{it}$ of 1.53×10^{11} eV^{-1}cm^{-2} @ E−E$_v$ = 0.36 eV is attained under the optimal Si cap method by preventing the formation of Hf-silicate/Hf-germanate and Si oxide from the reaction HfO$_2$ and Si$_{0.7}$Ge$_{0.3}$ substrate.

Author Contributions: Conceptualization, X.M., Y.W. and Y.L.; methodology, Q.Y., X.M., H.W., Y.W., G.W., J.Z., W.L., X.W., J.Y., Y.L. and W.W.; investigation, Q.Y., X.M., Y.W. and Y.L.; data curation, Q.Y., X.M., H.W., Y.W. and Y.L.; writing original draft preparation, Q.Y., X.M., Y.W. and Y.L.; writing review and editing, X.M., Y.W., Y.L. and J.Z.; supervision, J.Y., Y.L. and W.W.; project administration, Y.L. and W.W.; funding acquisition, W.W. All authors have read and agreed to the published version of the manuscript.

Funding: This research was funded in part by the Science and Technology Program of Beijing Municipal Science and Technology Commission (Grant no. Z201100004220001), in part by the CAS Pioneer Hundred Talents Program, in part by Beijing Municipal Natural Science Foundation (Grant no. 4202078), in part by National Natural Science Foundation of China (Grant no. 62074160) and in part by Scientific Research Startup Foundation of North China University of Technology.

Institutional Review Board Statement: Not applicable.

Informed Consent Statement: Not applicable.

Data Availability Statement: Not applicable.

Acknowledgments: We thank the Integrated Circuit Advanced Process Center (ICAC) at the Institute of Microelectronics of the Chinese Academy of Sciences for the devices fabricated on their advanced 200 mm CMOS platform.

Conflicts of Interest: The authors declare no conflict of interest.

References

1. Mertens, H.; Ritzenthaler, R.; Arimura, H.; Franco, J.; Sebaai, F.; Hikavyy, B.A.; Pawlak, J.; Machkaoutsan, V.; Devriendt, K.; Tsvetanova, D.; et al. Si-cap-free SiGe p-Channel FinFETs and Gate-All-Around Transistors in a Replacement Metal Gate Process: Interface Trap Density Reduction and Performance Improvement by High-Pressure Deuterium Anneal. In Proceedings of the 2015 Symposium on VLSI Technology, Kyoto, Japan, 16–18 June 2015.
2. Hashemi, P.; Balakrishnan, K.; Engelmann, S.U.; Ott, J.A.; Khakifirooz, A.; Baraskar, A.; Hopstaken, M.; Newbury, J.S.; Chan, K.K.; Leobandung, E.; et al. First Demonstration of High-Ge-Content Strained-Si1-xGex(x = 0.5) on Insulator PMOS FinFETs with High Hole Mobility and Aggressively Scaled Fin Dimensions and Gate Lengths for High-Performance Applications. In Proceedings of the 2014 IEEE International Electron Devices Meeting, San Francisco, CA, USA, 15–17 December 2014.
3. Hashemi, P.; Ando, T.; Balakrishnan, K.; Bruley, J.; Engelmann, S.; Ott, J.A.; Narayanan, V.; Park, D.-G.; Mo, R.T.; and Leobandung, E. High-Mobility High-Ge-Content Si1-xGex-OI PMOS FinFETs with Fins Formed Using 3D Germanium Condensation with Ge Fraction up to x~0.7, Scaled EOT~8.5Å and ~10nm Fin Width. In Proceedings of the 2015 Symposium on VLSI Circuits, Kyoto, Japan, 17–19 June 2015.
4. Han, J.; Zhang, R.; Osada, T.; Hata, M.; Takenaka, M.; Takagi, S. Impact of plasma post-nitridation on $HfO_2/Al_2O_3/SiGe$ gate stacks toward EOT scaling. *Microelectron. Eng.* **2013**, *109*, 266–269. [CrossRef]
5. Sardashti, K.; Hu, K.-T.; Tang, K.; Madisetti, S.; McIntyre, P.; Oktyabrsky, S.; Siddiqui, S.; Sahu, B.; Yoshida, N.; Kachian, J.; et al. Nitride passivation of the interface between high-k dielectrics and SiGe. *Appl. Phys. Lett.* **2016**, *108*, 011604. [CrossRef]
6. Sardashtia, K.; Hua, K.T.; Tang, K.; Parka, S.; Kim, H.; Madisetti, S.; Oktyabrsky, S.; Siddiqui, S.; Sahu, B.; Yoshida, N.; et al. Sulfur passivation for the formation of Si-terminated $Al_2O_3/SiGe(0\,0\,1)$ interfaces. *Appl. Surf. Sci.* **2016**, *366*, 455–463. [CrossRef]
7. Hellberg, P.-E.; Zhang, S.-L.; d'Heurle, F.M.; Petersso, C.S. Oxidation of silicon–germanium alloys. I. An experimental study. *J. Appl. Phys.* **1997**, *82*, 5773–5778.
8. Masanori, T.; Tatsuo, O.; Taizoh, S.; Miyao, M. Comprehensive study of low temperature (<1000 °C) oxidation process in SiGe/SOI structures. *Thin Solid Films* **2008**, *517*, 251–253.
9. Song, Y.J.; Mheen, B.; Kang, J.Y.; Lee, Y.S.; Lee, N.E.; Kim, J.H.; Song, J.I.; Shim, K.-H. A low-temperature and high-quality radical-assisted oxidation process utilizing a remote ultraviolet ozone source for high-performance SiGe/Si MOSFETs. *Semicond. Sci. Technol.* **2004**, *19*, 792–797. [CrossRef]
10. Ando, T.; Hashemi, P.; Bruley, J.; Rozen, J.; Ogawa, Y.; Koswatta, S.; Chan, K.K.; Cartier, E.A.; Mo, R.; Narayanan, V. High Mobility High-Ge-Content SiGe PMOSFETs Using Al_2O_3/HfO_2 Stacks with In-Situ O_3 Treatment. *IEEE Electron. Device Lett.* **2017**, *38*, 303–305. [CrossRef]
11. Ma, X.L.; Xiang, J.J.; Zhou, L.X.; Wang, X.L.; Li, Y.L.; Yang, H.; Zhang, J.; Zhao, C.; Yin, H.X.; Wang, W.W.; et al. Comprehensive Study and Design of High-k/SiGe Gate Stacks with Interface-Engineering by Ozone Oxidation. *ECS J. Solid-State Sci. Technol* **2019**, *8*, N100–N105. [CrossRef]
12. Ma, X.L.; Xiang, J.J.; Zhou, L.X.; Xu, H.; Wang, X.L.; Yang, H.; Li, Y.L.; Yin, H.X.; Wang, W.W. Understanding mechanisms impacting interface states of ozone-treated high-k/SiGe interfaces. *Semicond. Sci. Technol.* **2020**, *35*, 055018. [CrossRef]
13. Yeh, W.K.; Chen, Y.T.; Huang, F.S.; Hsu, C.W.; Chen, C.Y.; Fang, Y.K.; Gan, K.J.; Chen, P.Y. The Improvement of High-k/Metal Gate pMOSFET Performance and Reliability Using Optimized Si Cap/SiGe Channel Structure. *IEEE Trans. Device Mater. Reliab.* **2011**, *11*, 7–12. [CrossRef]
14. Tsutsui, G.; Durfee, C.; Wang, M.M.; Konar, A.; Wu, H.; Mochizuki, S.; Bao, R.Q.; Bedell, S.; Li, J.T.; Zhou, H.M.; et al. Leakage Aware Si/SiGe CMOS FinFET for Low Power Applications. In Proceedings of the 2008 IEEE Symposium on VLSI Technology, Honolulu, HI, USA, 18–22 June 2018.
15. Ma, X.L.; Zhou, L.X.; Xiang, J.J.; Yang, H.; Wang, X.L.; Li, Y.L.; Zhang, J.; Zhao, C.; Yin, H.X.; Wang, W.W.; et al. Identification of a suitable passivation route for high-k/SiGe interface based on ozone oxidation. *Appl. Surf. Sci.* **2019**, *493*, 478–484. [CrossRef]

16. Nicollian, E.H.; Brews, J.R. *MOS (Metal Oxide Semiconductor) Physics and Technology*; John Wiley & Sons Inc.: Hoboken, NJ, USA, 1982; p. 176.
17. Ma, X.L.; Wang, X.L.; Zhou, L.X.; Xu, H.; Zhang, Y.Y.; Duan, J.H.; Xiang, J.J.; Yang, H.; Li, J.J.; Li, Y.L.; et al. Experimental study of the ultrathin oxides on SiGe alloy formed by low-temperature ozone oxidation. *Mater. Sci. Semicond. Process.* **2020**, *107*, 104832. [CrossRef]
18. Gusev, E.P.; Shang, H.; Copel, M.; Gribelyuk, M.; D'Emic, C.; Kozlowski, P.; Zabel, T.; Gusev, E.P.; Shang, H.; Copel, M.; et al. Microstructure and thermal stability of HfO_2 gate dielectric deposited on Ge(100). *Appl. Phys. Lett.* **2004**, *85*, 2334. [CrossRef]
19. Chang, C.T.; Toriumi, A. Preferential Oxidation of Si in SiGe for Shaping Ge-Rich SiGe Gate Stacks. In Proceedings of the 2015 IEEE International Electron Devices Meeting (IEDM), Washington, DC, USA, 7–9 December 2015.
20. Shirley, D.A. High-Resolution X-ray Photoemission Spectrum of the Valence Bands of Gold. *Phys. Review B* **1972**, *5*, 4709. [CrossRef]

Article

Epitaxial Growth of Ordered In-Plane Si and Ge Nanowires on Si (001)

Jian-Huan Wang [1,2], Ting Wang [1,3] and Jian-Jun Zhang [1,3,*]

1. Beijing National Laboratory for Condensed Matter Physics and Institute of Physics, Chinese Academy of Sciences, Beijing 100190, China; jhwang1@iphy.ac.cn (J.-H.W.); wangting@iphy.ac.cn (T.W.)
2. School of Physical Sciences, University of Chinese Academy of Sciences, Beijing 100190, China
3. Songshan Lake Materials Laboratory, Dongguan 523808, China
* Correspondence: jjzhang@iphy.ac.cn

Abstract: Controllable growth of wafer-scale in-plane nanowires (NWs) is a prerequisite for achieving addressable and scalable NW-based quantum devices. Here, by introducing molecular beam epitaxy on patterned Si structures, we demonstrate the wafer-scale epitaxial growth of site-controlled in-plane Si, SiGe, and Ge/Si core/shell NW arrays on Si (001) substrate. The epitaxially grown Si, SiGe, and Ge/Si core/shell NW are highly homogeneous with well-defined facets. Suspended Si NWs with four {111} facets and a side width of about 25 nm are observed. Characterizations including high resolution transmission electron microscopy (HRTEM) confirm the high quality of these epitaxial NWs.

Keywords: in-plane nanowire; site-controlled; epitaxial growth; silicon; germanium; nanowire-based quantum devices

1. Introduction

Si and Ge nanowires (NWs) have potential applications for high-performance transistors [1,2] and for disruptively quantum computation technology [3–6]. The controllable growth of NW arrays in wafer-scale remains the major challenge for large scale integration. The top-down method by patterning and etching can precisely fabricate NWs in wafer-scale but also induce additional defects during the nanofabrications. For instance, IMEC has previously reported the vertically stacked horizontal Si NWs with selective etching of Si/SiGe multilayer fin structures [1]. Moreover, by selectively etching Si, stacked SiGe NWs were obtained to improve the channel mobility [7]. However, the top-down fabrication introduces atomic surface roughness and damages, which deteriorate the carrier mobility of the NWs [8].

Alternatively, the self-assembled growth of NWs via a vapor-liquid-solid (VLS) mechanism can form high quality NWs with a sharp interface [9,10]. A mobility of 730 cm^2(Vs)$^{-1}$ [11] and a ballistic conduction up to several hundred nanometers [12] were reported in such {111}-oriented Ge/Si core/shell NWs. Compared to the {111}-oriented NWs, {110}-oriented Ge/Si core/shell NWs have substantially enhanced hole mobility as high as 4200 cm^2(Vs)$^{-1}$ at 4 K [13]. Although, the VLS-grown Si and Ge NWs have recently presented single crystalline with controllable orientation [14–18], the out-of-planar geometry has not been compatible with the well-established planar device processing technology. Ex-situ assembly methods such as contact printing and capillary assembly have been developed to align the NWs on a target substrate [19,20], however, for such VLS-grown NWs, the precise positioning at a large scale is a challenge. Another challenge of the VLS-grown NWs is the poor size-controllability (including both length and diameter), which is considered to reduce the collective properties of NWs.

Combining top-down nanofabrication and bottom-up self-assembly, we have recently demonstrated site-controlled growth of Ge hut wires on trench-patterned Si (001) substrate [21]. The Ge hut wires have a height of 3.8 nm with sharp {105} facets specifically

oriented along <100> directions with high scalability. They are grown under a relatively high growth temperature where Si and Ge intermixing leads to a reduced Ge composition in the wires. Therefore, it is desirable to obtain epitaxial Si and Ge NWs with controllable size, orientation, and composition. In this work, we epitaxially grow {110}-orientated in-plane Si, SiGe, and Ge NWs on pre-patterned Si NW arrays. The pre-patterned Si NWs with an inverted trapezoidal structure are obtained through nanofabrications. On such pre-patterned Si NWs, homogeneous Si NWs with controllable sizes are epitaxially grown by molecular beam epitaxy. Furthermore, we demonstrate the formation of the conformal SiGe NWs and Ge NWs with {113} facets on the diamond-shaped Si NWs with {111} facets and truncated Si NWs. By transmission electron microscopy (TEM) characterizations, we investigate the material properties of the NWs mentioned above, which exhibit a high quality.

2. Materials and Methods

A CMOS-compatible top-down method was explored here to define the planar Si NW arrays on 200 mm Si (001) wafers. Figure 1 describes the fabrication process: a SiO_2 grating structure is firstly prepared along <110> direction on Si wafer by plasma enhanced vapor deposition, deep ultraviolet lithography, and reactive ion etching. Such SiO_2 grating structure is used as a hard mask for the subsequent wet etching of Si. The SiO_2 grating structure has periods that range from 360 to 440 nm with a constant duty cycle of nearly 1:1 and a depth of 150 nm. After dipping for 5 s in a buffered HF solution (7:1) to remove the native oxide on the exposed Si, a diluted tetramethylammonium hydroxide aqueous solution (TMAH 5%) is used to create the planar Si NWs at 75 °C. The SiO_2 hard mask is finally removed in diluted HF solution.

Figure 1. Schematic of process flow for the trapezoidal Si nanowire (NW) array.

By obtaining these pre-patterned Si NWs, we then studied the direct epitaxial growth of Si NWs, SiGe NWs, and Ge/Si core/shell NWs inside a SiGe molecular beam epitaxy system (Octoplus 500 EBV, MBE-Komponenten, Weil der Stadt, Germany). The patterned wafer was cleaved into 10×10 mm^2 small samples before dipping in a diluted HF solution for deoxidation and hydrogen passivation. To reduce the thermal instability of these tiny pre-patterned NWs, a low-temperature dehydrogenation was performed at 500 °C. The Si epitaxial NWs were obtained after homoepitaxial growth of Si at growth temperatures from 380 °C to 480 °C with a growth rate of 1 Å/s.

The SiGe NWs and Ge NWs were grown on the Si epitaxial NW after deposition 20 nm Si layer at 450 °C and 380 °C, respectively. The SiGe NWs were obtained by depositing 10 nm $Si_{0.5}Ge_{0.5}$ and 10 nm Si at 350 °C, where the growth rate of Si and Ge was 0.5 Å/s. The Ge NWs were obtained by depositing 2 nm Ge at 300 °C with a growth rate of 0.3 Å/s. The Ge/Si core/shell NWs were further formed after the deposition of 3 nm Si capping layer at 300 °C.

Focus ion-beam (FIB) system (NanoLab Helios 600i, FEI, Hillsboro, USA) equipped with high-resolution field-emission scanning electron microscope (SEM) was employed to elucidate the morphology of NWs and prepare the TEM lamellae. Before the FIB-milling, the NW sample was coated with 5 nm Ti and 50 nm Au for protection. TEM was performed to verify the quality of these epitaxial NWs, using a JEOL 2100 plus, operating at 200 kV.

3. Results and Discussion

3.1. Planar Trapezoidal Si NW Arrays

TMAH solution provides anisotropic wet etching for Si, with selectivity more than 1:10 between the Si {111} and Si {100} planes [22]. Therefore, {111}-faceted V-grooves were fabricated along the <110> direction, as shown in the SEM images (Figure 2a,b). In Figure 2a, on the tips of the Si V-grooves, we observed a Si hourglass figure with inverted {111} facets contributing to the SiO_2 hard mask. With optimized etching conditions, the formation of Si NWs with an inverted triangular or trapezoidal shape are achieved. Multiple widths of Si NWs ranging from 20 nm to 40 nm can be fabricated simultaneously on 200 mm Si (001) wafer by varying the pattern sizes. Figure 2a shows trapezoidal Si NWs with a minimum width of approximately 20 nm, while still preserving good uniformity, as confirmed by the surface SEM images, as shown in Figure 2b. The average width of the neck is approximately 3 nm, as shown in the inset of Figure 2a, expected to be facilely isolated by thermal oxidation [23,24]. The lengths of NW arrays are defined ranging from 2 μm up to 2 mm, suggesting a large aspect ratio (length: width) of nearly 10^5.

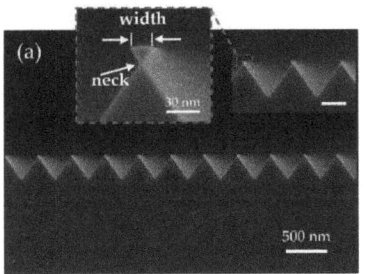 

Figure 2. (a) Cross-sectional view and (b) top view SEM images of the Si NW array with an average wire width of 19 nm. Insets: zoom-in SEM images of the NWs. Scale bar of insets: 300 nm.

3.2. Homoepitaxy of Si NWs

Figure 3a presents a typical NW array by homoepitaxially grown Si on pre-patterned trapezoidal Si NWs. They are highly uniform. The width of these epitaxial Si NWs can be tuned from 30 nm to 50 nm by simply changing the growth conditions. Figure 3b shows the cross-sectional SEM image of epitaxial NWs obtained after the deposition of 20 nm Si layer on 30 nm wide pre-patterned Si NWs at 380 °C. Although only 20 nm Si were deposited at 380 °C, the epitaxial NW evolved rapidly toward the {111}-faceted morphology and a small Si (001) terrace with a width less than 10 nm on the top was left, driven by the reduction of surface energy. We observe a truncated {111}-faceted Si NW with a Si (001) terrace on the top (Figure 3b). By depositing 20 nm Si at 380 °C on a 40 nm wide pre-patterned NW array, a Si (001) terrace with enlarged width of approximately 17 nm was obtained (Figure 3c). If we increase the growth temperature to 450 °C, the Si (001) terrace will evolve into two symmetric Si (111) facets (Figure 4a), which leads to a 33 nm wide diamond-shaped NW. By keeping the growth temperature at 450 °C, when the Si layer is increased to 50 nm, the average width of diamond-shape Si NWs enlarges to approximately 48 nm (Figure 4b). The NWs are characterized by high-resolution TEMs (HRTEMs). Figure 4c provides a cross-sectional HRTEM image of a single Si NW obtained at the identical growth conditions to those in Figure 4b. The green dashed line in Figure 4c represents the interface between the epitaxial layer and the initial hourglass structure (pre-patterned trapezoidal Si NW). The inset of Figure 4c provides a zoom-in HRTEM image of the epitaxial interface labeled in Figure 4c, showing a perfect arrangement of Si atoms. Atoms deposited on the Si hourglass structure diffuse upwards to the shoulder areas to reduce the surface area, as illustrated by black arrows.

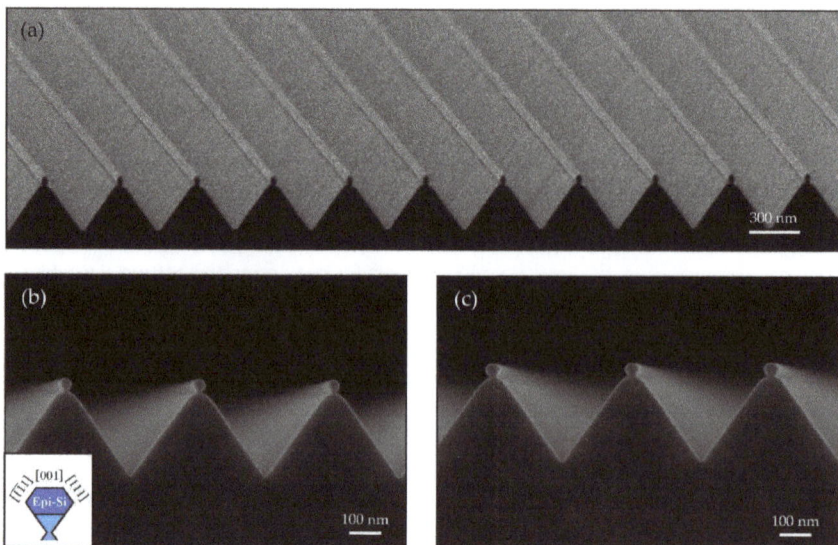

Figure 3. (**a**) Tilted SEM image showing the NW array of epitaxial Si on pre-patterned trapezoidal Si NWs. SEM images of epitaxial Si NWs obtained after the deposition of 20 nm Si at 380 °C on 30 nm wide pre-patterned trapezoidal NWs (**b**) and on 40 nm wide pre-patterned NWs at 380 °C (**c**). Inset of (**b**) schematically shows the truncated {111}-faceted cross-section.

Although the pre-patterned trapezoidal NWs are thermally stable at the aforementioned low-temperature epitaxy, we note that a high-temperature dehydrogenation process at more than 600 °C will deform the pre-patterned Si NWs. The thermal instability becomes remarkable for Si NWs with smaller dimensions [25,26], as we find that the pre-patterned NWs with a size of about 20 nm in Figure 2a deform into discrete Si beads only after 500 °C dehydrogenation. Similar phenomena have been previously reported on an isolated Si NW as Plateau–Rayleigh instability (PRI) [27,28], while the critical temperature reported is much higher at 775 °C for a Si NW with 100 nm diameter. In our case, the root causes of thermal instability are not just dominated by PRI, also strongly influenced by the fragile narrow Si necks as well as the surface diffusion between NWs and patterned V-grooves.

The supporting Si neck of the hourglass structure can significantly affect the thermal instability of the NW growth with small dimensions. Here, we then study the possibility of creating suspended NWs. Figure 5a shows a typical 2 µm long suspended Si trapezoidal NW with a sub-20 nm average width. The supporting Si necks are removed by similar fabrication method mentioned above with over-etched conditions. Absence of the neck, such suspended structure can avoid the diffusion between the NW and the V-groove more effectively. After the growth of the 20 nm Si layer, although the gap between the NW and the pre-patterned V-groove appears to be unclear in the SEM picture (Figure 5b), TEM characterization in Figure 5c has verified that still retains the suspended configuration and forms {111}-faceted diamond-shaped NW with a side width of about 25 nm. Overall, the suspended Si NW exhibit enhanced thermal stability and homogeneity with four {111} facets at small dimensions, which can be considered as an ideal isolated one-dimensional NW system. But these suspended Si NWs are limited to a few micrometers in length, due to insufficient mechanical strength.

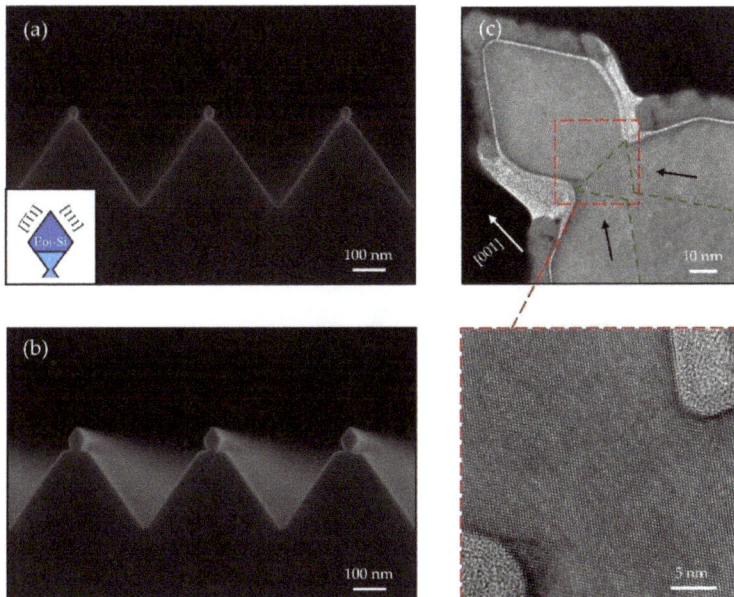

Figure 4. Cross-sectional SEM images of epitaxial Si NWs obtained after the deposition of 20 nm (**a**) and 50 nm (**b**) of Si at 450 °C on 30 nm wide pre-patterned trapezoidal NWs. Inset of (**a**) schematically shows the fully {111}-faceted cross-section. (**c**) Cross-sectional transmission electron microscopy (TEM) image of an epitaxial NW in (**b**), projected toward <110> direction. The interface of epitaxially formed Si NW and initial hourglass structure (pre-patterned Si NW) is sketched in green dashed line. The two shoulder areas marked in black are obtained by atomic diffusion during deposition. Inset of (**c**) shows a zoom-in high resolution transmission electron microscopy (HRTEM) confirming the perfect interface.

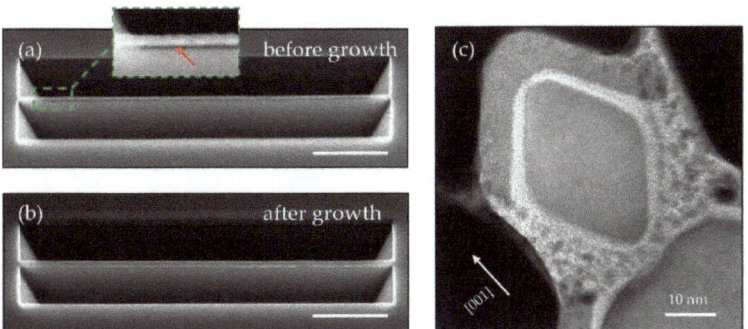

Figure 5. Suspended Si NW before (**a**) and after epitaxy (**b**). They both have a straight structure without distortion. Scale bar: 400 nm. (**c**) Cross-sectional HRTEM showing the high-quality diamond with four {111} facets after epitaxy. The red arrow in the inset of (**a**) highlights the suspended structure.

The size distributions of both the pre-patterned trapezoidal NW and epitaxial NWs were investigated. Figure 6a–c presents the SEM images of the pre-patterned NW and the epitaxial NWs obtained after the deposition of 20 and 50 nm-thick Si, respectively. After epitaxial growth, the rough surface of the pre-patterned NW has been significantly modified by forming atomic {111} facets. As illustrated in Figure 6d, the average width of pre-patterned trapezoidal NWs is 29.8 nm with relative standard deviation of 6.4%. By depositing a 20 nm (50 nm)-thick Si layer, the epitaxially formed Si NWs exhibit average widths of 35.8 nm (46.0 nm), and the relative standard deviation of the width distribution is reduced to 3.9% (2.9%).

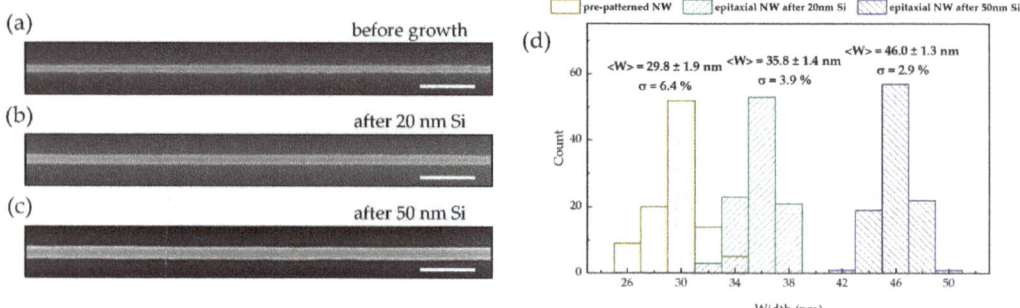

Figure 6. (**a**–**c**) SEM images of a 30 nm wide pre-patterned trapezoidal NW, epitaxial NWs after the deposition of 20 nm Si and 50 nm Si, respectively. Scale bar: 200 nm. (**d**) Statistical histogram showing the width distribution of 30 nm wide pre-patterned NWs, epitaxial NWs after the deposition of 20 nm and 50 nm Si layer. The average width <W> and relative standard deviation σ of the NWs are quoted.

3.3. Epitaxy of SiGe NWs

The epitaxial Si NWs provide platform for the subsequent growth of SiGe and Ge NWs. As mentioned, the SiGe NWs are obtained after the deposition of 10 nm $Si_{0.5}Ge_{0.5}$ and 10 nm Si layer at 350 °C on the {111}-faceted Si NW. We should note that all the thicknesses of the epitaxial layer mentioned in this work are referred to as-grown layer thickness on flat substrate. Here, the actual $Si_{0.5}Ge_{0.5}$ thickness that was deposited on the {111} facets should be 5.77 nm. The SEM images in cross-sectional view (Figure 7a,b) and top view (Figure 7c,d) indicate that these SiGe NWs are highly uniform. Attributed to the high Ge content in the SiGe layer, we can directly distinguish the SiGe layer in the magnified SEM image as shown in Figure 7b, where the SiGe layer has a brighter contrast.

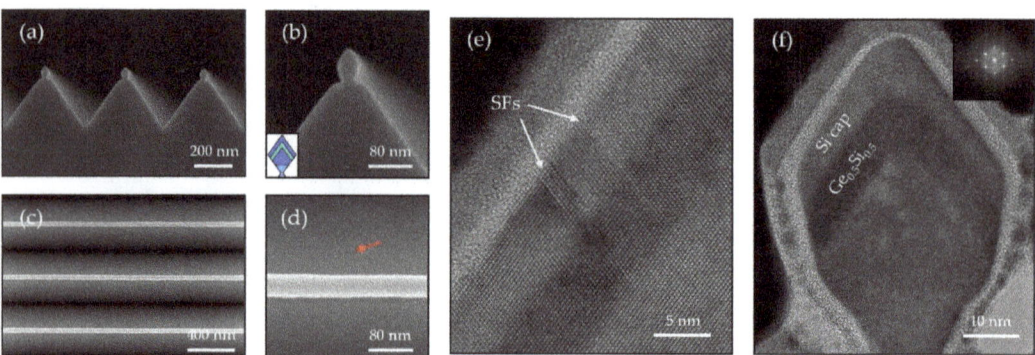

Figure 7. (**a**) Cross-sectional and (**c**) top view SEM images of the $Si_{0.5}Ge_{0.5}$ NW array and (**b**,**d**) the corresponding zoom-in images. The brighter contrast presenting in (**b**) results from the $Si_{0.5}Ge_{0.5}$ layer, where highlights in green in the schematic inset. The red arrow in (**d**) points to a strain-induced defect at the V-groove area. (**e**) Cross-sectional TEM image of SiGe at the V-groove area, showing that stacking faults (SFs) are generated from the interface and penetrate to the surface along the {111} gliding plane. (**f**) Cross-sectional HRTEM of a $Si_{0.5}Ge_{0.5}$ NW. Inset: FFT analysis of the SiGe/Si NW, showing a single set of spots indicating the SiGe is under fully strained condition.

Due to 2.1% lattice mismatch between $Si_{0.5}Ge_{0.5}$ and Si, misfit dislocations will generate if the SiGe film reaches the critical thickness for pseudomorphic growth. From the magnified planar SEM image (Figure 7d), the red arrow indicates strain-induced defects generated at the Si V-groove, indicating the excessive deposition of the SiGe layer. Figure 7e is a cross-sectional TEM image at the Si V-groove, showing that stacking faults (SFs) have generated from the interface and penetrated to the surface along the {111} gliding plane. In addition, we also observed other types of defects including SFs in parallel to the side-wall, attributed to plastic relaxation [29].

The situation is different for the SiGe NW. Figure 7f is a HRTEM image of directly grown in-plane SiGe/Si NW, with absence of defects, indicating the high crystal quality and conformal growth of the SiGe NW. The inset in Figure 7f is the fast Fourier transform (FFT) pattern of the SiGe/Si NW, showing only a single set of diffraction spots without distinct splitting. The FFT pattern is in-line with the spatial measurement result, indicating the SiGe NW is fully strained on Si NW.

3.4. Epitaxy of Ge/Si Core/Shell NWs

Despite a 4.2% lattice-mismatch between Ge and Si, we have further demonstrated Ge NW growth on the truncated {111}-faceted Si NW, where the average width of the Si (001) terrace is about 17 nm. As mentioned, the Ge NWs are obtained after the deposition of 2 nm Ge with a growth rate of 0.3 Å/s. In order to suppress the intermixing between Ge and Si, the growth is performed at a relatively low temperature of 300 °C [30]. Following a 3 nm Si capping layer deposited at 300 °C, Ge/Si core/shell NW is obtained, which can provide a high-performance one-dimensional hole gas system for exploring hole spin qubits [3,4,21]. Figure 8a,b shows cross-sectional and top view SEM images of the Ge/Si core/shell NW arrays, presenting a uniform morphology and smooth surface of NWs. To note, there are also numbers of strain-induced Ge islands formed on the Si V-grooves.

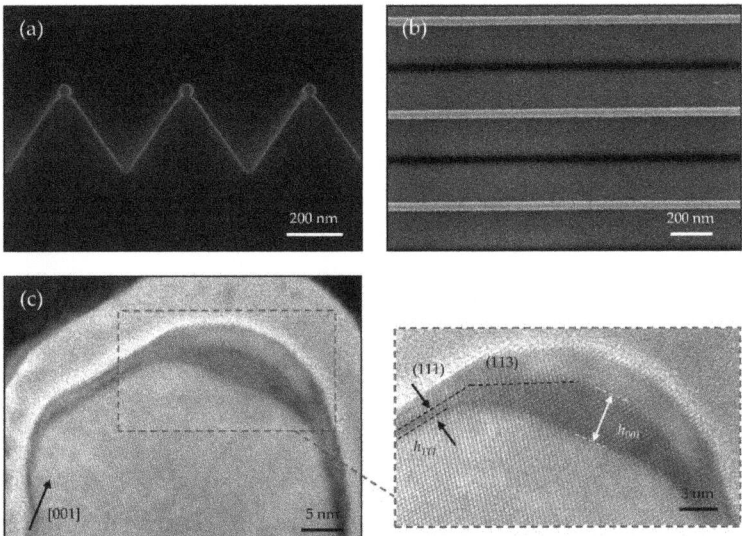

Figure 8. (a) Cross-sectional and (b) top view SEM images of the Ge/Si core/shell NW arrays. (c) Cross-sectional HRTEM image of a Ge/Si core/shell NW. Inset: a zoom-in HRTEM shows the two {113} side facets and the flat (001) top surface.

HRTEM micrograph in Figure 8c shows a typical cross-section of the Ge/Si core/shell NW. The Ge NW is grown on the <110>-oriented Si (001) terrace of the truncated {111}-faceted Si NW. The zoom-in HRTEM image in the inset of Figure 8c presents a trapezoidal geometry of the Ge NW composed of two (113) side facets and a flat (001) top surface. The formation of Ge (113) facets is attributed to the low surface energy, which has been reported in previous works [31–33]. Compared with <100>-oriented Ge hut wires [21,34], these <110>-oriented Ge NWs exhibit a larger aspect ratio of more than 0.2, where the height and width of the Ge NW are about 4 nm and 18 nm, respectively. Comparing the height of the Ge NW on the Si (001) terrace $h_{001} \approx 39.4$ Å and the thickness of the Ge wetting layer on (111) side facets $h_{111} \approx 6.3$ Å, we conclude that there is a significant Ge diffusion from the (111) facet towards the (001) facet. In terms of thermodynamics, Si (001) features higher surface energy than Si (111) [35,36], thus such Ge diffusion toward (001) facet is energetically favored.

Considering the low growth temperature, the intermixing of Ge and Si is strongly suppressed, thus we can expect an almost pure Ge-core in such Ge/Si NWs. Furthermore, atomically sharp interfaces between the Ge-core and the Si-shell are observed in the inset of Figure 8c, which further confirms the negligible intermixing between Ge and Si.

4. Conclusions and Perspectives

In summary, homogenous planar diamond-shaped Si NW arrays (30–50 nm in width) have been achieved on pre-patterned {111}-faceted Si arrays via direct epitaxial growth. Morphologies and dimensions of these NWs are controllable, while they can also be tuned under certain growth conditions. Suspended Si NWs exhibit diamond-shaped cross-section with four Si {111} facets. Furthermore, the SiGe NWs can be conformally grown on the {111}-faceted Si NWs. Additionally, {113}-faceted Ge NWs along [110] direction are also obtained after the deposition of 2 nm Ge on the truncated Si NWs. HRTEMs reveal the high quality of these epitaxial NWs.

The in-plane and site-controllable epitaxial NWs hold promise as the platform for the next generation of devices that require addressability and scalability. The Si and SiGe NWs have potential applications for high-perform transistors [7,23]. Moreover, the [110]-oriented Ge/Si core/shell NWs are expected to have a high mobility and a strong spin-orbit coupling [37,38] for the manipulation of hole spin qubits. Additionally, we believe this method is also applicable to obtain planar nanowires in other material systems with controllable size and orientation, such as III–V compound materials. However, the large V-groove poses a challenge for device fabrication, which needs to be addressed in future research work.

Author Contributions: Conceptualization, J.-J.Z.; methodology, J.-H.W. and J.-J.Z.; formal analysis, J.-H.W., T.W., and J.-J.Z.; investigation, J.-H.W., T.W., and J.-J.Z.; resources, J.-J.Z.; data curation, J.-H.W.; writing—original draft preparation, J.-H.W.; writing—review and editing, T.W. and J.-J.Z.; project administration, J.-J.Z.; funding acquisition, J.-J.Z. All authors have read and agreed to the published version of the manuscript.

Funding: This work was supported by the National Key R&D Program of China (Grant Nos. 2016YFA0301701), the NSFC (Grant Nos. 11574356, 11434010, and 11404252), the Strategic Priority Research Program of CAS (Grant No. XDB30000000).

Data Availability Statement: Data available on request.

Conflicts of Interest: The authors declare no conflict of interest.

References

1. Mertens, H.; Ritzenthaler, R.; Hikavyy, A.; Kim, M.S.; Tao, Z.; Wostyn, K.; Chew, S.A.; Keersgieter, A.D.; Mannaert, G.; Rosseel, E.; et al. Gate-all-around MOSFETs based on vertically stacked horizontal Si nanowires in a replacement metal gate process on bulk Si substrates. In Proceedings of the 2016 IEEE Symposium on VLSI Technology, Honolulu, HI, USA, 14–16 June 2016; pp. 1–2.
2. Yin, X.; Yang, H.; Xie, L.; Ai, X.Z.; Zhang, Y.B.; Jia, K.P.; Wu, Z.; Ma, X.; Zhang, Q.Z.; Mao, S.; et al. Vertical Sandwich Gate-All-Around Field-Effect Transistors with Self-Aligned High-k Metal Gates and Small Effective-Gate-Length Variation. *IEEE Electron Device Lett.* **2019**, *41*, 8–11. [CrossRef]
3. Scappucci, G.; Kloeffel, C.; Zwanenburg, F.A.; Loss, D.; Myronov, M.; Zhang, J.J.; Franceschi, S.D.; Katsaros, G.; Veldhorst, M. The germanium quantum information route. *Nat. Rev. Mater.* **2020**, *479*, 1–18.
4. Watzinger, H.; Kukučka, J.; Vukušić, L.; Gao, F.; Wang, T.; Schäffler, F.; Zhang, J.J.; Katsaros, G. A germanium hole spin qubit. *Nat. Commun.* **2018**, *9*, 3902. [CrossRef]
5. Gonzalez-Zalba, M.F.; Shevchenko, S.N.; Barraud, S.; Johansson, J.R.; Ferguson, A.J.; Nori, F.; Betz, A.C. Gate-sensing coherent charge oscillations in a silicon field-effect transistor. *Nano Lett.* **2016**, *16*, 1614–1619. [CrossRef]
6. Froning, F.N.M.; Camenzind, L.C.; van der Molen, O.A.H.; Li, A.; Bakkers, E.P.A.M.; Zumbühl, D.M.; Braakman, F.R. Ultrafast hole spin qubit with gate-tunable spin-orbit switch functionality. *Nat. Nanotechnol.* **2021**. [CrossRef]
7. Cheng, X.; Li, Y.; Liu, H.; Zan, Y.; Lu, Y.; Zhang, Q.; Li, J.; Du, A.; Wu, Z.; Luo, J.; et al. Selective wet etching in fabricating SiGe nanowires with TMAH solution for gate-all-around MOSFETs. *J. Mater. Sci. Mater. Electron.* **2020**, *31*, 22478–22486. [CrossRef]
8. Poli, S.; Pala, M.G.; Poiroux, T.; Deleonibus, S.; Baccarani, G. Size Dependence of Surface-Roughness-Limited Mobility in Silicon Nanowire FETs. *IEEE Trans. Electron Devices* **2008**, *55*, 2968–2976. [CrossRef]
9. Goldthorpe, I.A.; Marshall, A.F.; McIntyre, P.C. Inhibiting strain-induced surface roughening: Dislocation-free Ge/Si and Ge/SiGe core−shell nanowires. *Nano Lett.* **2009**, *9*, 3715–3719. [CrossRef]
10. Lauhon, L.J.; Gudiksen, M.S.; Wang, D.; Lieber, C.M. Epitaxial Core–Shell and Core–Multishell Nanowire Heterostructures. *Nature* **2002**, *420*, 57–61. [CrossRef]
11. Xiang, J.; Lu, W.; Hu, Y.; Wu, Y.; Yan, H.; Lieber, C.M. Ge/Si Nanowire Heterostructures as High-Performance Field Effect Transistors. *Nature* **2006**, *441*, 489–493. [CrossRef]
12. Lu, W.; Xiang, J.; Timko, B.P.; Wu, Y.; Lieber, C.M. One-dimensional hole gas in germanium/silicon nanowire heterostructures. *Proc. Natl. Acad. Sci. USA* **2005**, *102*, 10046. [CrossRef] [PubMed]
13. Conesa-Boj, S.; Li, A.; Koelling, S.; Brauns, M.; Ridderbos, J.; Nguyen, T.T.; Verheijen, M.A.; Koenraad, P.M.; Zwanenburg, F.A.; Bakkers, E.P.A.M. Boosting Hole Mobility in Coherently Strained [110]-Oriented Ge-Si Core-Shell Nanowires. *Nano Lett.* **2017**, *17*, 2259–2264. [CrossRef] [PubMed]
14. Adhikari, H.; Marshall, A.F.; Chidsey, C.E.D.; McIntyre, P.C. Germanium nanowire epitaxy: Shape and orientation control. *Nano Lett.* **2006**, *6*, 318–323. [CrossRef] [PubMed]
15. Constantinou, M.; Rigas, G.P.; Castro, F.A.; Stolojan, V.; Hoettges, K.F.; Hughes, M.P.; Adkins, E.; Korgel, B.A.; Shkunov, M. Simultaneous Tunable Selection and Self-Assembly of Si Nanowires from Heterogeneous Feedstock. *ACS Nano* **2016**, *10*, 4384–4394. [CrossRef]
16. Fortuna, S.A.; Li, X. Metal-catalyzed semiconductor nanowires: A review on the control of growth directions. *Semicond. Sci. Technol.* **2010**, *25*, 024005. [CrossRef]
17. Toko, K.; Nakata, M.; Jevasuwan, W.; Fukata, N.; Suemasu, T. Vertically Aligned Ge Nanowires on Flexible Plastic Films Synthesized by (111)-Oriented Ge Seeded Vapor–Liquid–Solid Growth. *ACS Appl. Mater. Interfaces* **2015**, *7*, 18120–18124. [CrossRef]
18. Seravalli, L.; Bosi, M.; Beretta, S.; Rossi, F.; Bersani, D.; Musayeva, N.; Ferrari, C. Extra-long and taper-free germanium nanowires: Use of an alternative Ge precursor for longer nanostructures. *Nanotechnology* **2019**, *30*, 415603. [CrossRef]
19. Yan, H.; Choe, H.S.; Nam, S.W.; Hu, Y.J.; Das, S.; Klemic, J.F.; Ellenbogen, J.C.; Lieber, C.M. Programmable Nanowire Circuits for Nanoprocessors. *Nature* **2011**, *470*, 240–244. [CrossRef]
20. Collet, M.; Salomon, S.; Klein, N.Y.; Seichepine, F.; Vieu, C.; Nicu, L.; Larrieu, G. Large-Scale Assembly of Single Nanowires Through Capillary-Assisted Dielectrophoresis. *Adv. Mater.* **2015**, *27*, 1268–1273. [CrossRef]
21. Gao, F.; Wang, J.H.; Watzinger, H.; Hu, H.; Rančić, M.J.; Zhang, J.Y.; Wang, T.; Yao, Y.; Wang, G.L.; Kukučka, J.; et al. Site-Controlled Uniform Ge/Si Hut Wires with Electrically Tunable Spin–Orbit Coupling. *Adv. Mater.* **2020**, *32*, 1906523. [CrossRef]
22. Tabata, O.; Asahi, R.; Funabashi, H.; Shimaoka, K.; Sugiyama, S. Anisotropic Etching of Silicon in TMAH Solutions. *Sens. Actuators A* **1992**, *34*, 51–57. [CrossRef]
23. Zhang, Q.Z.; Yin, H.X.; Meng, L.K.; Yao, J.X.; Li, J.J.; Wang, G.L.; Li, Y.D.; Wu, Z.H.; Xiong, W.J.; Yang, H.; et al. Novel GAA Si Nanowire p-MOSFETs With Excellent Short-Channel Effect Immunity via an Advanced Forming Process. *IEEE Electron Device Lett.* **2018**, *39*, 464–467. [CrossRef]
24. Gu, J.; Zhang, Q.; Wu, Z.; Yao, J.; Zhang, Z.; Zhu, X.; Wang, G.; Li, J.; Zhang, Y.; Cai, Y.; et al. Cryogenic Transport Characteristics of P-Type Gate-All-Around Silicon Nanowire MOSFETs. *Nanomaterials* **2021**, *11*, 309. [CrossRef] [PubMed]
25. Nanda, K.K.; Sahu, S.N.; Behera, S.N. Liquid-Drop Model for the Size-Dependent Melting of Low-Dimensional Systems. *Phys. Rev. A* **2002**, *66*, 013208. [CrossRef]
26. Shin, H.S.; Yu, J.; Song, J.Y. Size-dependent thermal instability and melting behavior of Sn nanowires. *Appl. Phys. Lett.* **2007**, *91*, 173106. [CrossRef]

27. Day, R.W.; Mankin, M.N.; Gao, R.; No, Y.S.; Kim, S.K.; Bell, D.C.; Park, H.G.; Lieber, C.M. Plateau–Rayleigh crystal growth of periodic shells on one-dimensional substrates. *Nat. Nanotechnol.* **2015**, *10*, 345–352. [CrossRef]
28. Day, R.W.; Mankin, M.N.; Lieber, C.M. Plateau–Rayleigh crystal growth of nanowire heterostructures: Strain-modified surface chemistry and morphological control in one, two, and three dimensions. *Nano Lett.* **2016**, *16*, 2830–2836. [CrossRef]
29. Lee, L.L.; Antoniadis, D.A.; Fitzgerald, E.A. Challenges in Epitaxial Growth of SiGe Buffers on Si(111), (110), and (112). *Thin Solid Films* **2006**, *508*, 136–139. [CrossRef]
30. Zhang, J.J.; Rastelli, A.; Schmidt, O.G.; Bauer, G. Role of the wetting layer for the SiGe Stranski-Krastanow island growth on planar and pit-patterned substrates. *Semicond. Sci. Technol.* **2011**, *26*, 014028. [CrossRef]
31. Gai, Z.; Yang, W.S.; Sakurai, T.; Zhao, R.G. Heteroepitaxy of germanium on Si (103) and stable surfaces of germanium. *Phys. Rev. B* **1999**, *59*, 13009–13013. [CrossRef]
32. Gai, Z.; Ji, H.; Gao, B.; Zhao, R.G.; Yang, W.S. Surface structure of the (3× 1) and (3× 2) reconstructions of Ge (113). *Phys. Rev. B* **1996**, *54*, 8593. [CrossRef]
33. Laracuente, A.; Erwin, S.C.; Whitman, L.J. Structure of Ge (113): Origin and Stability of Surface Self-Interstitials. *Phys. Rev. Lett.* **1998**, *81*, 5177. [CrossRef]
34. Zhang, J.J.; Katsaros, G.; Montalenti, F.; Scopece, D.; Rezaev, R.O.; Mickel, C.; Rellinghaus, B.; Miglio, L.; De Franceschi, S.; Rastelli, A.; et al. Monolithic growth of ultrathin Ge nanowires on Si (001). *Phys. Rev. Lett.* **2012**, *109*, 085502. [CrossRef]
35. Eaglesham, D.J.; White, A.E.; Feldman, L.C.; Moriya, N.; Jacobson, D.C. Equilibrium shape of Si. *Phys. Rev. Lett.* **1993**, *70*, 1643–1646. [CrossRef]
36. Lu, G.H.; Huang, M.; Cuma, M.; Liu, F. Relative stability of Si surfaces: A first-principles study. *Surf. Sci. Rep.* **2005**, *588*, 61–69. [CrossRef]
37. De Vries, F.K.; Shen, J.; Skolasinski, R.J.; Nowak, M.P.; Varjas, D.; Wang, L.; Wimmer, M.; Ridderbos, J.; Zwanenburg, F.A.; Li, A.; et al. Spin-Orbit Interaction and Induced Superconductivity in a One-Dimensional Hole Gas. *Nano Lett.* **2018**, *18*, 6483–6488. [CrossRef] [PubMed]
38. Kloeffel, C.; Trif, M.; Loss, D. Strong spin-orbit interaction and helical hole states in Ge/Si nanowires. *Phys. Rev. B.* **2011**, *84*, 195314. [CrossRef]

Article

Investigation on $Ge_{0.8}Si_{0.2}$-Selective Atomic Layer Wet-Etching of Ge for Vertical Gate-All-Around Nanodevice

Lu Xie [1,2], Huilong Zhu [1,*], Yongkui Zhang [1], Xuezheng Ai [1], Junjie Li [1], Guilei Wang [1,2,3,*], Anyan Du [1], Zhenzhen Kong [1,2,3], Qi Wang [1], Shunshun Lu [1], Chen Li [1,2], Yangyang Li [1,2], Weixing Huang [1,2] and Henry H. Radamson [1,2,3,*]

[1] Key Laboratory of Microelectronics Devices & Integrated Technology, Institute of Microelectronics, Chinese Academy of Sciences, Beijing 100029, China; xielu@ime.ac.cn (L.X.); zhangyongkui@ime.ac.cn (Y.Z.); aixuezheng@ime.ac.cn (X.A.); lijunjie@ime.ac.cn (J.L.); duanyan@ime.ac.cn (A.D.); kongzhenzhen@ime.ac.cn (Z.K.); wangqi@ime.ac.cn (Q.W.); lushunshun@ime.ac.cn (S.L.); lichen2017@ime.ac.cn (C.L.); liyangyang@ime.ac.cn (Y.L.); huangweixing@ime.ac.cn (W.H.)
[2] Microelectronics Institute, University of Chinese Academy of Sciences, Beijing 100049, China
[3] Research and Development Center of Optoelectronic Hybrid IC, Guangdong Greater Bay Area Institute of Integrated Circuit and System, Guangdong 510535, China
* Correspondence: zhuhuilong@ime.ac.cn (H.Z.); wangguilei@ime.ac.cn (G.W.); rad@ime.ac.cn (H.H.R.)

Abstract: For the formation of nano-scale Ge channels in vertical Gate-all-around field-effect transistors (vGAAFETs), the selective isotropic etching of Ge selective to $Ge_{0.8}Si_{0.2}$ was considered. In this work, a dual-selective atomic layer etching (ALE), including $Ge_{0.8}Si_{0.2}$-selective etching of Ge and crystal-orientation selectivity of Ge oxidation, has been developed to control the etch rate and the size of the Ge nanowires. The ALE of Ge in p^+-$Ge_{0.8}Si_{0.2}$/Ge stacks with 70% HNO_3 as oxidizer and deionized (DI) water as oxide-removal was investigated in detail. The saturated relative etched amount per cycle (REPC) and selectivity at different HNO_3 temperatures between Ge and p^+-$Ge_{0.8}Si_{0.2}$ were obtained. In p^+-$Ge_{0.8}Si_{0.2}$/Ge stacks with (110) sidewalls, the REPC of Ge was 3.1 nm and the saturated etching selectivity was 6.5 at HNO_3 temperature of 20 °C. The etch rate and the selectivity were affected by HNO_3 temperatures. As the HNO_3 temperature decreased to 10 °C, the REPC of Ge was decreased to 2 nm and the selectivity remained at about 7.4. Finally, the application of ALE in the formation of Ge nanowires in vGAAFETs was demonstrated where the preliminary I_d–V_{ds} output characteristic curves of Ge vGAAFET were provided.

Keywords: vertical Gate-all-around (vGAA); p^+-$Ge_{0.8}Si_{0.2}$/Ge stack; dual-selective wet etching; atomic layer etching (ALE)

1. Introduction

As the continuous scaling down of complementary metal-oxide-semiconductor (CMOS) technology nodes, novel device designs and high-mobility channel materials have been under investigation [1–8]. Vertical nanowire GAAFET is a powerful candidate for the 3 nm process, since its superiority in the short channel effects (SCEs) control [9,10] and can greatly reduce the gate pitch and increase the device integration density [11,12]. In addition, Ge is one of the most promising channel materials for pMOS due to its high carrier mobility and excellent bandgap [13]. Therefore, vertical GAAFETs with Ge as channel material have become an ideal choice for next era CMOS technology.

For vertical GAA devices, a new structure of vertical sandwich GAAFETs (VSAFETs) with Si source/drain and SiGe channel has been proposed [14,15]. The main process flow of VSAFETs is shown in Figure 1, the selective etching of the channel is a key step in the formation of vertical nanostructures. And the dimension of the channel is determined by selective etching. For the formation of Ge vertical nanowires, several selective etching methods have been reported, including dry etching with Cl_2 or CF_4 RF plasma [16–18] or mixtures of Cl_2/HBr [19] and wet etching with H_2O_2 (HNO_3) [20] or TMAH [21] or other alkaline solutions [22] or mixtures of HF/H_2O_2/CH_3COOH [20,23]. SiGe/Ge multilayer structures have been used to release with SiGe sacrificial layer etching to fabricate Ge nanowires in the lateral GAA device [19,24]. However, the above-mentioned methods for forming Ge nanowires are all continuous etching methods, and the etching depth is time-dependent. They achieve high selectivity and high etch rate at the cost of repeatability. At and beyond the 3 nm technology node, the gate, and channel of vGAAFETs need to be precisely aligned and the size controlled at the atomic-scale to achieve good device performance. In order to form Ge channels with self-aligned gate structure in VSAFETs, Ge was required to be laterally released to nano-scale size with source/drain-selective etching of Ge. Atomic layer etching (ALE) [25–27] is a promising technology that can remove ultra-thin materials through at least one self-limiting reaction step to achieve lower atomic-scale process variation. At present, the ALE for isotropic selective etching of SiGe to Si [28–30] has been reported, ALE for isotropic selective etching of Ge to GeSi has not been extensively reported. However, the ALE is not universal, and the selective ALE method depends on the recipe and it is difficult to achieve.

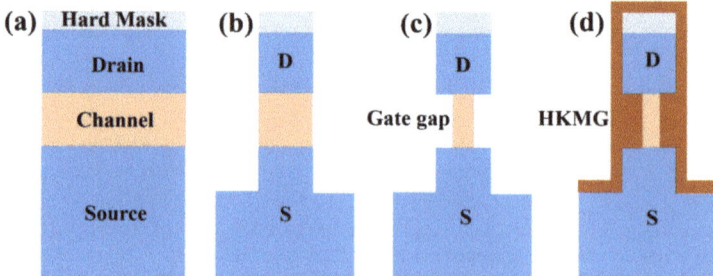

Figure 1. Schematic diagram of the basic flow of vertical Gate-all-around FET (vGAAFETs). (**a**) Sandwich structure and hard mask growth, (**b**) lithographic patterning and plasma anisotropic etching, (**c**) channel isotropic selective etching to form laterally depressed channels and gate gaps, (**d**) gate gaps filling with high-k metal gate (HKMG).

In this work, a developed wet dual-selective ALE process with selective etching of Ge and crystal-orientation selectivity of Ge oxidation was proposed. Based on the principle of atomic layer etching (ALE) and the oxidation–removal reaction of HNO_3 and deionized (DI) water, the characteristics of $Ge_{0.8}Si_{0.2}$-selective ALE of Ge at different temperatures were investigated systematically. The ALE process with a focus on the selective etching of Ge in p^+-$Ge_{0.8}Si_{0.2}$/Ge multilayers with 70% HNO_3 as oxidizer and DI water as oxide removal. The saturated relative etched amount per cycle (REPC) and selectivity at different HNO_3 temperatures between Ge and p^+-$Ge_{0.8}Si_{0.2}$ were investigated in detail. The application of ALE in the formation of Ge nanowires in VSAFETs was demonstrated.

2. Materials and Methods

The samples were performed on 200-mm p-type Si (100) wafers with a resistivity of 8–12 Ohm·cm. The high-quality epitaxial p^+-$Ge_{0.8}Si_{0.2}$/Ge vertical heterostructure multilayers started with a Ge buffer layer growth by ASM E2000 (ASM, Munich, Germany) plus RPCVD on Si wafers [20]. Dichlorosilane (SiH_2Cl_2), germane (10% GeH_4 in H_2), and diborane (1% B_2H_6 in H_2) were utilized as gas precursors for Si, Ge, and B, respectively. $Ge_{0.8}Si_{0.2}$ layers were in-situ doped with boron (concentration: 1.0×10^{19} cm^{-3}). The growth parameters and boron content have been carefully optimized to avoid boron precipitates in the $Ge_{0.8}Si_{0.2}$ layers [31]. The Ge buffer layer was grown with a two-step growth of low-high temperature (400 °C and 650 °C), and then a post-growth in situ annealing was applied at 820 °C in H_2 ambient [32]. Then, the p^+-$Ge_{0.8}Si_{0.2}$/Ge stacks were grown at 500 °C using an adjusted gas source with H_2 as a carrier gas. Then, a hard mask with 30 nm SiN and 50 nm SiO$_2$ was deposited with plasma enhanced chemical vapor deposition (PECVD) on the epitaxial p$^+$-Ge$_{0.8}$Si$_{0.2}$/Ge stack layers. Finally, the p$^+$-Ge$_{0.8}$Si$_{0.2}$/Ge stack fins were patterned by I-line optical lithography and fabricated by using HBr-based dry anisotropic etching. Afterward, the samples were cut into small slices to facilitate the etching experiments.

There were three kinds of samples to estimate the etch rate and selectivity between Ge and p$^+$-Ge$_{0.8}$Si$_{0.2}$, as shown in Figure 2a–c. The etch rate of Ge (100 nm) and p$^+$-Ge$_{0.8}$Si$_{0.2}$ (300 nm) films with (100) flat surfaces were expressed by the etch rate per cycle (EPC). In order to measure the relative etch rate and selectivity of Ge and p$^+$-Ge$_{0.8}$Si$_{0.2}$ with ALE, a structure with (110) sidewall was fabricated and keep p$^+$-Ge$_{0.8}$Si$_{0.2}$/Ge/p$^+$-Ge$_{0.8}$Si$_{0.2}$/Ge/p$^+$-Ge$_{0.8}$Si$_{0.2}$ at 120 nm/50 nm/75 nm/50 nm/75 nm, and p$^+$-Ge$_{0.8}$Si$_{0.2}$ with boron dopant concentration of 1.0×10^{19} cm^{-3}. And the thickness and composition of the samples were kept constant among repetitive experiments. The prepared p$^+$-Ge$_{0.8}$Si$_{0.2}$/Ge stack structure is shown in Figure 2c. The relative total etched amount (tunnel depth) and GeSi loss are shown in the insert view, selective etching of Ge at (111) planes result in a lateral angle of 54.7°. The etch selectivity between p$^+$-Ge$_{0.8}$Si$_{0.2}$/Ge stacks with (110) sidewall is estimated by (tunnel depth + GeSi loss)/GeSi loss [28]. The relative etched amount per cycle (REPC) was calculated as the tunnel depth divided by the number of etching cycles. The flow diagram of ALE is shown in Figure 2d, the steps in the dashed frame are one cycle of ALE, including 70% HNO$_3$ oxidation, deionized (DI) water rinsing, and repeating the number of cycles until the required etching amount was reached. Before etching experiments, the samples were immersed in diluted BOE (dBOE, 49 wt% HF and 40 wt% NH$_4$F with volume ratio of 1:7) for 5 min to remove the natural oxide. During the experiments, the volume of the nitric acid solution was constant at 2 L, and the DI water was with overflow rinsing. As high-concentration nitric acid is easier to decompose, the experimental process requires a high-precision density meter to monitor the concentration of HNO$_3$ solution. In view of the different oxidation mechanism of nitric acid (HNO$_3$) concentration on Ge, the nitric acid concentration must be maintained at 70% with 5% variation in the ALE experiments compared to the continuous etching of Ge with nitric acid at low concentration. The experimental temperature was kept at 20 °C ± 0.5 °C, and the temperature with variation ±0.5 °C of the HNO$_3$ solution was controlled by the water bath method of the cryostat at low temperature. The oxidation time t_{ox} of the control recipe is 30 s and the DI water rinsing time (oxide remove time) is set as 1 min to make sure that oxide is removed totally.

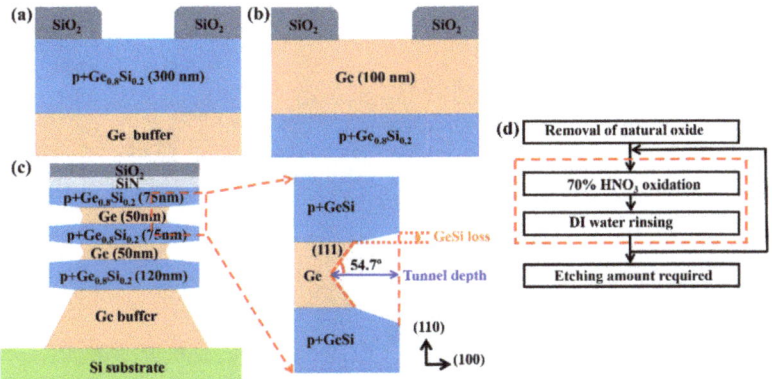

Figure 2. (a) Scheme of structure for p$^+$-Ge$_{0.8}$Si$_{0.2}$ (a) and Ge (b) with (100) flat surface; (c) scheme of p$^+$-Ge$_{0.8}$Si$_{0.2}$/Ge stacks with SiO$_2$ and SiN as hard mask. The tunnel depth and GeSi loss are indicated in the insert view, selective etching of Ge at (111) planes result in a lateral angle of 54.7°. (d) Flow diagram of HNO$_3$-DI water for p$^+$-Ge$_{0.8}$Si$_{0.2}$-selective etching of Ge with ALE.

Scanning electron microscopy (SEM) was used to examine the morphology and etching depth of the samples. Transmission electron microscopy (TEM) characterized the sample to determine the layer profile and evaluate the results of wet etching. Energy dispersive spectroscopy (EDS) was employed to determine the elemental analysis of the etched layers. High-resolution X-ray diffraction (HRXRD) was used to measure the strain relaxation and examine the epitaxial quality. Atomic force microscopy (AFM) measured surface roughness.

3. Results and Discussion

3.1. Dual-Selective Etching Ge to p^+-$Ge_{0.8}Si_{0.2}$ with ALE

In order to form Ge channels with self-aligned gate structure in pVSAFETs, Ge was required to be laterally released to nano-scale size with ALE. First, we tried the ALE recipe of selective SiGe to Si in the previous work [28,29], using H_2O_2 and diluted HNO_3 solution as oxidants to oxidize p^+-$Ge_{0.8}Si_{0.2}$/Ge stacks. However, due to the water-soluble of GeO_2, GeO_2 generated by the reaction of Ge with H_2O_2 and diluted HNO_3 solution was dissolved in water immediately, resulting in uncontrolled continuous etching [20], these recipes cannot be used as ALE recipes for Ge. Then, we used O_2 plasma and O_3 as oxidants, combined with the etchant for Ge selective etching experiments, but failed with non-selectivity (not shown). Finally, we found that (1) only a high concentration of HNO_3 (70% with 5% variation) as oxidant and DI water as etchant can achieve the selective ALE of Ge to p^+-$Ge_{0.8}Si_{0.2}$, and (2) 70% concentrated HNO_3 has crystal-orientation selectivity for the oxidation of Ge, but not for p^+-$Ge_{0.8}Si_{0.2}$. The oxidation rate of HNO_3 on Ge surface is inversely proportional to the atomic density of the crystallographic plane, which determines the slower oxidation rate on Ge (111) planes [33,34]. Therefore, the self-saturated selective etching of Ge at (111) planes result in a lateral angle of 54.7°, as shown in the insert view of Figure 3a. Within the experimental error, this angle is equal to the theoretical angle between the (100) and (111) set of planes. A developed dual-selective ALE method, including material selectivity and crystal-orientation selective oxidation, is suitable for the selective etching of Ge in p^+-$Ge_{0.8}Si_{0.2}$/Ge stacks.

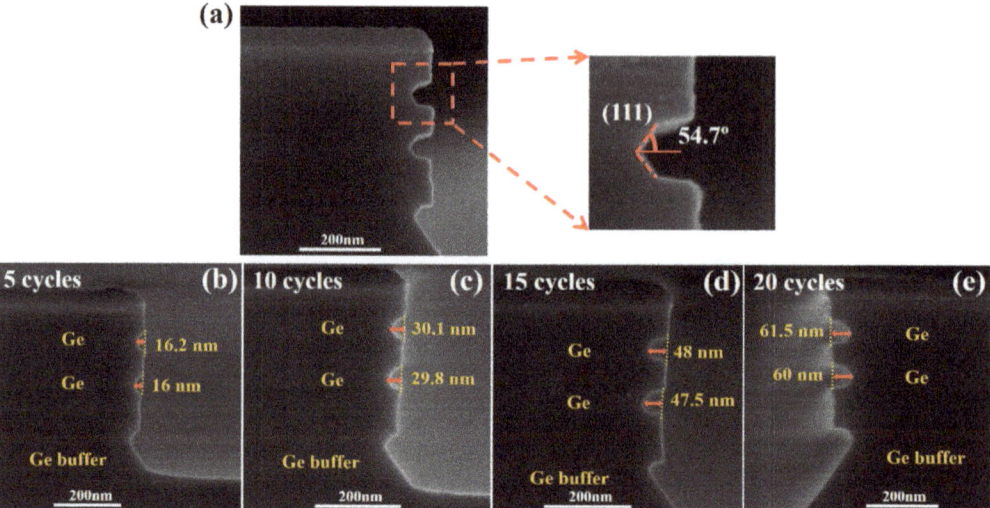

Figure 3. (a) The dual-selective etch of Ge at (111) planes result in a lateral angle of 54.7°. The SEM cross-section images of p^+-$Ge_{0.8}Si_{0.2}$/Ge multilayers after 70% HNO_3-DI water ALE at (b) 5 cycles, (c) 10 cycles, (d) 15 cycles, and (e) 20 cycles. (Oxidation time t_{ox} = 30 s; temperature of HNO_3 = 20 °C).

Due to the ALE separated into two parts: oxidation and oxide remover. The amount of etching is determined according to the amount of oxidation, so the oxidation step is the key to control the etching rate. First, 70% HNO$_3$ was used to periodically oxidize the surface of the sample for time t$_{ox}$, showing a self-limiting surface passivation reaction, and then deionized (DI) water was used to directly remove the oxide. The oxidation of Ge in high-concentration nitric acid can be described as follows [20]:

$$Ge + 4HNO_3 \rightarrow GeO_2 \cdot H_2O + 4NO_2 \uparrow + H_2O \qquad (1)$$

In the reaction of high-concentration nitric acid to germanium, the reaction product is GeO$_2$·H$_2$O. In the whole reaction process, GeO$_2$ formed at the germanium-liquid interface diffuses slowly, and enough GeO$_2$ covers the entire germanium surface and produces passivation to prevent further corrosion of germanium [35]. Compared with ALE of SiGe to Si, the ALE of Ge to p$^+$-Ge$_{0.8}$Si$_{0.2}$ did not need another etchant (hydrofluoric acid or dBOE) but only DI water can directly remove oxides. To ensure the stability of the nitric acid concentration during the ALE experiment, the nitric acid concentration needs to be monitored with a high-precision density meter every 5 min. It was found that the concentration of HNO$_3$ changes by 5% within 1 h during the ALE experiment. When the nitric acid concentration changed more than 5%, that is, the nitric acid concentration is less than 65% (original nitric acid concentration is 70%), the p$^+$-Ge$_{0.8}$Si$_{0.2}$/Ge stacks became continuous etching in HNO$_3$ solution. In order to maintain the stability and repeatability of the ALE experiment, the nitric acid solution was changed every 1 h. It can be seen that the concentration of nitric acid had a great influence on the experimental results. Since nitric acid is exothermic and volatilized when exposed to water, the sample must be dried before the next cycle in nitric acid. In order to verify whether germanium oxide is completely removed in DI water, three comparative experiments were taken at the same time: (1) 70% HNO$_3$ 30 s + DI water cleaning for 1 min, (2) 70% HNO$_3$ 30 s + DI water cleaning for 2 min and (3) 70% HNO$_3$ 30 s + DI water cleaning for 1 min + dBOE immersing for 1 min + DI water cleaning for 1 min. Excluding measurement errors, the Ge etching amounts of these three cleaning conditions were almost the same (not shown), which proved that 1 min of DI water cleaning was sufficient to remove germanium oxide without adding other etching agents. It was also proved that enough water molecules in DI water can pass through GeO$_2$ to reach the Ge-GeO$_2$ interface and strip the germanium oxide.

Samples were immersed in concentrated HNO$_3$ (70%) with oxidation time t$_{ox}$ = 30 s at HNO$_3$ temperature of 20 °C, and rinsed in DI water for at least 1 min to remove GeO$_x$ effectively. Figure 3b–e shows the SEM cross-section images of p$^+$-Ge$_{0.8}$Si$_{0.2}$/Ge stacks after etching with 5 cycles, 10 cycles, 15 cycles, and 20 cycles, respectively. Within the measurement error, the mean relative etching amounts of Ge in p$^+$-Ge$_{0.8}$Si$_{0.2}$/Ge stacks were 16.1 nm, 30 nm, 47.8 nm, and 60.8 nm, respectively. The relative etching amount is linearly proportional to the number of etching cycles, and the etch rate per cycle (EPC) is independent of the etching cycle number with the mean value of 3.1 nm.

In p$^+$-Ge$_{0.8}$Si$_{0.2}$/Ge stacks, the selective ALE of Ge can be performed smoothly in a 70% HNO$_3$-DI water system. One is that the Ge layers (atoms) were oxidized preferentially by HNO$_3$ owing to the weaker Ge-Ge bond energy than those of Ge-Si and Si-Si [36–39]. Second, GeO$_2$ generated in the Ge layer is soluble in water, while SiO$_2$ generated by oxidation of the p$^+$-Ge$_{0.8}$Si$_{0.2}$ layer is stable and hydrophobic, and GeO$_2$ far away from the surface cannot be directly dissolved in water because of the passivation of the SiO$_2$ layer.

The EPC was calculated by dividing the relative etching amounts by the number of etching cycles and this experiment was carried out at HNO$_3$ temperature of 20 °C. The EPC for Ge and p$^+$-Ge$_{0.8}$Si$_{0.2}$ as a function of the oxidation time is shown in Figure 4. The EPC increased with the increase of oxidation time and gradually saturates when the oxidation time exceeds 30 s, indicating that the oxidation of Ge and p$^+$-Ge$_{0.8}$Si$_{0.2}$ in concentrated HNO$_3$ (70%) is quasi-self-limiting. When the oxidation time is 30 s, the etching selectivity between Ge and p$^+$-Ge$_{0.8}$Si$_{0.2}$ in (100) flat surface is 6.5, as determined from the two oxidation curves in Figure 4. The EPC of Ge and p$^+$-Ge$_{0.8}$Si$_{0.2}$ were 3.7 nm and 0.5 nm, respectively. The etch rate of lateral selective etching of Ge on p$^+$-Ge$_{0.8}$Si$_{0.2}$/Ge stacks with (110) sidewall, which is more important in vertical device fabrication, is defined as the relative etched amount per cycle (REPC). Similarly, as shown in Figure 4, the REPC of Ge in the vertical structure of p^+-$Ge_{0.8}Si_{0.2}$/Ge stacks is fixed at ~3.1 nm. The estimated data in these experiments in Figure 4 show very small etching errors. Therefore, it can be verified that the ALE method is repeatable if the measurement error is taken out.

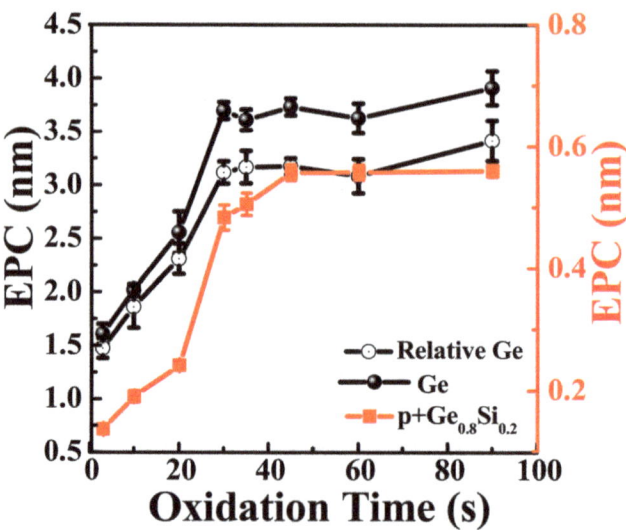

Figure 4. EPCs of ALE with different oxidation time for Ge and p^+-$Ge_{0.8}Si_{0.2}$ (100) planes and Ge selectively etched in p^+-$Ge_{0.8}Si_{0.2}$/Ge stacks. (30 cycles; Temperature of HNO_3 = 20 °C; Experimental data of average values and error bars are shown in the graph).

3.2. Effect of HNO_3 Temperature on Ge ALE

At room temperature (20 °C), the REPC of Ge (3.1 nm) is about 8 times higher than that of REPC of SiGe (0.4 nm) [29]. The etching rate is relatively fast. Since oxidation plays an important role in ALE, the effect of lowering the temperature of the nitric acid solution on the oxidation rate was studied to obtain a lower oxidation rate. Due to the strong corrosiveness of nitric acid, the temperature of the nitric acid solution is cooled by the water bath of the cryostat. In order to explore the effect of low-temperature on the REPC and selectivity between Ge and p^+-$Ge_{0.8}Si_{0.2}$, experiments were carried out at HNO_3 temperature of 5 °C, 10 °C, 15 °C, and 20 °C, respectively. The low-temperature etching morphology was characterized by SEM and TEM. As shown in Figure 5, Ge ALE at 20 °C and 5 °C both exhibit dual-selectivity (material and crystal-orientation selectivity). The TEM results show the clear layering of the p^+-$Ge_{0.8}Si_{0.2}$/Ge stacks and the etching morphology corresponding to the SEM characterization. The oxidation time was t_{ox} = 30 s, where the cycles of 20 °C and 5 °C were 15 cycles and 20 cycles, respectively, and the etching amount of Ge is 50 nm and 36 nm, respectively. Experiments have proved that low temperature can reduce the oxidation rate of nitric acid without changing the etching morphology, which is an effective method to reduce REPC of Ge.

The REPC curves of ALE at different HNO_3 temperatures (20 °C, 15 °C, 10 °C, and 5 °C) in p^+-$Ge_{0.8}Si_{0.2}$/Ge multilayers are shown in Figure 6a. As the temperature of HNO_3 decreased from 20 °C to 5 °C, REPC decreased from 3.1 nm to 1.8 nm. With the HNO_3 temperature reducing to 10 °C, the REPC of Ge was reduced to 2 nm. Similarly, the REPCs at all four temperatures reached saturation when the oxidation time exceeded 30 s, which is the quasi-self-limiting process. Due to the temperature sensitivity of the Ge oxidation reaction, the saturated oxide will be thinner with the lower HNO_3 temperature. The temperature with variation ±0.5 °C was controlled by the water bath of the low-constant temp tank. According to the corrosion model proposed by Seidei [40], its point of view was to attribute chemical reactions to differences in energy. When the temperature of the HNO$_3$ decreased, the energy difference between its molecular kinetic energy and the surface activation energy of the sample became smaller, so the oxidation rate became slower.

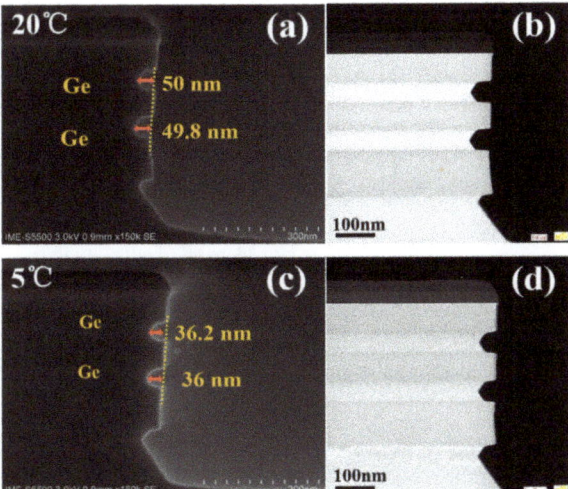

Figure 5. The SEM and TEM images of etching profile of p$^+$-Ge$_{0.8}$Si$_{0.2}$/Ge multilayers after ALE at (**a**,**b**) 15 cycles (temperature of HNO$_3$ = 20 °C); (**c**,**d**) 20 cycles (temperature of HNO$_3$ = 5 °C). The oxidation time t_{ox} = 30 s.

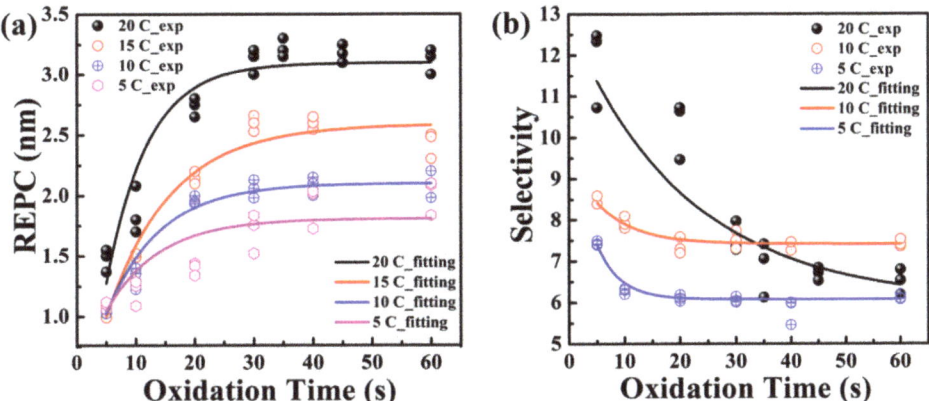

Figure 6. (**a**) The REPC of p$^+$-Ge$_{0.8}$Si$_{0.2}$/Ge at different HNO$_3$ temperatures (20 °C, 15 °C, 10 °C, 5 °C) as a function of oxidation time. (**b**) The selectivity between p$^+$-Ge$_{0.8}$Si$_{0.2}$ and Ge at different HNO$_3$ temperatures (20 °C, 10 °C, and 5 °C) as a function of oxidation time. The scatters and solid lines represent the experimental data and fitting curves of the experimental data.

Figure 6b shows the etching selectivity of p$^+$-Ge$_{0.8}$Si$_{0.2}$/Ge stacks at different HNO$_3$ temperatures (5 °C, 10 °C, and 20 °C). The selectivity is defined as [28]:

$$\text{Selectivity} = \frac{\text{GeSiloss} + \text{Tuneldepth}}{\text{GeSiloss}} \quad (2)$$

where GeSi loss is vertical etching amount of GeSi, which is equal to the horizontal etching amount, and the tunnel depth is the relative total etched amount of Ge to Ge$_{0.8}$Si$_{0.2}$, as shown in Figure 2c. The experimental data were obtained by measuring the SEM images of the etching profile in the p$^+$-Ge$_{0.8}$Si$_{0.2}$/Ge stacks with (110) sidewalls. The mean values of p$^+$-Ge$_{0.8}$Si$_{0.2}$/Ge selectivity at HNO$_3$ temperatures of 20 °C, 10 °C, and 5 °C were 6.5, 7.4, and 6.1, respectively. The results show that Ge ALE can achieve a high selectivity of Ge to Ge$_{0.8}$Si$_{0.2}$, independent of HNO$_3$ temperature. At HNO$_3$ temperatures of 20 °C, when the oxidation time increased, there was enough time for HNO$_3$ to destroy and oxidize the Si-Si bonds and Ge-Si bonds. After 30 cycles of ALE, the etching amount of Ge$_{0.8}$Si$_{0.2}$ became larger, and the EPC became larger accordingly. Thereby reducing the selectivity of Ge to Ge$_{0.8}$Si$_{0.2}$, and reached saturation when the oxidation time was 45 s. At low temperatures, due to the decrease of the molecular kinetic energy of HNO$_3$, the selectivity of Ge to Ge$_{0.8}$Si$_{0.2}$ does not change much with time and reached saturation in about 20 s of oxidation time.

3.3. Structure Characterization and Material Quality Analysis

In order to more accurately characterize the Ge ALE results in this study, the quality of the epitaxial layers and the etching morphology and element analysis of the p$^+$-Ge$_{0.8}$Si$_{0.2}$/Ge stack structure were characterized by HRTEM and EDS. Figure 7 shows the cross-section TEM micrograph and elemental analysis of the etched area of p$^+$-Ge$_{0.8}$Si$_{0.2}$/Ge stacks with (110) sidewalls with Ge ALE. The results show that the Ge content in the GeSi layer maintains at 80%, and there is no mixing between the layers. The relative etching amount is the same as that characterized by SEM. It was further verified that the selective etching of Ge on (111) planes resulted in a lateral angle of 54.7°. The EDS mapping of Figure 8 shows that the boundaries of the layers are obvious, the thickness of the film is consistent with the design, and there is no obvious element diffusion between the layers. The germanium content of the Ge layer is almost 100%. Si element is distributed in the layers of SiO$_2$, SIN, and Ge$_{0.8}$Si$_{0.2}$, O mainly exists in the hard mask of SiO$_2$.

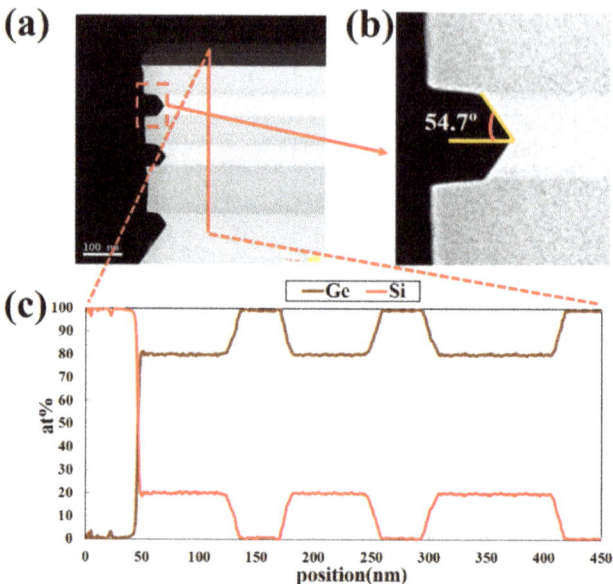

Figure 7. (**a**,**b**) TEM images of etching profile, and (**c**) EDS analysis with line scanning of Si and Ge in vertical orientation of the Ge buffer/p$^+$-Ge$_{0.8}$Si$_{0.2}$/Ge stack structure. The selective etching of Ge at (111) planes results in a lateral angle of 54.7°. (20 cycles; temperature of HNO$_3$ = 15 °C; oxidation time t$_{ox}$ = 30 s).

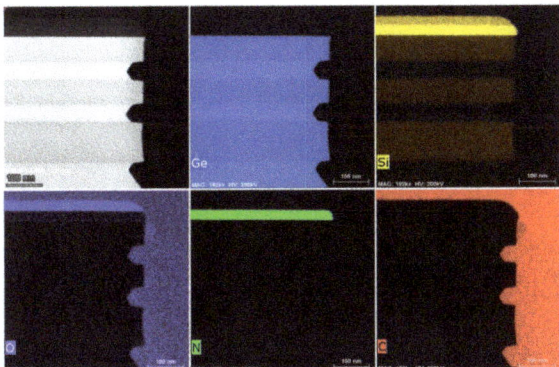

Figure 8. EDS mapping near etching regions with elements Ge, Si, O, N, and C, the sample etching for 15 cycles (oxidation time t_{ox} = 30 s; temperature of HNO_3 = 20 °C).

Crystallinity and strain have a strong impact on channel carrier mobility and device performance, therefore, crystal quality and strain relaxation of the sample before and after the etching steps has been studied. HRXRD is a technique that is widely used for the detection of defects in crystal materials [41]. Figure 9 displays rocking curves (RCs) measured around the (004) reflection on stack samples: as-grown, after vertical stack etch, after ALE at 20 °C, and after ALE at 10 °C. For the as-grown sample, the $Ge_{0.8}Si_{0.2}$ and Ge peaks were intense with low Full-width-of-half-Maximum (FWHM) showing good crystalline quality of $Ge_{0.8}Si_{0.2}$/Ge stack. Since most of the film layer was removed by etching, the intensity of the $Ge_{0.8}Si_{0.2}$ peak was weaker, while the Ge peak was still strong due to the presence of the Ge buffer layer. Compared with the as-grown sample, the $Ge_{0.8}Si_{0.2}$ peak shifts towards the Ge peak after stack etching. This indicates that (tensile-strained) $Ge_{0.8}Si_{0.2}$ is partially relaxed after the vertical etching. Moreover, the Ge peak became asymmetric after vertical etch and the amount of strain in the Ge was minor. No further shift of $Ge_{0.8}Si_{0.2}$ or Ge peak was detected after ALE etching, which indicates that there was no further strain relaxation after lateral etching at 20 °C and 10 °C.

Since Ge will be used as the channel material in vertical GAAFETs, the etched germanium surface can be used as the channel interface. Due to the scattering of the surface roughness, the channel surface roughness will cause gate oxide integrity degradation and mobility degradation. Therefore, it is necessary to measure the surface roughness of Ge that has been etched many times by DI water. Figure 10 shows the AFM images of the flat (100) Ge surface with as-grown epi-Ge, after etching with ALE at HNO_3 temperatures of 20 °C (20 cycles), after etching with ALE at HNO_3 temperatures of 5 °C (20 cycles), and after etching with $HF:HNO_3:CH_3COOH$ mixtures. It was found that the root mean square (RMS) roughness of the ALE process at 20 °C HNO_3 temperatures is 0.85 nm, which is similar to that of the as-grown sample (RMS of 0.67 nm). However, the RMS increases as the temperature of HNO_3 decreases. The roughness is very poor etching with $HF:HNO_3:CH_3COOH$ mixtures. Table 1 shows the comparison of the RMS of flat (100) Ge surfaces of as-grown and different etching processes. It is demonstrated that the surface roughness of ALE is better than chemical continuous etching. The smoothing effect can be explained by a model, where the depressions on the surface asperity are preferentially oxidized, and the protrusions on the asperity are preferentially etched. When the temperature of nitric acid decreases, the oxidizing ability decreases, resulting in uneven surface oxidation. Finally, the height difference between the depressions and the protrusions was increased during the etching, so the surface after the low-temperature treatment is rougher. In summary, we need to make a trade-off between the etching rate and the surface roughness and choose a suitable temperature for ALE etching. Because of the equipment, the preparation process of our devices was mainly carried out at room temperature.

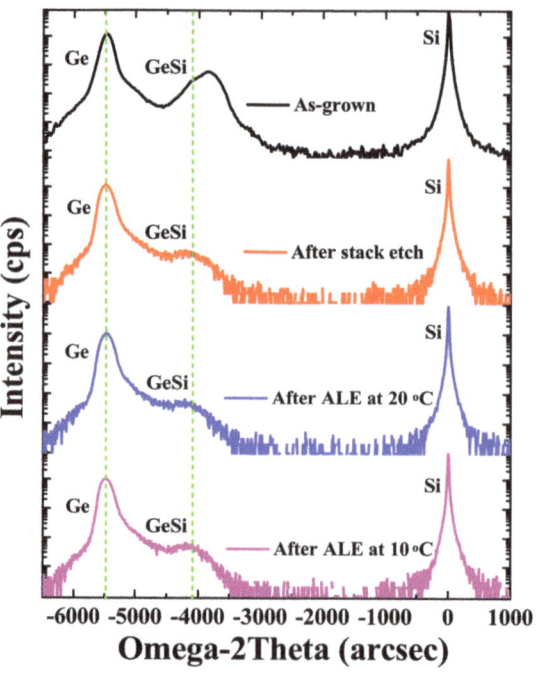

Figure 9. HRXRD rocking curves (RCs) around (004) reflection of stack samples of as-grown, after vertical stack etch, after ALE at 20 °C, and after ALE at 10 °C with 20 cycles.

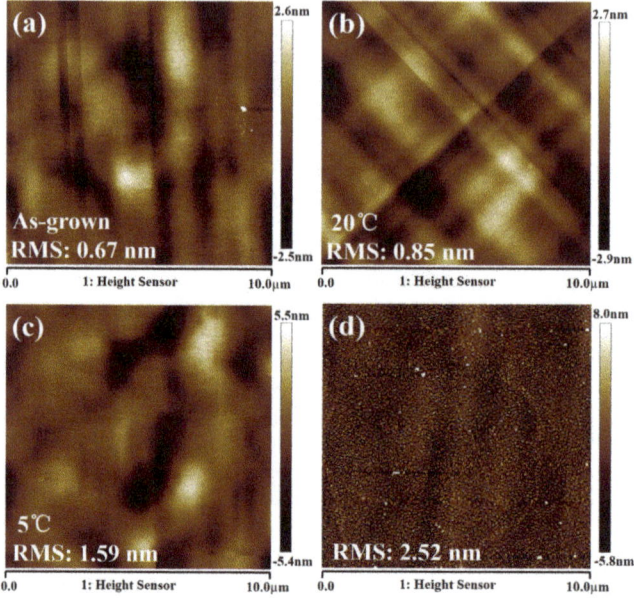

Figure 10. Typical AFM images (10 × 10 μm^2) of flat (100) Ge surfaces before and after the etching process: (**a**) as–grown; (**b**) ALE with 20 cycles (temperature of HNO_3 = 20 °C); (**c**) ALE with 20 cycles (temperature of HNO_3 = 5 °C); (**d**) HF:HNO_3:CH_3COOH mixtures.

Table 1. RMS of flat (100) Ge surfaces with as-grown film and different etching processes.

	As-Grown	20 °C ALE	15 °C ALE	10 °C ALE	5 °C ALE	HF:HNO_3:CH_3COOH
RMS (nm)	0.67	0.85	1.12	1.39	1.59	2.52

3.4. Application of ALE for Ge Vertical Sandwich GAAFETs (VSAFETs)

The ALE of Ge selective to p^+-$Ge_{0.8}Si_{0.2}$ will be adopted in the vertical sandwich GAAFETs (VSAFETs) to form Ge channel nanowire, as mentioned in the introduction. The cross-sectional SEM images of the p^+-$Ge_{0.8}Si_{0.2}$/Ge/p^+-$Ge_{0.8}Si_{0.2}$ sandwich structure forming vertical nanowire are shown in Figure 11a,b. Figure 11a,b respectively show a 40 nm Ge channel with 15 cycles of ALE and a 15 nm Ge channel with 20 cycles of ALE, implying the well-controlled Ge channel size with ALE. Figure 11c,d show the TEM top view for NS with perimeter 185 nm and NW with perimeter 143 nm formed by 30 cycles of ALE, respectively. The TEM top views were cut at the top drain, the brighter part is the channel, and the black part is metal. The vertical nanosheet (NS) and nanowire (NW) shown in Figure 11c,d were obtained on the same wafer, and the channel size and shape are determined by initial dimensions (defined by electron-beam lithography) and the dual-selective ALE.

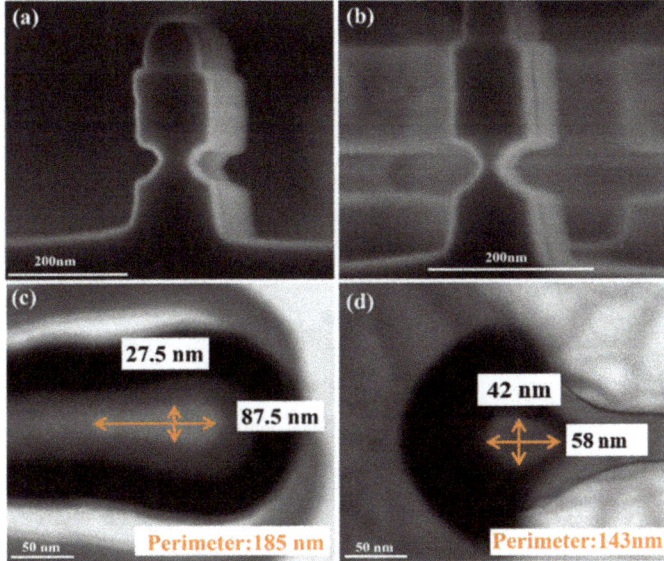

Figure 11. Cross-sectional SEM images of Ge selective etching with ALE, (**a**) 40 nm Ge channel with 15 cycles of ALE, and (**b**) 15 nm Ge channel with 20 cycles of ALE. Cross-sectional TEM top views of Ge VSAFETs: (**c**) NS with perimeter 185 nm, (**d**) NW with a square cross-section with perimeter 143 nm. Oxidation time t_{ox} = 30 s and temperature of HNO_3 = 20 °C.

Figure 12a,b show the cross-sectional SEM and TEM images of the filled high-k metal gate (HKMG) of the gate gap formed by selective etching of the channel with ALE, respectively. Figure 12c shows the TEM image of the gate stack on the side-wall of the hourglass-shaped Ge channel. As shown in Figure 12d–f, the EDX mapping of elements Ge, Si, and W shows sharp contours, proving the absence of element intermixing. The self-saturated dual-selective etch of Ge at (111) planes result in an hourglass-shape with a lateral angle of 54.7°. The current transports along the (111) planes of the hourglass-shaped Ge channel.

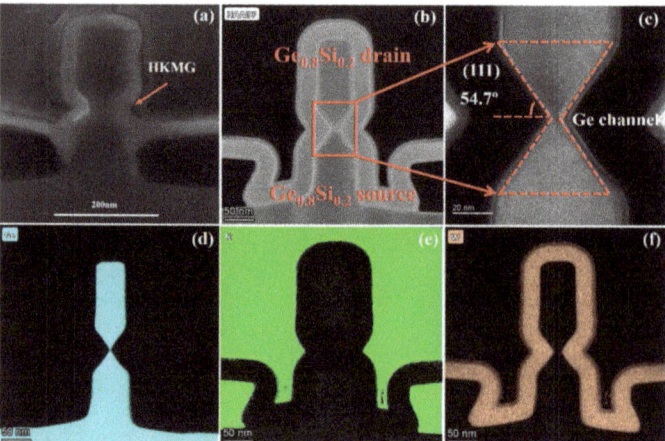

Figure 12. (**a**) SEM and (**b**) TEM images of Ge pVSAFET with gate gaps filling high-k metal gate (HKMG). (**c**) The self-saturated dual-selective etch of Ge at (111) planes result in an hourglass shape with a lateral angle of 54.7°. Energy-dispersive Xray spectroscopy (EDX) mapping of (**d**) Ge (cyan), (**e**) Si (cyan), and (**f**) W (yellow) atoms with sharp contours.

In this stage, a vertical sandwich GAAFET (VSAFET) was processed by ALE method when a Ge nanosheet (NS) with thickness of ~27.5 nm was the channel material as shown in Figure 11c. The process flow of the transistor includes sandwich structure growth, lithographic patterning and plasma anisotropic etching, channel selective etching to form channels, and filling of the high-k metal gate (HKMG) by ALD [14]. Figure 13 illustrates the I_d–V_{ds} output characteristic curve when the I_{on} is 141 µA/um (I_d@V_{ov} = V_{gs} − V_t = −0.6 V, V_{ds} = −1.0 V). It is important to emphasize that these results are preliminary for Ge pVSAFETs, and the device performance will be further studied in the future.

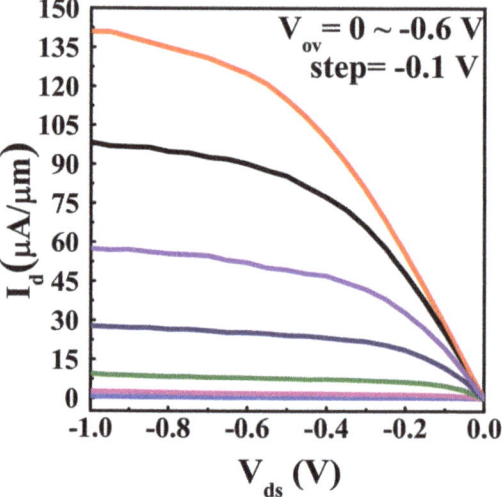

Figure 13. I_d–V_{ds} output characteristic curves for Ge pVSAFET with NS thickness of ~27.5 nm, V_{ov} from 0 to −0.6 V, showing characteristic current saturation behavior. I_d normalizes with the perimeter of NS.

4. Conclusions

In this work, a developed wet dual-selective ALE process with selective etching of Ge and crystal-orientation selectivity of Ge oxidation was proposed. With the oxidation–removal reaction of 70% HNO_3 and deionized (DI) water, the characteristics of $Ge_{0.8}Si_{0.2}$-selective ALE of Ge at different HNO_3 temperatures were investigated systematically. In p^+-$Ge_{0.8}Si_{0.2}$/Ge stacks with (110) sidewalls, the saturated relative etched amount per cycle (REPC) of Ge was 3.1 nm and the saturated etching selectivity was 6.5 at an HNO_3 temperature of 20 °C. The etch rate and the selectivity were affected by HNO_3 temperatures. As the HNO_3 temperature decreased to 10 °C, the REPC of Ge was decreased to 2 nm and the selectivity remained at about 7.4. The Ge channel size in the VSAFETs was well-controlled by ALE. The hourglass-shaped channel of the VSAFETs is formed by the dual-selective ALE of Ge, narrow in the middle and wide close to S/D. Finally, the preliminary I_d–V_{ds} output characteristic curve of Ge pVSAFET was demonstrated.

Author Contributions: Conceptualization, L.X., H.Z., G.W. and H.H.R.; Data curation, L.X.; Funding acquisition, H.Z.; Methodology, L.X., Y.Z., X.A., J.L., A.D., Z.K., Q.W., S.L., C.L., Y.L., and W.H.; Project administration, H.Z.; Supervision, H.Z., G.W., and H.H.R.; Writing—original draft, L.X.; Writing—review & editing, L.X., H.Z., G.W., and H.H.R. All authors have read and agreed to the published version of the manuscript.

Funding: This work was supported by the Academy of Integrated Circuit Innovation under Grant No. Y7YC01X001 and E0YC03X001, the National Key Research and Development Program of China (Grant No. 2016YFA0301701), the Youth Innovation Promotion Association of CAS (Grant No. Y2020037), the National Natural Science Foundation of China (Grant No. 92064002), the Guangdong Greater Bay Area Institute of Integrated Circuit and System (Grant No. 2019B090909006) and the projects of the construction of new research and development institutions (Grant No. 2019B090904015).

Data Availability Statement: The data presented in this study are available on request from the corresponding author.

Conflicts of Interest: The authors declare no conflict of interest.

References

1. Radamson, H.H.; Zhu, H.L.; Wu, Z.H.; He, X.B.; Lin, H.X.; Liu, J.B.; Xiang, J.J.; Kong, Z.Z.; Wang, G.L.; Li, J.; et al. State of the Art and Future Perspectives in Advanced CMOS Technology. *Nanomaterials* **2020**, *10*, 1555. [CrossRef] [PubMed]
2. Radamson, H.H. *Monolithic Nanoscale Photonics-Electronics Integration in Silicon and Other Group IV Elements*; Academic Press: Cambridge, MA, USA, 2014; ISBN 978-0124199750.
3. Liu, M.S.; Schlykow, V.; Hartmann, J.M.; Knoch, J.; Grützmacher, D.; Buca, D.; Zhao, Q.T. Vertical Heterojunction $Ge_{0.92}Sn_{0.08}$/Ge GAA Nanowire pMOSFETs: Low SS of 67 mV/dec, Small DIBL of 24 mV/V and Highest $G_{m, ext}$ of 870 µS/µm. In Proceedings of the IEEE Symposium on VLSI Technology, Honolulu, HI, USA, 16–19 June 2020; pp. 1–2. [CrossRef]
4. Loubet, N.; Hook, T.; Montanini, P.; Yeung, C.-W.; Kanakasabapathy, S.; Guillom, M.; Yamashita, T.; Zhang, J.; Miao, X.; Wang, J.; et al. Stacked nanosheet gate-all-around transistor to enable scaling beyond FinFET. *Symp. VLSI Technol.* **2017**, T230–T231. [CrossRef]
5. Radamson, H.H.; Zhang, Y.B.; He, X.B.; Cui, S.H.; Li, J.J.; Xiang, J.J.; Liu, J.B.; Gu, S.H.; Wang, G.L. The Challenges of Advanced CMOS Process from 2D to 3D. *Appl. Sci.* **2017**, *7*, 1047. [CrossRef]
6. Gu, J.J.; Wang, X.W.; Shao, J.; Neal, A.T.; Manfra, M.J.; Gordon, R.G.; Ye, P.D. III-V gate-all-around nanowire MOSFET process technology: From 3D to 4D. *Int. Electron Devices Meet. (IEDM)* **2012**, 23.7.1–23.7.4. [CrossRef]
7. Radamson, H.H.; Simoen, E.; Luo, J.; Zhao, C. *Past, Present and Future of CMOS*; Woodhead Publishing: Cambridge, UK, 2018; pp. 95–114. ISBN 978-008-102-139-2.
8. Singh, N.; Buddharaju, K.D.; Manhas, S.K.; Agarwal, A.; Rustagi, S.C.; Lo, G.Q.; Balasubramanian, N.; Kwong, D.L. Si, SiGe nanowire devices by top–down technology and their applications. *IEEE Trans. Electron. Devices* **2008**, *55*, 3107–3118. [CrossRef]
9. Xu, W.J.; Wong, H.; Kakushima, K.; Iwai, H. Quasi-analytical model of ballistic cylindrical surrounding gate nanowire MOSFET. *Microelectron. Eng.* **2015**, *138*, 111–117. [CrossRef]
10. Bae, G.; Bae, D.I.; Kang, M.; Hwang, S.M.; Kim, S.S.; Seo, B.; Kwon, T.Y.; Lee, T.J.; Moon, C.; Choi, Y.M.; et al. 3 nm GAA technology featuring multi-bridge-channel FET for low power and high-performance applications. In Proceedings of the 2018 IEEE International Electron Devices Meeting (IEDM), San Francisco, CA, USA, 1–5 December 2018; pp. 28.7.1–28.7.4. [CrossRef]
11. Veloso, A.; Altamirano-Sánchez, E.; Brus, S.; Chan, B.T.; Cupak, M.; Dehan, M.; Delvaux, C.; Devriendt, K.; Eneman, G.; Ercken, M.; et al. Vertical Nanowire FET Integration and Device Aspects. *ECS Trans.* **2016**, *72*, 31–42. [CrossRef]

12. Radamson, H.H.; He, X.B.; Zhang, Q.Z.; Liu, J.B.; Cui, H.S.; Xiang, J.J.; Kong, Z.Z.; Xiong, W.; Li, J.; Gao, J.; et al. Miniaturization of CMOS. *Micromachines* **2019**, *10*, 293. [CrossRef]
13. Toriumi, A.; Tabata, T.; Lee, C.H.; Nishimura, T.; Kita, K.; Nagashio, K. Opportunities and challenges for Ge CMOS-Control of interfacing field on Ge is a key. *Microelectron. Eng.* **2009**, *86*, 1571–1576. [CrossRef]
14. Yin, X.G.; Zhang, Y.K.; Zhu, H.L.; Wang, G.L.; Li, J.J.; Du, A.Y.; Li, C.; Zhao, L.H.; Huang, W.X.; Yang, H.; et al. Vertical Sandwich Gate-All-Around Field-Effect Transistors with Self-Aligned High-k Metal Gates and Small Effective-Gate-Length Variation. *IEEE Electron Device Lett.* **2020**, *41*, 8–11. [CrossRef]
15. Zhang, Y.K.; Ai, X.Z.; Yin, X.G.; Zhu, H.L.; Yang, H.; Wang, G.L.; Li, J.J.; Du, A.Y.; Li, C.; Huang, W.X.; et al. Vertical Sandwich GAA FETs with Self Aligned High-k Metal Gate Made by Quasi Atomic Layer Etching Process. *IEEE Trans. Electron Devices* **2021**. [CrossRef]
16. Porret, C.; Vohra, A.; Sebaai, F.; Douhard, B.; Hikavyy, A.; Loo, R. A New Method to Fabricate Ge Nanowires: Selective Lateral Etching of GeSn:P/Ge Multi-Stacks. *Solid State Phenom.* **2018**, *282*, 113–118. [CrossRef]
17. Fischer, A.C.; Belova, L.M.; Rikers, Y.G.M.; Malm, B.G.; Radamson, H.H.; Kolahdouz, M.; Gylfason, K.B.; Stemme, G.; Niklaus, F. 3D free-form patterning of silicon by ion implantation, silicon deposition and selective silicon etching. *Adv. Funct. Mater.* **2012**, *22*, 4004–4008. [CrossRef]
18. Gupta, S.; Chen, R.; Huang, Y.C.; Kim, Y.; Sanchez, E.; Harris, J.S.; Saraswat, K.C. Highly Selective Dry Etching of Germanium over Germanium–Tin ($Ge_{1-x}Sn_x$): A Novel Route for $Ge_{1-x}Sn_x$ Nanostructure Fabrication. *Nano Lett.* **2013**, *13*, 3783–3790. [CrossRef]
19. Lee, Y.J.; Hou, F.J.; Chuang, S.S.; Hsueh, F.K.; Kao, K.H.; Sung, P.J.; Yuan, W.Y.; Yao, J.Y.; Lu, Y.C.; Lin, K.L.; et al. Diamond-shaped Ge and $Ge_{0.9}Si_{0.1}$ Gate-All-Around Nanowire FETs with Four {111} Facets by Dry Etch Technology. In Proceedings of the 2015 IEEE International Electron Devices Meeting (IEDM), Washington, DC, USA, 7–9 December 2015; pp. 15.1.1–15.1.4. [CrossRef]
20. Xie, L.; Zhu, H.L.; Zhang, Y.K.; Ai, X.Z.; Wang, G.L.; Li, J.J.; Du, A.Y.; Kong, Z.Z.; Yin, X.G.; Li, C.; et al. Strained $Si_{0.2}Ge_{0.8}$/Ge multilayer Stacks Epitaxially Grown on a Low-/High-Temperature Ge Buffer Layer and Selective Wet-Etching of Germanium. *Nanomaterials* **2020**, *10*, 1715. [CrossRef]
21. Sebaai, F.; Witters, L.; Holsteyns, F.; Wostyn, K.; Rip, J.; Yukifumi, Y.; Lieten, R.R.; Bilodeau, S.; Cooper, E. Wet Selective SiGe Etch to Enable Ge Nanowire Formation. *Solid State Phenom.* **2016**, *255*, 3–7. [CrossRef]
22. Liu, W.D.; Lee, Y.C.; Sekiguchi, R.; Yoshida, Y.; Komori, K.; Wostyn, K.; Sebaai, F.; Holsteyns, F. Selective Wet Etching in Fabricating SiGe and Ge Nanowires for Gate-All-Around MOSFETs. *Solid State Phenom.* **2018**, *282*, 101–106. [CrossRef]
23. Holländer, B.; Buca, D.; Mantl, S.; Hartmann, J.M. Wet Chemical Etching of Si, $Si_{1-x}Ge_x$, and Ge in HF: H_2O_2:CH_3COOH. *J. Electrochem. Soc.* **2010**, *157*, H643–H646. [CrossRef]
24. Witters, L.; Arimura, H.; Sebaai, F.; Hikavyy, A.; Milenin, A.P.; Loo, R.; De Keersgieter, A.; Eneman, G.; Schram, T.; Wostyn, K.; et al. Strained Germanium Gate-All-Around pMOS Device Demonstration Using Selective Wire Release Etch Prior to Replacement Metal Gate Deposition. *IEEE Trans. Electron Devices* **2017**, *64*, 4587–4593. [CrossRef]
25. Kanarika, K.J.; Tan, S.; Yang, W.; Kim, T.; Lill, T.; Kabansky, A.; Hudson, E.A.; Ohba, T.; Nojiri, K.; Yu, J.; et al. Predicting synergy in atomic layer etching. *J. Vac. Sci. Technol. A* **2017**, *35*, 05C302. [CrossRef]
26. Kanarika, K.J.; Lill, T.; Hudson, E.A.; Sriraman, S.; Tan, S.; Marks, J.; Vahedi, V.; Gottscho, R.A. Overview of atomic layer etching in the semiconductor industry. *J. Vac. Sci. Technol. A* **2015**, *33*, 020802. [CrossRef]
27. Ikeda, H.; Imai, S.; Matsumura, M. Atomic layer etching of germanium. *Appl. Surf. Sci.* **1997**, *112*, 87–91. [CrossRef]
28. Li, C.; Zhu, H.L.; Zhang, Y.K.; Yin, X.G.; Jia, K.P.; Li, J.J.; Wang, G.L.; Kong, Z.Z.; Du, A.Y.; Yang, T.Z.; et al. Selective Digital Etching of Silicon–Germanium Using Nitric and Hydrofluoric Acids. *ACS Appl. Mater. Interfaces* **2020**, *12*, 48170–48178. [CrossRef] [PubMed]
29. Yin, X.G.; Zhu, H.L.; Zhao, L.H.; Wang, G.L.; Li, C.; Huang, W.X.; Zhang, Y.K.; Jia, K.P.; Li, J.J.; Radamson, H.H. Study of Isotropic and Si-Selective Quasi Atomic Layer Etching of $Si_{1-x}Ge_x$. *ECS J. Solid-State Sci. Technol.* **2020**. [CrossRef]
30. Li, Y.Y.; Zhu, H.L.; Kong, Z.Z.; Zhang, Y.K.; Ai, X.Z.; Wang, G.L.; Wang, Q.; Liu, Z.Y.; Lu, S.S.; Xie, L.; et al. The Effect of Doping on the Digital Etching of Silicon-Selective Silicon–Germanium Using Nitric Acids. *Nanomaterials* **2021**, *11*, 1209. [CrossRef]
31. Radamson, H.H.; Joelsson, K.B.; Ni, W.-X.; Hultman, L.; Hansson, G.V. Characterization of highly boron-doped Si, $Si_{1-x}Ge_x$ and Ge layers by high-resolution transmission electron microscopy. *J. Cryst. Growth* **1995**, *157*, 80–84. [CrossRef]
32. Du, Y.; Kong, Z.Z.; Toprak, M.S.; Wang, G.L.; Miao, Y.H.; Xu, B.Q.; Yu, J.H.; Li, B.; Lin, H.X.; Han, J.H.; et al. Investigation of the Heteroepitaxial Process Optimization of Ge Layers on Si (001) by RPCVD. *Nanomaterials* **2021**, *11*, 928. [CrossRef]
33. Prabhakaran, K.; Ogino, T. Oxidation of Ge (100) and Ge (111) surfaces: An UPS and XPS study. *Surf. Sci.* **1995**, *325*, 263–271. [CrossRef]
34. Wostyn, K.; Sebaai, F.; Rip, J.; Mertens, H.; Witters, L.; Loo, R.; Hikavyy, A.; Milenin, A.; Horiguchi, N.; Collaert, N.; et al. Selective Etch of Si and SiGe for Gate All-Around Device Architecture. *ECS Trans.* **2015**, *69*, 147–152. [CrossRef]
35. Cretella, M.C.; Gatos, H.C. The Reaction of Germanium with Nitric Acid Solutions. *J. Electrochem. Soc.* **1958**, *105*, 487. [CrossRef]
36. Xue, Z.Y.; Wei, X.; Liu, L.J.; Chen, D.; Zhang, B.; Zhang, M.; Wang, X. Etch characteristics of $Si_{1-x}Ge_x$ films in HNO_3: H_2O: HF. *Sci. China Technol. Sci.* **2011**, *54*, 2802. [CrossRef]
37. Koyama, K.; Hiroi, M.; Tatsumi, T.; Hirayama, H. Etching characteristics of $Si_{1-x}Ge_x$ alloy in ammoniac wet cleaning. *Appl. Phys. Lett.* **1990**, *57*, 2202–2204. [CrossRef]

38. Spadafora, M.; Privitera, G.; Terrasi, A. Oxidation rate enhancement of SiGe epitaxial films oxidized in dry ambient. *Appl. Phys. Lett.* **2003**, *83*, 3713–3715. [CrossRef]
39. Yeo, C.C.; Cho, B.J.; Gao, F.; Lee, S.J.; Lee, M.H.; Yu, C.Y.; Liu, C.W.; Tang, L.J.; Lee, T.W. Electron mobility enhancement using ultrathin pure Ge on Si substrate. *IEEE Electron Device Lett.* **2005**, *26*, 761–763. [CrossRef]
40. Seidel, H.; Csepregi, L.; Heuberger, A.; Baumgartel, H. Anisotropic etching of crystalline silicon in alkaline solutions. *J. Electrochem. Soc.* **1990**, *137*, 3612.
41. Radamson, H.H.; Hallstedt, J. Application of high-resolution X-ray diffraction for detecting defects in SiGe (C) materials. *J. Phys. Condens. Matter* **2005**, *17*, S2315–S2322. [CrossRef]

Article

The Effect of Doping on the Digital Etching of Silicon-Selective Silicon–Germanium Using Nitric Acids

Yangyang Li [1,2], Huilong Zhu [1,*], Zhenzhen Kong [1], Yongkui Zhang [1], Xuezheng Ai [1], Guilei Wang [1,2,3], Qi Wang [1], Ziyi Liu [1,2], Shunshun Lu [1], Lu Xie [1,2], Weixing Huang [1,2], Yongbo Liu [1,2], Chen Li [1,2], Junjie Li [1], Hongxiao Lin [1,3], Jiale Su [1], Chuanbin Zeng [4] and Henry H. Radamson [1,2,3,*]

1 Key Laboratory of Microelectronics Devices & Integrated Technology, Institute of Microelectronics, Chinese Academy of Sciences, Beijing 100029, China; liyangyang@ime.ac.cn (Y.L.); kongzhenzhen@ime.ac.cn (Z.K.); zhangyongkui@ime.ac.cn (Y.Z.); aixuezheng@ime.ac.cn (X.A.); wangguilei@ime.ac.cn (G.W.); wangqi@ime.ac.cn (Q.W.); liuziyi@ime.ac.cn (Z.L.); lushunshun@ime.ac.cn (S.L.); xielu@ime.ac.cn (L.X.); huangweixing@ime.ac.cn (W.H.); liuyongbo@ime.ac.cn (Y.L.); lichen2017@ime.ac.cn (C.L.); lijunjie@ime.ac.cn (J.L.); linhongxiao@ime.ac.cn (H.L.); sujiale@ime.ac.cn (J.S.)
2 University of Chinese Academy of Sciences, Beijing 100049, China
3 Research and Development Center of Optoelectronic Hybrid IC, Guangdong Greater Bay Area Institute of Integrated Circuit and System, Guangzhou 510535, China
4 Institute of Microelectronics, Chinese Academy of Sciences, Beijing 100029, China; chbzeng@ime.ac.cn
* Correspondence: zhuhuilong@ime.ac.cn (H.Z.); rad@ime.ac.cn (H.H.R.)

Abstract: Gate-all-around (GAA) field-effect transistors have been proposed as one of the most important developments for CMOS logic devices at the 3 nm technology node and beyond. Isotropic etching of silicon–germanium (SiGe) for the definition of nano-scale channels in vertical GAA CMOS and tunneling FETs has attracted more and more attention. In this work, the effect of doping on the digital etching of Si-selective SiGe with alternative nitric acids (HNO_3) and buffered oxide etching (BOE) was investigated in detail. It was found that the HNO_3 digital etching of SiGe was selective to n^+-Si, p^+-Si, and intrinsic Si. Extensive studies were performed. It turned out that the selectivity of SiGe/Si was dependent on the doped types of silicon and the HNO_3 concentration. As a result, at 31.5% HNO_3 concentration, the relative etched amount per cycle (REPC) and the etching selectivity of $Si_{0.72}Ge_{0.28}$ for n^+-Si was identical to that for p^+-Si. This is particularly important for applications of vertical GAA CMOS and tunneling FETs, which have to expose both the n^+ and p^+ sources/drains at the same time. In addition, the values of the REPC and selectivity were obtained. A controllable etching rate and atomically smooth surface could be achieved, which enhanced carrier mobility.

Keywords: vertical gate-all-around (vGAA); digital etch; quasi-atomic-layer etching (q-ALE); selective wet etching; HNO_3 concentration; doping effect

1. Introduction

Gate-all-around (GAA) nanowire transistors are ideal candidates for various CMOS applications due to their outstanding gate control, excellent performance, immunity to short-channel effects, and scalability [1–3]. Tunneling field-effect transistors (TFETs) have arisen as promising devices with emerging device concepts by breaking through the subthreshold swing limit of 60 mV/dec for low-power applications [4–6]. GAA nanowire TFETs have become candidates for substitutes for conventional MOS technology, especially in terms of their energy efficiency and scaling due to the better electrostatic control of the tunneling carriers provided by their nanowire structure [7–10]. SiGe channel materials have been introduced due to their excellent bandgap, high mobility, high density of states, and high compatibility with existing CMOS technology [11,12]. In order to precisely define the nanowire diameter and effective gate length, SiGe materials need to be selectively etched with accurate etching depth control and high selectivity for both n^+-Si and p^+-Si for

CMOS and TFET applications, which have to expose both the n$^+$ and p$^+$ sources/drains at the same time.

Several techniques have been proposed for selective etching of SiGe, such as mixtures of HNO_3, HF, and H_2O [13–15], as well as solutions of H_2O_2, HF, and CH_3COOH [16,17]. Unfortunately, the wet etching of mixtures is not appropriate for small-sized features due to the high etching rate [18,19]. Vapor etching using gaseous HCl in a chemical vapor deposition (CVD) reactor is also limited because of its high-temperature process, which degrades the sharpness of the junction [20]. Moreover, dry etching using CF_4-based plasma has been extensively researched [21–23]. The disadvantage is that the plasma equipment is more complex and the loading effect is serious [24]. The etching techniques mentioned above involve continuous etching that is controlled by the etching time. Therefore, they do not meet the requirements of nano-scale transistors for process control. Atomic-layer etching (ALE) draws has significantly attracted researchers and the industrial community due to its self-limiting characteristics. The superiority of ALE techniques over other methods is due to the controllable etching rate and excellent variation control [25,26]. It has been employed for the etching of dielectrics [26,27], some nitrides [28], and metals [29,30]. Recently, an isotropic and quasi-ALE (q-ALE) method for Si-selective SiGe was proposed and reported by our group [31,32]. This q-ALE method is based on a cyclic oxidation–etching process in which hydrogen peroxide (H_2O_2) [32] or nitric acid (HNO_3) [31] and buffered oxide etchants (BOEs) are separately used as an oxidant and an oxide remover agent, which is also called digital etching. The experimental etching rate of about 5 Å (approximately four monolayers) per cycle accounted for the quasi-self-limited behavior in our q-ALE process [31,32]. This was explained and understood from the perspective of the activation energy, which was extracted by fitting the experimental data with the proposed oxidation model [31]. The works mentioned above mainly focused on the digital etching characteristics of SiGe that is selective of p-type doped Si. However, the digital etching of SiGe that is selective of n-type doped Si and intrinsic Si has not been studied.

In this work, the effect of doping on digital wet etching of SiGe that is selective of Si was investigated systematically. The digital etching was based on a combination of HNO_3 and buffered oxide etchants (BOEs) as an oxidant and an oxide remover agent, respectively. The selectivity characteristics of SiGe for n$^+$-Si were demonstrated. The effects of different parameters on the selectivity of the etching of germanium–silicon, such as for Si doping, HNO_3 concentration, and SiGe doping, were examined and discussed in detail.

2. Materials and Methods

The substrates were 8 inch p-type Si (100) wafers with a resistivity of 8–12 ohm·cm. The p$^+$-Si/SiGe/n$^+$-Si stack layers were grown in an ASM E2000 plus RPCVD reactor (ASM, Munich, Germany). First, after a standard pre-epitaxial cleaning, the wafers were baked at 900 °C in ambient H_2 with a pressure of 20 Torr for 5 min, achieving a pure and smooth silicon surface [32]. Then, the p$^+$-Si/SiGe/n$^+$-Si stack layers and p$^+$-Si/SiGe/i-Si stack layers were grown at 650 °C using an adjusted gas source with H_2 as a carrier gas. Dichlorosilane (SiH_2Cl_2), germane (GeH_4), diborane (B_2H_6), and phosphine (PH_3) were utilized as gas precursors of Si, Ge, B, and P, respectively. The Ge incorporation, P concentration, and B concentration in silicon were achieved by tuning the gas flow and gas pressure. Finally, the epitaxial stack layers were fabricated. Then, a hard mask was deposited on the epitaxial stacked layers, and the pattern was formatted with an optical lithography with an I-line. The Si/SiGe stack layers were etched using hydrogen bromide (HBr)-based dry anisotropic etching. The details of the sample preparation can be found in [32]. Afterwards, the prepared samples were cut into same-sized slices of about 3×3 cm^2 to facilitate the etching experiments.

There were five kinds of Si/SiGe stack layer structures, as shown in Figure 1a–c. Sample I was a laminated structure in which ~300 nm p-type doped Si with a boron dopant concentration of 1.0×10^{20} cm^{-3}, 55 nm intrinsic $Si_{0.72}Ge_{0.28}$, and 120 nm n-type doped Si with a phosphorus dopant concentration of 1.7×10^{19} cm^{-3} were epitaxially grown in sequence. Bottom p-type doped Si was etched to ~120 nm. The structural diagram is shown in Figure 1a. Sample II was a laminated structure in which ~300 nm p-type doped Si with a boron dopant concentration of 9×10^{19} cm^{-3}, 55 nm intrinsic Si$_{0.72}$Ge$_{0.28}$, and 120 nm intrinsic Si were grown in situ and in sequence. The structural diagram is shown in Figure 1b. In Sample III, arsenic (As) ion implantation with the energy of 30 keV and dose of 4×10^{15} cm^{-2} was performed on the top intrinsic Si, and then 900 °C spike annealing was carried out to activate arsenic in the top Si. The structural diagram is shown in Figure 1c. This sample was employed to demonstrate the digital etching characteristics of n-type doped Si with the implantation of As. Sample IV was a laminated structure with nine Si$_{0.72}$Ge$_{0.28}$ layers, as shown in Figure 1d. The n$^+$-SiGe layers with in situ phosphorus included SiGe layer 1, SiGe layer 2, and SiGe layer 3. The concentrations were 2×10^{19}, 1.3×10^{19}, and 2×10^{19} cm^{-3}, respectively. The intrinsic SiGe layers consisted of SiGe layer 4, SiGe layer 5, and SiGe layer 6. The p$^+$-SiGe layers with in situ boron included SiGe layer 7 and SiGe layer 8 with a concentration of 4×10^{19} cm^{-3}. SiGe layer 9 was doped with boron with a concentration of 4×10^{19} cm^{-3} and arsenic with a concentration of 4×10^{19} cm^{-3}. The thickness of the SiGe was ~35 nm. The thickness of the Si was ~50 nm. The sample was used for the investigation of the digital etching characteristics of doped SiGe. Sample V was a laminated structure with ~30 nm n$^+$-Si layers and ~35 nm n$^+$-SiGe layers with a varying Ge fraction; these layers were alternated, and the sample was used to examine the influence of the Ge mole fraction.

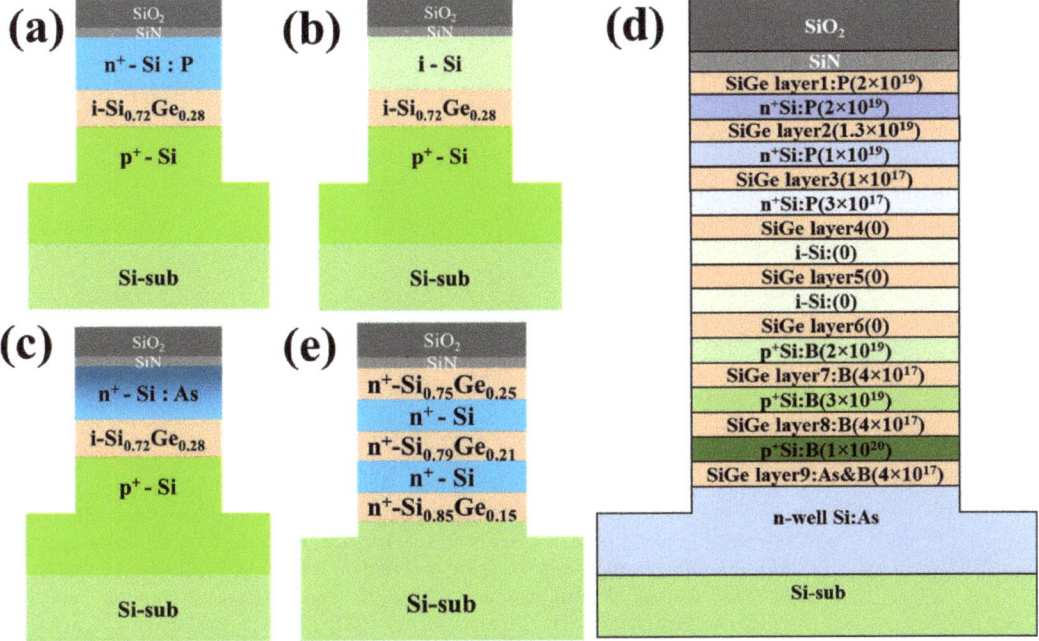

Figure 1. Scheme of the laminated structure with a lateral opening: (**a**) in situ n$^+$-Si/i-SiGe/p$^+$-Si; (**b**) i-Si/i-SiGe/p$^+$-Si; (**c**) implanted n$^+$-Si/i-SiGe/p$^+$-Si; (**d**) SiGe/Si multilayers with different doping types; (**e**) SiGe/Si multilayers with different Ge fractions.

The q-ALE process of digital wet etching, including oxidation, deionized (DI) water rinsing, oxide removal, and DI water rinsing, and the investigations on the self-limiting behavior of SiGe etching were described previously [31,32]. The flow is shown in Figure 2. The diluted BOE solutions were utilized for sample pretreatment. The steps within dotted border, including HNO₃ oxidation, DI water rinsing, oxide removal, and DI water rinsing, were repeated for many cycles until the desired etching amount was reached. The HNO₃ solutions in the experiments were prepared by adjusting the volume of analytical-grade nitric acid (70% (wt/wt)) and the volume of deionized water under the condition that the total volume of the solution was kept constant (2 L). The concentrations of the HNO₃ solutions were monitored with a high-precision density meter. The values of concentrations mentioned in this paper represent mass fractions. The nitric acid solutions were employed for oxidation, and were then cooled to room temperature before use. The oxidation time was set to 60 s. It was long enough to reach saturation with an oxidation time of 27.6 s [31]. The BOE solutions used in the experiments were prepared by diluting the original BOEs (NH₄F 34.8%, HF 6.23%) 50 times with deionized water, and the total volumes of the BOE solutions were kept at 2 L. The BOE solutions were used for oxide removal. HF/BOE concentrations that were too high would damage the Si or SiGe layers. The oxide removal time and DI water rinsing time were fixed at 60 s, which ensured the complete removal of oxides and the non-existent cross-contamination of solutions. The temperature of the control recipe was kept at room temperature (20.5 ± 0.5 °C). Unless otherwise specified, the oxidation–etching procedure in every experiment was performed for 50 cycles repeatedly. Additionally, the H_2O_2 (30% (wt/wt)) solutions were prepared for the H_2O_2 q-ALE experiments as a comparison with the HNO₃ q-ALE experiments.

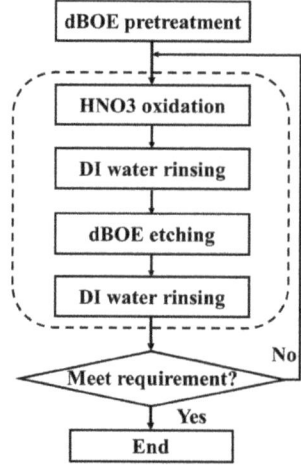

Figure 2. Flow diagram of the main process of digital etching.

The etched morphology of the samples and the etched depth were examined with scanning electron microscopy (SEM) (Hitachi, Tokyo, Japan). Secondary ion mass spectroscopy (SIMS) was used to analyze the doping and mole fraction. Atomic force microscopy (AFM) (Dimension Icon AFM, Bruker, Billerica, MA, USA) was used to measure the surface roughness. High-resolution X-ray diffraction (HRXRD) (Delta-X, Bruker, Billerica, MA, USA) was used to determine the crystallinity and strain relaxation of the Si/SiGe/Si structures.

3. Results and Discussion

3.1. n-Type Doped Si Selectivity with H_2O_2 or HNO₃ q-ALE

The digital etching of SiGe and selectivity of SiGe for p⁺-Si were previously identified with H_2O_2-dBOE q-ALE and HNO₃-dBOE q-ALE [31,32]. We chose the above two q-ALE processes to investigate the selectivity of n-type silicon. Figure 3a,b show the SEM cross-section images of Sample I with n⁺-Si and in situ phosphorus after etching for 40 cycles with 30% H_2O_2 q-ALE and 40 cycles with 31.5% HNO₃ q-ALE. The relative etching amounts (REAs) of SiGe/p⁺-Si with H_2O_2 and with HNO₃ were 18.4 nm (see Figure 3a) and 24.8 nm (see Figure 3b), respectively. The REPC was calculated by dividing the REA by the number of etching cycles. The REPC with H_2O_2 was 0.46 nm, which was almost identical to the previous results [32]. The REPC with HNO_3 was 0.62 nm, 20% higher than the previous value [31]. This may have been caused by the increase in nitric acid concentration.

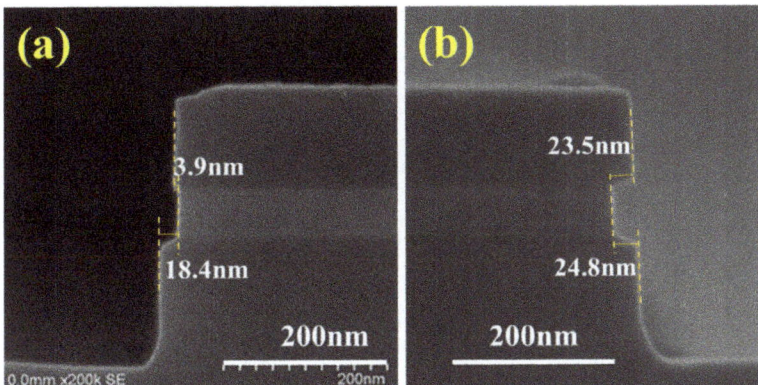

Figure 3. The SEM cross-section images of Sample I after digital etching at 40 cycles with (**a**) 30% H_2O_2-dBOE q-ALE and (**b**) 31.5% HNO_3-dBOE q-ALE.

It is shown in Figure 3a that the REA of SiGe/n$^+$-Si with H_2O_2 was just 3.9 nm, which is smaller than that of SiGe/p$^+$-Si, indicating poor selectivity for n$^+$-Si and the high reactivity of n-Si. Therefore, H_2O_2-dBOE q-ALE was not suitable for p$^+$-Si/SiGe/n$^+$-Si structure etching. The differences between n$^+$-Si and p$^+$-Si in terms of selectivity and etching rate might be related to the types of carriers or the dopant types. Sample III with the arsenic ion implantation was assessed using 50 cycles with the H_2O_2-dBOE and HNO_3-dBOE q-ALE process. The SEM cross-section images of Sample III are shown in Figure 4a,b. The results are almost consistent. As shown in Figure 4a, with H_2O_2 q-ALE, Sample III exhibited weak selectivity for n$^+$-Si formed by As implantation. It was demonstrated that the carrier type—instead of dopant type—enhanced the etching rate of n$^+$-Si in the H_2O_2 q-ALE process. The high concentration of electrons in n$^+$-Si might accelerate oxide growth in H_2O_2 solutions, which could be explained by the improved relativity of Si-Si back bonds [33].

As shown in Figure 3a,b, the REA of SiGe/n$^+$-Si with HNO_3 was obviously larger than with H_2O_2, and was close to that of p$^+$-Si. Similar results are shown in Figure 4a,b. This indicates the excellent selectivity for n$^+$-Si with the HNO_3 q-ALE process compared with the H_2O_2 q-ALE process, regardless of if in situ doped Si or implanted Si is used. In addition, the etched notch on top of Sample III shown in Figure 4b is assumed to be the result of high dose implantation. Figure 4c shows the SIMS data of boron/arsenic doping and the Ge/Si fraction. The results showed that the dopant concentration was above 1×10^{20} cm^{-3} within a depth of about 100 nm. Such high arsenic doping might lead to local polycrystalline or even amorphous characteristics, which enhance the etching reaction. In addition, a phenomenon that was not easy to observe was that the etching rate of SiGe near the n-type Si was slightly faster than that near the p-type Si. In the arsenic doping profile shown in Figure 4c, the arsenic was distributed in the SiGe. This might have been caused by arsenic implantations. However, the boron distribution in SiGe was negligible. It was considered that the digital etching of SiGe is dependent on the doping of SiGe. We will perform an in-depth study in the third part.

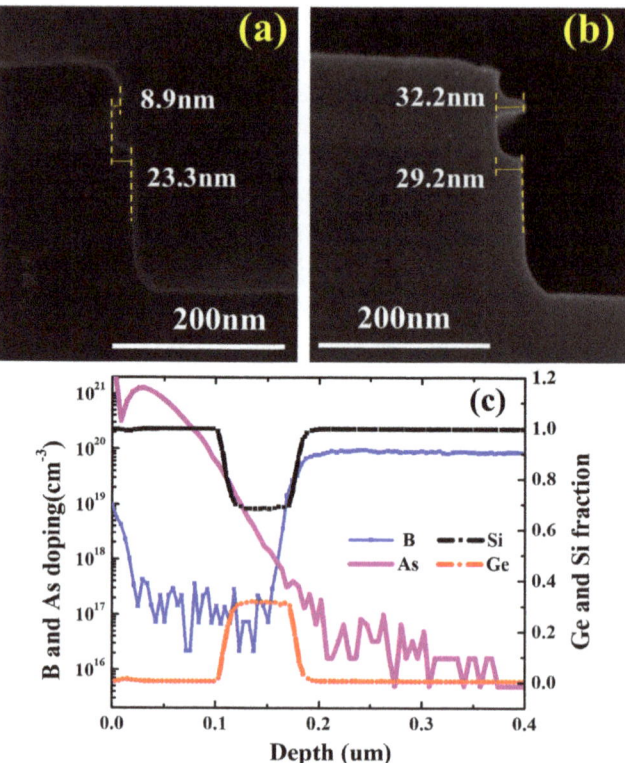

Figure 4. The SEM cross-section images of Sample III after digital etching at 50 cycles with (**a**) 30% H_2O_2-dBOE q-ALE and (**b**) 31.5% HNO_3-dBOE q-ALE. (**c**) SIMS data of boron/arsenic and the Ge/Si mole fraction in Sample III. An abrupt B profile was formed by in situ doped epi, as the profile exhibits a large diffusion into SiGe.

3.2. Effect of Doped Si and HNO_3 Concentration Dependence

In order to explore the effect of doping in silicon on the selectivity of SiGe etching, 31.5% HNO_3 q-ALE experiments were carried out with Sample I and Sample II. The structures of SiGe/n^+-Si, SiGe/p^+-Si, and SiGe/i-Si were included and could be investigated. All of the samples were processed together. Groups of samples including Sample I and Sample II were taken out every 50 cycles. For samples with different doping conditions between the top silicon and bottom silicon, the etching morphologies of the top and bottom silicon might be different. For example, the SEM images of Sample II shown in Figure S1 exhibited different REA values for SiGe/i-Si from those of SiGe/p^+-Si and different i-Si losses with p^+-Si loss.

The structural diagram of the etching morphology is shown in Figure 5. The dashed line in Figure 5 represents the initial envelope lines of the fresh sample. The solid boxes are the envelope lines as they were etched. The angle between the etching slope at the surface of the etched Si and the horizontal direction is θ. Silicon was etched in the vertical and lateral directions. The etching amounts are described as the vertical Si loss (Si loss_v) and lateral Si loss (Si loss_l). As discussed in a previous work [31], the influence of crystal planes on the etching rate was ignored, that is, Si loss_v was almost equal to Si loss_l. The etching amount in the vertical direction could be directly measured. Therefore, the etching amount in the vertical direction is usually regarded as the Si etching amount (Si loss) in the following section. Silicon–germanium was only etched laterally. The etching amount can be described as the sum of the REA and Si loss_l. The selectivity can be expressed as the ratio of SiGe loss to Si loss, as described in Equation (1).

$$selectivity = \frac{(REA + Si\ loss)}{Si\ loss} = 1 + \cot\theta \qquad (1)$$

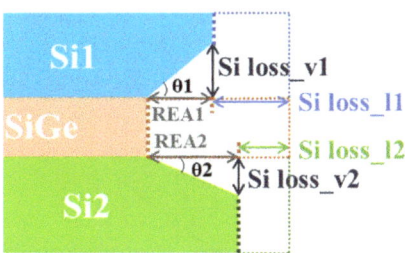

Figure 5. The structural diagram of the etching morphology. The dashed lines represent the initial envelope lines of the fresh sample. The solid boxes are the envelope lines as they were etched.

In addition, in Sample I and Sample II, the diffusion of impurities from silicon to silicon–germanium was negligible, and SiGe could be regarded as intrinsic. The germanium component was fixed in the whole SiGe layer. Therefore, the q-ALE etching of SiGe/Si1 and the q-ALE etching of SiGe/Si2 were independent of each other. The selectivity could be separately calculated by using the Equation (1). According to the angle calculation method and the length calculation method, the values of the selection ratios were very close. This was proved by our experimental data.

Figure 6 shows the dependence of the REA and Si loss on the number of etching cycles for SiGe/n^+-Si, SiGe/p^+-Si, and SiGe/i-Si. The scatters in Figure 6 are the data points obtained through the experiments, and the lines are the curves fitted linearly according to the experimental data. It is shown that the REA of SiGe/n^+-Si, REA of SiGe/p^+-Si, and REA of SiGe/i-Si were highly linear dependent on the number of etching cycles, which was confirmed by the R_square up to 0.975. The Si losses of n^+-Si, p^+-Si, and i-Si were also linearly related to the number of cycles. In the table embedded in Figure 6, the fitting slopes of the REA curves and the Si loss curves represent the REPC and silicon etching amount of each cycle (EPC). It was shown that the REPC of SiGe/p^+-Si was 0.6079 nm, which was close to the REPC value of SiGe/n^+-Si (0.6389 nm). The EPC of p^+-Si was 0.2262 nm, which was also close to the EPC value of n^+-Si (0.2255 nm). The results indicate that the 31.5% HNO_3 concentration had the same etching rate for p^+-Si and n^+-Si. The concentration is expected to be used for the digital etching of p^+-Si/SiGe/n^+-Si stack structures, such as GAA CMOS and TFET applications. Moreover, the slopes in the fitting curves of SiGe/i-Si REA are lower than that of doped Si, suggesting its poor selectivity for i-Si. The EPC of i-Si was 0.3732 nm. It was demonstrated that the etching rate was larger than that of doped Si. It was considered that the doping of silicon contributed to the better selectivity for silicon with the nitric acid etching of SiGe.

Figure 7 shows the selectivity of SiGe/n^+-Si, SiGe/p^+-Si, and SiGe/i-Si. The selectivity was calculated with Equation (1). The experimental data were obtained by measuring the SEM images of Sample I and Sample II with the 31.5% HNO_3 q-ALE process. We carried out the experiments six times on Sample I and Sample II, and six sets of data were obtained. The mean values of the SiGe/n^+-Si, SiGe/p^+-Si, and SiGe/i-Si selectivity were 3.59, 3.68, and 2.56, respectively. The values of the standard deviations were 0.0759, 0.1228, and 0.2512, respectively. The results show the significant improvements in selectivity for doped Si compared with intrinsic Si. The selectivity of SiGe/n^+-Si and SiGe/p^+-Si was similar—40% larger than that of intrinsic silicon. It was demonstrated that doped Si was more difficult to etch in the process of digital etching, which might be due to the oxidation difficulty in the HNO_3 solutions. Moreover, it was observed that the variation in SiGe/n^+-Si was larger than that in SiGe/p$^+$-Si. This indicates that it is more susceptible to process factors, such as concentration monitoring and solution preparation. The selectivity might be sensitive to the actual HNO$_3$ concentration.

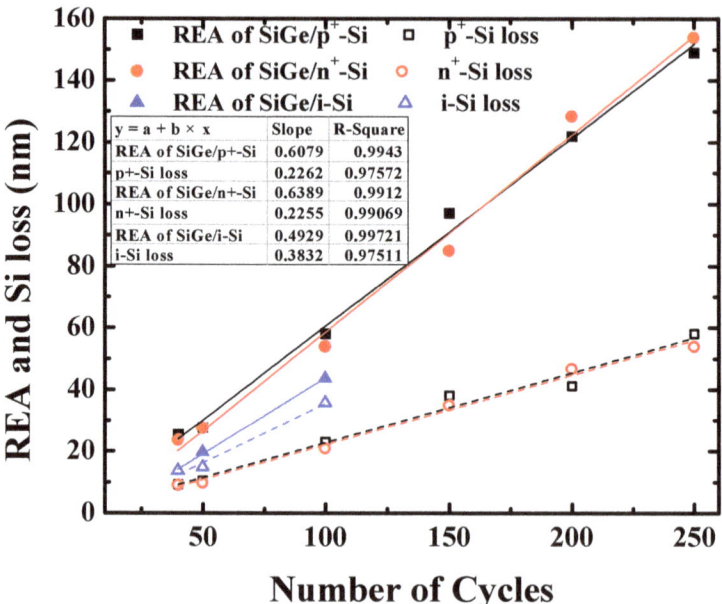

Figure 6. Dependence of the REA and Si loss on the number of etching cycles for SiGe/n$^+$-Si, SiGe/p$^+$-Si, and SiGe/i-Si. The scatters are the experimental data, and the lines are the linear fitting curves of the experimental data. The slopes represent the REPC and silicon etching amounts for each cycle (EPC).

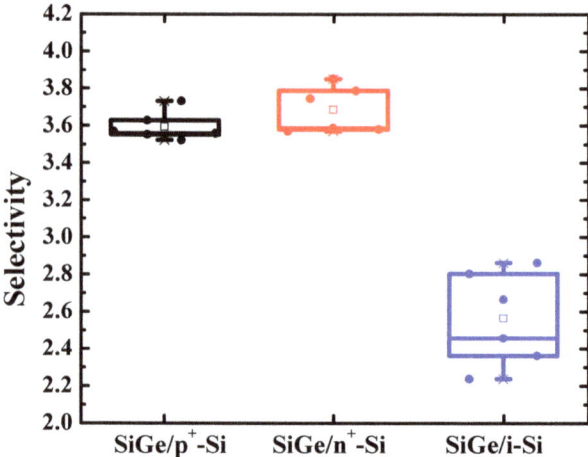

Figure 7. Box plot of the selectivity of SiGe/n$^+$-Si, SiGe/p$^+$-Si, and SiGe/i-Si. The means and the standard deviations are 3.59, 3.68, and 2.56 and 0.0759, 0.1228, and 0.2512, respectively. Significant improvements in the selectivity for doped Si were observed. The selectivity of SiGe/n$^+$-Si and SiGe/p$^+$-Si was similar, but the variation in SiGe/n$^+$-Si was larger.

In order to explore the effect of nitric acid concentration, we carried out digital etching experiments on Sample I and Sample II with different HNO_3 concentrations. Figure 8 shows the REPC of SiGe/n^+-Si, SiGe/p^+-Si, and SiGe/i-Si as a function of HNO_3 concentration. As described in Figure 8, with the increase in HNO_3 concentration, the REPC of SiGe/p^+-Si increased and became saturated at 0.61 nm/cycle with the 29.5% HNO_3 concentration. The appearance of saturation is helpful for the stability of the process. However, to achieve accurate etching control for small-sized devices, a controllable etching rate is expected. Moreover, when the concentration was lower than 26.5%, the etched surface of SiGe was very rough, as shown in (Supporting Information, Figure S2a). In the case of high concentrations, damage occurred on the etched surface, as shown in (Supporting Information, Figure S2c). The critical concentration might be between 47.5% and 52%, and the 52% concentration exceeded this limit, resulting in etching damage as shown in (Supporting Information, Figure S2c,d). Therefore, there is a tradeoff between the controllable etching rate, high etching control, small process variations, and excellent etched surfaces when choosing a HNO_3 concentration for the digital etching of SiGe that is selective of p^+-Si.

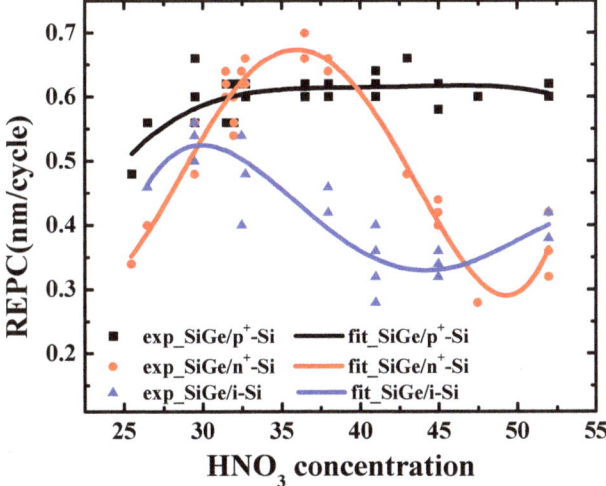

Figure 8. REPC of SiGe/n^+-Si, SiGe/p^+-Si, and SiGe/i-Si as a function of HNO_3 concentration. The dots in the figure are the experimental data, and the lines are the fitting curves of the experimental data. The slopes represent the relative etching amount per cycle (REPC) and the etching amount per cycle (EPC) of silicon.

As discussed above, the concentration range from 26.5% to 47.5% led to an etching morphology with a smooth surface that was free of damage, showing that the study on HNO_3 concentration was meaningful. In this concentration range, it was observed that the REPC of SiGe/n^+-Si had a trend of first increasing and then decreasing. At the concentration of 36.5%, the REPC reached the maximum. The etching rate might be the least influenced by concentration fluctuations. The HNO_3 concentration might be used for the fabrication of GAA transistors due to the small process variations. At the concentrations of 31.5% and 40%, the fitting curves of SiGe/n^+-Si and SiGe/p^+-Si intersected. Despite its identical relative etching rate, the 40% HNO_3 concentration requires a greater nitric acid concentration, thus increasing the cost. It was considered that 31.5% is the most suitable concentration for the digital etching of p^+-Si/SiGe/n^+-Si stack structures, such as for GAA CMOS and TFET applications, which must expose both the n^+ and p^+ sources/drains at the same time.

For the digital etching of SiGe/i-Si, the REPC first increased and then decreased with the increase in HNO_3 concentration. The REPC of SiGe/i-Si reached the maximum, which was equal to that of SiGe/n^+-Si at the 30% HNO_3 concentration. However, there was a large process variation, which is a burden in the control of the etching process. Through many repeated experiments with fine concentration intervals, the HNO_3 concentration relationship can be further verified.

3.3. Effect of Doped SiGe and Ge Fraction Dependence

In the first part, it was observed that the diffusion of arsenic into SiGe might enhance the etching rate of SiGe. Sample IV with in situ doping in SiGe was treated with the 31.5% HNO_3 q-ALE process. To make the effect of doping more obvious and easier to observe, 300 cycles of etching were performed. Figure 9 shows the SEM cross-section images of Sample IV after digital etching at 40, 100, 200, and 300 cycles with 31.5% HNO_3-dBOE q-ALE. It was shown that SiGe layer 1 disappeared at 100 cycles, which might have been due to the etching from the top. The SiGe layer 2 was penetrated horizontally at 200 cycles. It was observed that the remaining SiGe layer 3 at 300 cycles was the lowest. The etching amounts of the intrinsic SiGe layers, including SiGe layer 4, SiGe layer 5, and SiGe layer 6, were almost equal—slightly more than in the p-type SiGe, such as in SiGe layer 7 and SiGe layer 8. The results demonstrate that the relationship of the etching rate with the doping type is: p-type < intrinsic < n-type. The etching rate increased with the increase in n-type dopant concentration. As shown in Figure 9, the remaining amount of SiGe layer 9 doped by almost equal concentrations of arsenic and boron was similar to that of the intrinsic SiGe layers. This indicates the dependence on the carrier type instead of the dopant type. Additionally, it is demonstrated that the etching selectivity between the same doped SiGe and Si always exists regardless of the doping type.

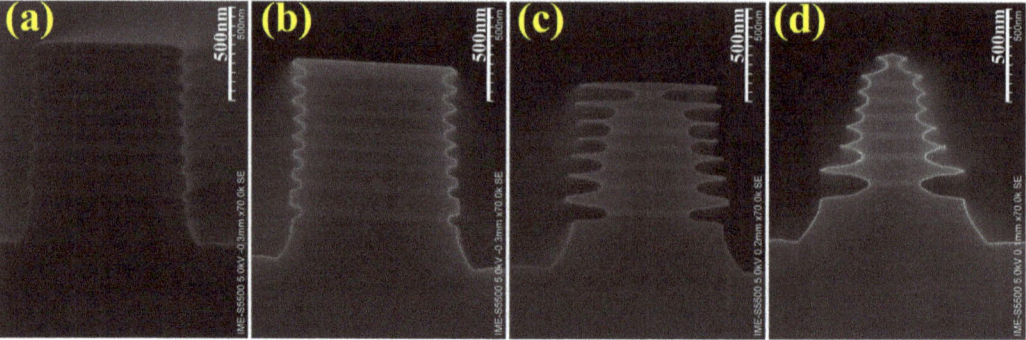

Figure 9. The SEM cross–section images of Sample IV after digital etching with 31.5% HNO_3-dBOE q-ALE at (**a**) 40 cycles, (**b**) 100 cycles, (**c**) 200 cycles, and (**d**) 300 cycles.

To investigate the influence of the Ge fraction on the selectivity and etching rate of n^+-SiGe/n^+-Si, Sample V was etched with 31.5% HNO_3-dBOE q-ALE for 100 cycles. Figure 10 shows the SEM cross-section images of Sample V after digital etching with 31.5% HNO_3-dBOE q-ALE for 100 cycles. As shown in Figure 10, there is a selectivity for n^+-Si in the n^+-SiGe digital etching. The top SiGe might have been etched from the top opening. The REA of the n^+-SiGe increased with the increase in the Ge fraction. This might have been due to the easier hole injection and larger valence band offset [18]. It was demonstrated that increasing the Ge fraction could increase the etching rate of n^+-SiGe and the selectivity of n^+-SiGe/n^+-Si.

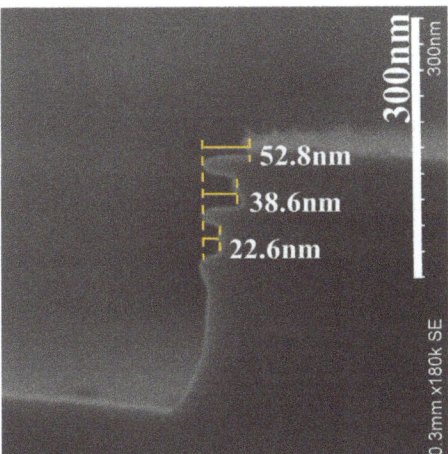

Figure 10. The SEM cross–section images of Sample V after digital etching with 31.5% HNO$_3$-dBOE q-ALE for 100 cycles.

3.4. Strain and Material Quality Analyses

In order to further determine the strain and material quality of the samples after the etching process, the HRXRD analysis scanning around the (004) diffraction order has been performed on p$^+$-Si/SiGe/n$^+$-Si stack layers as grown, after vertical stack etch, and after SiGe q-ALE with 31.5% HNO$_3$. The HRXRD rocking curves are shown in Figure 11. For the epitaxial growth sample, the SiGe signal was intense, and many fringes were observed around the SiGe peak due to X-ray interference at the SiGe/Si interface, which indicated a high-quality SiGe/Si interface. Therefore, the subsequent etching experiments could be implemented based on the high-quality epitaxial film.

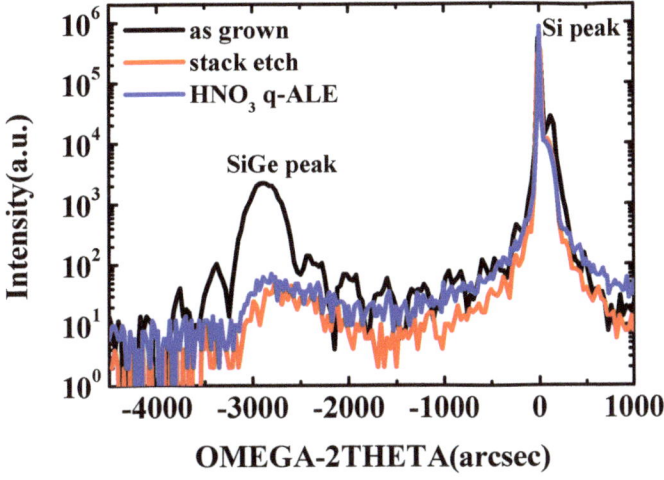

Figure 11. HRXRD rocking curves around the (004) reflection of the as–grown p$^+$-Si/SiGe/n$^+$-Si stack layers after vertical stack etching and after 31.5% HNO$_3$ q-ALE with 50 cycles.

A high full-width at half-maximum (FWHM) is a characteristic of a material's quality [34]. Compared with the epitaxial growth sample, the intensity of the SiGe peak after the etching process was weaker, which might be due to the reduction of SiGe material into chips after etching. There was also a slight shift of the SiGe peak towards the Si peak, which is an indicator of strain in the SiGe layer. As shown, the SiGe peak of the stack-etched sample was shifted toward the Si substrate peak compared to the SiGe peak of the as-grown sample. This was a result of a strain relaxation induced by the stack-etching process. No continued shift of SiGe peak was detected after SiGe q-ALE etching, indicating that there was no further strain relaxation. This is important point out in the SiGe channel because the energy band and carrier mobility are dependent on the strain.

In vertical GAA CMOS and TFET applications, SiGe is often used as a channel material, and the etched surface can be used as a channel interface. It is necessary to check the surface roughness after it is etched. Figure 12 shows the AFM morphology of the SiGe surface on as-grown epi-SiGe after etched with HNO_3:HF:H_2O mixtures and after etched with q-ALE. It was found that the root mean square (RMS) roughness of the q-ALE process still maintained a comparatively low value after many cycles. The RMS was 0.418 nm at 50 cycles and 0.474 nm at 30 cycles. AFM measurements were performed at many sites. The RMS was always in the range of 0.40 to 0.50 nm. It turned out that the RMS variation was due to differences in the test sites, and there was no dependence on the number of cycles. It was demonstrated that the surface roughness after the HNO_3-dBOE q-ALE process stayed in the range of 0.40 to 0.50 nm and was better than dry [35] and wet chemical continuous etching.

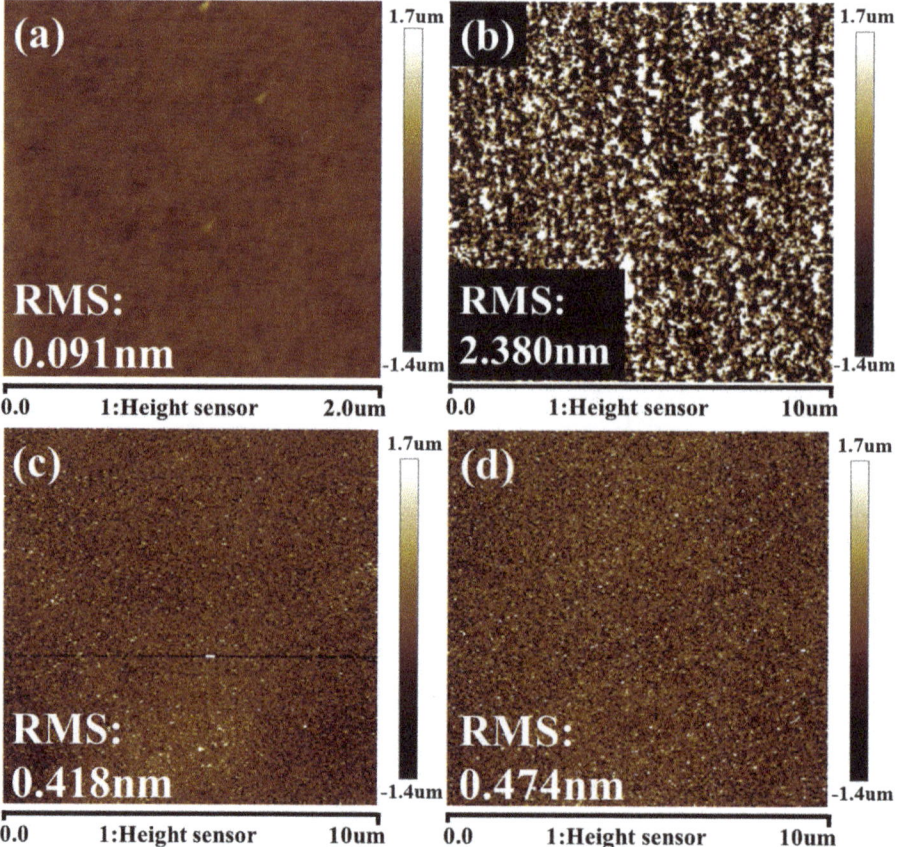

Figure 12. Typical AFM images of flat (100) $Si_{0.72}Ge_{0.28}$ surfaces before and after the etching process: (**a**) as–grown; (**b**) HNO_3:HF:H_2O mixtures; (**c**) q-ALE with 50 cycles; (**d**) q-ALE with 30 cycles.

4. Conclusions

The HNO_3-dBOE q-ALE process consists of alternative HNO_3 oxidation and dBOE oxide removal. Compared with the H_2O_2-dBOE q-ALE process, excellent selectivity for n-type doped Si could be found with HNO_3-dBOE q-ALE. Doping plays an important role in the selective etching of SiGe. The selectivity of SiGe/Si was enhanced by doped Si. In addition, the selectivity for n-type doped Si had a strong dependence on the HNO_3 concentration. The relative etching of n^+-Si reached a maximum at 36.5% HNO_3 concentration, and p^+-Si was saturated at 29.5% HNO_3 concentration. It was found that at 31.5% HNO_3 concentration, identical selectivity levels for p^+-Si and n^+-Si could be achieved. The REPC was 0.6 nm. The etching selectivity was 3.6–40% higher than that of intrinsic Si. The most suitable concentration for digital etching of p^+-Si/SiGe/n^+-Si stack structures, such as for GAA CMOS and TFET applications, which have to expose both the n^+ and p^+ sources/drains at the same time, is considered to be 31.5%. The relationship between the etching rate of doped SiGe and the doping type is: p-type < intrinsic < n-type. The etching rate of doped SiGe could be improved by the Ge fraction. Finally, this technique is a promising process for the fabrication of GAA CMOS transistors and TFETs due to its perfectly controllable etching rate and the resulting atomically smooth surface roughness.

Supplementary Materials: The following are available online at https://www.mdpi.com/article/10.3390/nano11051209/s1, Figure S1: The SEM cross-section images of Sample II after digital etching at 50 cycles: (a) in 30% H_2O_2-dBOE q-ALE (b) 31.5% HNO_3-dBOE q-ALE, Figure S2: The SEM cross-section images of Sample I after HNO_3 digital etching at 50 cycles with varying HNO_3 concentrations: (a) 25.5% HNO_3 concentration (b) 36.5% HNO_3 concentration (c) 52% HNO_3 concentration, the etch damage is marked in the yellow dotted line. (d) significant etch damage at 52% HNO_3 concentration.

Author Contributions: Conceptualization, Y.L. (Yangyang Li), H.Z. and H.H.R.; Data curation, Y.L. (Yangyang Li); Funding acquisition, H.Z.; Methodology, Y.L. (Yangyang Li), Z.K., Y.Z., X.A., G.W., Q.W., Z.L., S.L., L.X., W.H., Y.L. (Yongbo Liu), C.L., J.L., H.L., J.S. and C.Z.; Project administration, H.Z.; Supervision, H.Z. and H.H.R.; Writing—original draft, Y.L. (Yangyang Li); Writing—review and editing, Y.L. (Yangyang Li), H.Z. and H.H.R. All authors have read and agreed to the published version of the manuscript.

Funding: This work was supported by the Academy of Integrated Circuit Innovation (Grant No. Y7YC01X001 and Grant No. E0YC03X001), the National Key Research and Development Program of China (Grant No. 2016YFA0301701), the Youth Innovation Promotion Association of CAS (Grant No. Y2020037), and the National Natural Science Foundation of China (Grant No. 92064002).

Data Availability Statement: The data presented in this study are available on request from the corresponding author.

Conflicts of Interest: The authors declare no conflict of interest.

References

1. Radamson, H.H.; Zhu, H.; Wu, Z.; He, X.; Lin, H.; Liu, J.; Xiang, J.; Kong, Z.; Xiong, W.; Li, J.; et al. State of the Art and Future Perspectives in Advanced CMOS Technology. *Nanomaterials* **2020**, *10*, 1555. [CrossRef]
2. Singh, N.; Buddharaju, K.D.; Manhas, S.K.; Agarwal, A.; Rustagi, S.C.; Lo, G.Q.; Balasubramanian, N.; Kwong, D.L. Si, SiGe nanowire devices by top–down technology and their applications. *IEEE Trans. Electron. Devices* **2008**, *55*, 3107–3118. [CrossRef]
3. Radamson, H.H.; He, X.; Zhang, Q.; Liu, J.; Cui, H.; Xiang, J.; Kong, Z.; Xiong, W.; Li, J.; Gao, J.; et al. Miniaturization of CMOS. *Micromachines* **2019**, *10*, 293. [CrossRef] [PubMed]
4. Radamson, H.H.E.; Luo, J.; Zhao, C. Past, Present and Future of CMOS. In *Woodhead Publishing Series in Electronic and Optical Materials*; Elsevier: Amsterdam, The Netherlands, 2018; pp. 95–114. [CrossRef]
5. Gandhi, R.; Chen, Z.; Singh, N.; Banerjee, K.; Lee, S. Vertical Si-Nanowire n-Type Tunneling FETs with Low Subthreshold Swing 50 mV/decade at Room Temperature. *IEEE Electron. Device Lett.* **2011**, *32*, 437–439. [CrossRef]
6. Bhuwalka, S.S.K.K.; Ludsteck, A.K.; Tolksdorf, C.; Schulze, J.; Eisele, I. Vertical Tunnel Field-Effect Transistor. *IEEE Trans. Electron. Devices* **2004**, *51*, 279–282. [CrossRef]
7. Saremi, M.; Afzali-Kusha, A.; Mohammadi, S. Ground plane fin-shaped field effect transistor (GP-FinFET): A FinFET for low leakage power circuits. *Microelectron. Eng.* **2012**, *95*, 74–82. [CrossRef]
8. Imenabadi, R.M.; Saremi, M.; Vandenberghe, W.G. A Novel PNPN-Like Z-Shaped Tunnel Field-Effect Transistor with Improved Ambipolar Behavior and RF Performance. *IEEE Trans. Electron. Devices* **2017**, *64*, 4752–4758. [CrossRef]

9. Abadi, R.M.I.; Saremi, M. A Resonant Tunneling Nanowire Field Effect Transistor with Physical Contractions: A Negative Differential Resistance Device for Low Power Very Large Scale Integration Applications. *J. Electron. Mater.* **2018**, *47*, 1091–1098. [CrossRef]
10. Sivieri, V.D.B.; Bordallo, C.C.M.; Der Agopian, P.G.; Martino, J.A.; Rooyackers, R.; Vandooren, A.; Simoen, E.; Thean, A.; Claeys, C. Vertical Nanowire TFET Diameter Influence on Intrinsic Voltage Gain for Different Inversion Conditions. *ECS Trans.* **2015**, *66*, 187–192. [CrossRef]
11. Bhuwalka, K.K.; Schulze, J.; Eisele, I. Scaling the vertical tunnel FET with tunnel bandgap modulation and gate workfunction engineering. *IEEE Trans. Electron. Devices* **2005**, *52*, 909–917. [CrossRef]
12. Yin, X.; Zhang, Y.; Zhu, H.; Wang, G.L.; Li, J.J.; Du, A.Y.; Li, C.; Zhao, L.H.; Huang, W.X.; Yang, H.; et al. Vertical Sandwich Gate-All-Around Field-Effect Transistors with Self-Aligned High-k Metal Gates and Small Effective-Gate-Length Variation. *IEEE Electron. Device Lett.* **2020**, *41*, 8–11. [CrossRef]
13. Acker, J.; Rietig, A.; Steinert, M.; Hoffmann, V. Mass and Electron Balance for the Oxidation of Silicon during the Wet Chemical Etching in HF/HNO_3 Mixtures. *J. Phys. Chem. C* **2012**, *116*, 20380–20388. [CrossRef]
14. Steinert, M.J.A.; Wetzig, K. New Aspects on the Reduction of Nitric Acid during Wet Chemical Etching of Silicon in Concentrated $HF-HNO_3$ Mixtures. *J. Phys. Chem. C* **2008**, *112*, 14139–14144. [CrossRef]
15. Steinert, M.J.A.; Oswald, S.; Wetzig, K. Study on the Mechanism of Silicon Etching in HNO_3-Rich HF HNO_3 Mixtures. *J. Phys. Chem. C* **2007**, *111*, 2133–2140. [CrossRef]
16. Baraissov, Z.; Pacco, A.; Koneti, S.; Bisht, G.; Panciera, F.; Holsteyns, F.; Mirsaidov, U. Selective Wet Etching of Silicon Germanium in Composite Vertical Nanowires. *ACS Appl. Mater. Interfaces* **2019**, *11*, 36839–36846. [CrossRef]
17. Gondek, C.; Lippold, M.; Röver, I.; Bohmhammel, K.; Kroke, E. Etching Silicon with $HF-H_2O_2$-Based Mixtures: Reactivity Studies and Surface Investigations. *J. Phys. Chem. C* **2014**, *118*, 2044–2051. [CrossRef]
18. Choi, Y.; Jang, H.; Byun, D.-s.; Ko, D.-H. Selective chemical wet etching of $Si_{1-x}Ge_x$ versus Si in single-layer and multi-layer with HNO_3/HF mixtures. *Thin Solid Films* **2020**, *709*, 138230. [CrossRef]
19. Cams, T.K.; Tanner, M.O.; Wang, K.L. Chemical Etching of $Si_{1-x}Ge_x$ in $HF:H_2O_2:CH_3COOH$. *J. Electrochem. Soc.* **1995**, *142*, 1260–1266. [CrossRef]
20. Loubet, N.; Kormann, T.; Chabanne, G.; Denorme, S.; Dutartre, D. Selective etching of $Si_{1-x}Ge_x$ versus Si with gaseous HCl for the formation of advanced CMOS devices. *Thin Solid Films* **2008**, *517*, 93–97. [CrossRef]
21. Caubet, V.; Beylier, C.; Borel, S.; Renault, O. Mechanisms of isotropic and selective etching between SiGe and Si. *J. Vac. Sci. Technol. B Microelectron. Nanometer Struct.* **2006**, *24*, 2748. [CrossRef]
22. Borel, S.; Arvet, C.; Bilde, J.; Caubet, V.; Louis, D. Control of Selectivity between SiGe and Si in Isotropic Etching Processes. *Jpn. J. Appl. Phys.* **2004**, *43*, 3964–3966. [CrossRef]
23. Ahles, C.F.; Choi, J.Y.; Wolf, S.; Kummel, A.C. Selective Etching of Silicon in Preference to Germanium and $Si_{0.5}Ge_{0.5}$. *ACS Appl. Mater. Interfaces* **2017**, *9*, 20947–20954. [CrossRef] [PubMed]
24. Oehrlein, G.S.; Tromp, R.M.; Lee, Y.H.; Petrillo, E.J. Study of silicon contamination and near-surface damage caused by CF_4/H_2 reactive ion etching. *Appl. Phys. Lett.* **1984**, *45*, 420–422. [CrossRef]
25. Cano, A.M.; Marquardt, A.E.; DuMont, J.W.; George, S.M. Effect of HF Pressure on Thermal Al_2O_3 Atomic Layer Etch Rates and Al_2O_3 Fluorination. *J. Phys. Chem. C* **2019**, *123*, 10346–10355. [CrossRef]
26. Min, K.S.; Kang, S.H.; Kim, J.K.; Jhon, Y.I.; Jhon, M.S.; Yeom, G.Y. Atomic layer etching of Al_2O_3 using BCl_3/Ar for the interface passivation layer of III–V MOS devices. *Microelectron. Eng.* **2013**, *110*, 457–460. [CrossRef]
27. DuMont, J.W.; Marquardt, A.E.; Cano, A.M.; George, S.M. Thermal Atomic Layer Etching of SiO_2 by a "Conversion-Etch" Mechanism Using Sequential Reactions of Trimethylaluminum and Hydrogen Fluoride. *ACS Appl. Mater. Interfaces* **2017**, *9*, 10296–10307. [CrossRef] [PubMed]
28. Lee, Y.; George, S.M. Thermal Atomic Layer Etching of Titanium Nitride Using Sequential, Self-Limiting Reactions: Oxidation to TiO_2 and Fluorination to Volatile TiF_4. *Chem. Mater.* **2017**, *29*, 8202–8210. [CrossRef]
29. Lu, W.; Lee, Y.; Murdzek, J.; Gertsch, J.; Vardi, A. First Transistor Demonstration of Thermal Atomic Layer Etching: InGaAs FinFETs with sub-5 nm Fin-width Featuring in situ ALE-ALD. In Proceedings of the 2018 IEEE International Electron Devices Meeting (IEDM), San Francisco, CA, USA, 1–5 December 2018.
30. Xie, W.; Lemaire, P.C.; Parsons, G.N. Thermally Driven Self-Limiting Atomic Layer Etching of Metallic Tungsten Using WF6 and O_2. *ACS Appl. Mater. Interfaces* **2018**, *10*, 9147–9154. [CrossRef]
31. Li, C.; Zhu, H.; Zhang, Y.; Yin, X.; Jia, K.; Li, J.; Wang, G.; Kong, Z.; Du, A.; Yang, T.; et al. Selective Digital Etching of Silicon-Germanium Using Nitric and Hydrofluoric Acids. *ACS Appl. Mater. Interfaces* **2020**, *12*, 48170–48178. [CrossRef]
32. Yin, X.; Zhu, H.; Zhao, L.; Wang, G.; Li, C.; Huang, W.; Zhang, Y.; Jia, K.; Li, J.; Radamson, H.H. Study of Isotropic and Si-Selective Quasi Atomic Layer Etching of $Si_{1-x}Ge_x$. *ECS J. Solid State Sci. Technol.* **2020**, *9*, 034012. [CrossRef]
33. Gokce, B.; Aspnes, D.E.; Lucovsky, G.; Gundogdua, K. Bond-specific reaction kinetics during the oxidation of (111) Si Effect of n-type doping. *Appl. Phys. Lett.* **2011**, *98*, 021904. [CrossRef]
34. Hansson, G.V.; Radamsson, H.H.; Ni, W.X. Strain and Relaxation in Si-Mbe Structures Studied by Reciprocal Space Mapping Using High-Resolution X-Ray-Diffraction. *J. Mater. Sci-Mater. El* **1995**, *6*, 292–297. [CrossRef]
35. Li, J.; Wang, W.; Li, Y.; Zhou, N.; Wang, G.; Kong, Z.; Fu, J.; Yin, X.; Li, C.; Wang, X.; et al. Study of selective isotropic etching $Si_{1-x}Ge_x$ in process of nanowire transistors. *J. Mater. Sci. Mater. Electron.* **2019**, *31*, 134–143. [CrossRef]

Review

Review of Si-Based GeSn CVD Growth and Optoelectronic Applications

Yuanhao Miao [1,2,*], Guilei Wang [1,2,3], Zhenzhen Kong [1,3], Buqing Xu [1,3], Xuewei Zhao [1,3], Xue Luo [2], Hongxiao Lin [2], Yan Dong [1], Bin Lu [2,4], Linpeng Dong [2,5], Jiuren Zhou [6], Jinbiao Liu [1] and Henry H. Radamson [1,2,3,*]

[1] Key Laboratory of Microelectronic Devices Integrated Technology, Institute of Microelectronics, Chinese Academy of Sciences, Beijing 100029, China; wangguilei@ime.ac.cn (G.W.); kongzhenzhen@ime.ac.cn (Z.K.); xubuqing@ime.ac.cn (B.X.); zhaoxuewei@ime.ac.cn (X.Z.); dongyan2019@ime.ac.cn (Y.D.); liujinbiao@ime.ac.cn (J.L.)
[2] Research and Development Center of Optoelectronic Hybrid IC, Guangdong Greater Bay Area Institute of Integrated Circuit and System, Guangzhou 510535, China; luoxue@giics.com.cn (X.L.); linhongxiao@giics.com.cn (H.L.); lubinsxnu@sina.com (B.L.); lpdong@xatu.edu.cn (L.D.)
[3] Institute of Microelectronics, University of Chinese Academy of Sciences, Beijing 100049, China
[4] School of Physics and Information Engineering, Shanxi Normal University, Linfen 041004, China
[5] Shaanxi Province Key Laboratory of Thin Films Technology Optical Test, Xi'an Technological University, Xi'an 710032, China
[6] Department of Electrical and Computer Engineering, National University of Singapore, Singapore 117576, Singapore; zhoujiuren@163.com
* Correspondence: miaoyuanhao@ime.ac.cn (Y.M.); rad@ime.ac.cn (H.H.R.); Tel.: +86-010-8299-5793 (H.H.R.)

Abstract: GeSn alloys have already attracted extensive attention due to their excellent properties and wide-ranging electronic and optoelectronic applications. Both theoretical and experimental results have shown that direct bandgap GeSn alloys are preferable for Si-based, high-efficiency light source applications. For the abovementioned purposes, molecular beam epitaxy (MBE), physical vapour deposition (PVD), and chemical vapor deposition (CVD) technologies have been extensively explored to grow high-quality GeSn alloys. However, CVD is the dominant growth method in the industry, and it is therefore more easily transferred. This review is focused on the recent progress in GeSn CVD growth (including ion implantation, in situ doping technology, and ohmic contacts), GeSn detectors, GeSn lasers, and GeSn transistors. These review results will provide huge advancements for the research and development of high-performance electronic and optoelectronic devices.

Keywords: GeSn; CVD; lasers; detectors; transistors

1. Introduction

Si-based integrated circuits (ICs), which are dominated by Si CMOS technology, have reached their physics limit. The influences of quantum effects, parasitic parameters, and process parameters on data transmission applications are also reaching their limits, as the rapid development of microelectronics has led to higher requirements for data transmission technology. For these reasons, scientists have proposed schemes to integrate optoelectronic devices with microelectronic devices [1–7]. However, Si-based on-chip integrated light source was lacking, and the light sources for existing optoelectronic integrated circuits (OEICs) were all externally coupled; though the coupling efficiency between the edge of the light source and grating coupler was high enough, the lack of an on-chip light source restricted OEICs' applications [8–10]. As such, many research programs started to pay more attention to Si-based monolithic OEIC technology [11–15], which has the following advantages over the baseline technology: (i) it is compatible with mature Si CMOS technology; (ii) has low costs; (iii) has larger wafer sizes and larger scale production; (iv) its partial electrical interconnection can be replaced by optical interconnection, which can realize high-efficiency, high-speed, and low loss data transmission. Si-based monolithic OEIC technology uses Si-compatible semiconductor technology to integrate optoelectronic devices into Si chips in order to improve chip performance, extend chip function, and reduce costs. Though Si-based photonic devices, such as optical waveguides [16,17], photodetectors [18–20], optical modulators [21–23], and optical switches [24,25], have been successfully developed, it is difficult to achieve high-efficiency emission due to the facts that Si is an indirect bandgap semiconductor and its light emission efficiency is about five orders of magnitude lower than that of direct band gap III–V compound semiconductors. Thus, the need for an Si-based high-efficiency light source represents an important technical bottleneck in the development of Si-based monolithic OEICs. Therefore, looking for a direct bandgap semiconductor material that is compatible with the Si CMOS process is of great significance in the creation of large scale Si-based monolithic OEICs [26–28].

Group IV materials are compatible with the traditional Si CMOS process, and Si, SiGe, and Ge are commonly used as indirect band gap semiconductors despite not being suitable for light emission. Fortunately, tensile strain engineering and Sn-alloying engineering have enabled Ge to become a quasi-direct bandgap or direct bandgap material due to the small bandgap difference between its two minima in conduction bands (only 136 meV). Experimental research has shown an optical gain of 0.24% for tensile-strained n^+-type Ge (the n-type doping level is 1×10^{19} cm^{-3}), which led to the creation of optically injected and electrically injected Ge lasers [29–32]. However, the threshold for a Ge laser is too high, which means that weak tensile-strained n^+-type Ge is not able to supply enough optical gain to achieve low-threshold lasing.

In recent decades, GeSn alloys have demonstrated novel indirect-to-direct bandgap transition, as well excellent carrier transport. Due to their tunable band structures, GeSn materials have become promising candidates to create Si-based OEICs with higher hole mobility, enhanced light absorption, etc. [33–37]. Growing high-quality GeSn layers with relatively high Sn contents has different challenges, e.g., Sn segregation during growth and the poor thermal stability of SnGe layers [38–41]. These issues root from the low solid solubility of Sn in Ge (<1%) and the large lattice mismatch between Si or Ge and GeSn. As early as 1995, the first growth of a GeSn/Ge superlattice was reported using a very low growth temperature in a molecular beam epitaxy (MBE) chamber. Such GeSn layers had an Sn content of 26% [42,43]. Based on these early pioneer works, other growth techniques, such as chemical vapor deposition (CVD) and magnetron sputtering, have been widely used to grow high-quality direct bandgap GeSn materials with high Sn contents [44–49]. Although MBE can grow GeSn materials well, its growth rate is extremely low, which makes it tough to manufacture on a large scale. To achieve a significant impact within the industry, it is very important to develop a commercially available tool to grow high-quality GeSn materials. At a very early development stage of GeSn growth via CVD, SnD_4 and Ge_2H_6 were chosen as the Sn precursors and Ge precursors, respectively. Although there were many foundational studies on GeSn growth via CVD, SnD_4 is a high-cost material with a short lifetime, which makes it incompatible with the industry. For this reason, other precursors such as $SnCl_4$ have been explored. The IMEC and KTH groups pioneered the growing of GeSn layers using commercially available reaction precursors ($SnCl_4/Ge_2H_6$) [50]. A major breakthrough was later demonstrated using the production of commercially available reaction precursors ($SnCl_4/GeH_4$) [51,52]. The limitations of incorporating Sn into Ge have been conquered, and two major breakthroughs for GeSn CVD growth have been reached: (i) a world record high Sn content (22.3%) in bulk GeSn materials with PL emission was observed at room temperature (indicating good material quality), and (ii) SiGeSn/GeSn/SiGeSn multiple quantum well (MQW) structure growth and low-temperature PL intensity were later able to be remarkable enhanced [53–56]. Furthermore, the low costs and widespread availability of these chemicals in large-scale fabrication makes them the best choice for GeSn-based optoelectronic integration into CMOS processing. To make GeSn an efficient N-type or P-type semiconductor material for optoelectronic device application, there is an urgent need to research and develop doping engineering for GeSn. Currently, doping technologies, such as ion implantation and in situ CVD doping, have been optimized regarding their target doping concentration and doping distributions.

After the successful growth of P^+-Si/i–GeSn/n–GeSn via CVD, Jay Mathews et al. demonstrated the world's first GeSn photodetector with a 2% Sn content in 2009 [57]. The wavelength cutoff was extended to be at least 1750 nm, which means that the GeSn photodetector with a 2% Sn content can cover the entire telecommunication band. Since then, GeSn photoconductor detectors [58–63], and p–GeSn/i–GeSn/n–GeSn heterostructure detectors [64–68] have been demonstrated. Advances in GeSn CVD growth technology have occurred alongside material quality and detector performance improvements, including: (i) the wavelength cutoff for the GeSn photodetector has been progressively broadened from 1800 nm to 2100, 2400, 2600, 2650, and 3650 nm [63]; (ii) based on wafer-bonding technology, the dark current for GeSn photodetector has been suppressed by more than two orders of magnitude [69]; (iii) peak specific detectivity values are now comparable to those of commercial extended-InGaAs detectors (4×10^{10} cm·$Hz^{1/2}$·W^{-1}) at the same wavelength range; (iv) a passivation technique was developed to enhance responsivity and peak specific detectivity [65]; and (v) mid-IR imaging was demonstrated with GeSn photodetectors, and the image quality of the GeSn photodetectors was found to be superior to that of a commercial PbSe detector [63].

Alongside the significant breakthroughs in GeSn growth and detectors, GeSn lasing had also developed to an advanced stage. Recently reported GeSn laser structures have all been grown via the CVD technique. Following the observation of a PL peak with narrowed line widths, a true direct bandgap GeSn material with an Sn content of up to 10% was experimentally demonstrated in 2014 [33]. Encouraged by this major technical breakthrough, researchers used the injection methods such as optical injection with a Ge laser to check the GeSn waveguide, and lasing behavior was clearly observed at a low temperature in 2015 [70]. Following this breakthrough, several types of GeSn lasers [71–82] were demonstrated, though they still suffer from the problems of low-temperature operation and high lasing thresholds. To overcome these difficulties, several methods have been proposed to improve performance, such as greater Sn incorporation into Ge [73,75,76], the use of SiGeSn/GeSn/SiGeSn heterostructures or SiGeSn/GeSn/SiGeSn MQWs as the gain medium [83–86], a modulation doping scheme in SiGeSn/GeSn/SiGeSn MQWs [87], defect management [80], and thermal management [81,82]. Considerable efforts in GeSn lasing research have led to an increased maximum lasing temperature of 270 K [76] due to the amazing discovery of strain relaxation growth mechanism [88]. Near-room-temperature lasing was also observed for a GeSn active medium with a 16% Sn content and high uniaxial tensile strain [77]. A breakthrough regarding the optical pumping threshold was reported in 2020, when a low-Sn-content GeSn material with a high uniaxial tensile strain was utilized as an active medium; continuous wave (CW) lasing was also achieved. However, the lasing temperature only reached 100 K due to the low directness of the active medium [80]. In the same year, electrically pumped GeSn/SiGeSn heterostructure lasers with operation temperatures of up to 100 K were demonstrated [89,90]; this was an essential achievement for Si-based electrically pumped group IV interband lasing.

As a group IV material, GeSn is compatible with Si and can realize the transition from indirect band gap to direct band gap by adjusting its Sn content, which makes it the best substitute for group IV materials in Si-based optoelectronic integration applications. GeSn has an extremely high carrier mobility, so it may also be an ideal materials for transistor applications. Due to the significant development of GeSn CVD growth technology, vertically stacked 3-GeSn-nanosheet pGAAFETs (gate-all-around FETs) [91], GeSn p-FinFETs [92,93], GeSn n-channel MOSFETs [94,95], GeSn/Ge vertical nanowire pFETs [96], GeSn GAA nanowire pFETs [97], and GeSn n-FinFETs [98] have been successfully demonstrated. Additionally, GeSn's direct band gap property was found to effectively improve the tunneling probability of electrons, making an excellent material for TFET preparation [99,100], this opening a new development direction for the integrated circuit after Moore's era. The discovery of this property has attracted considerable research interest in recent years. Since Sn naturally has low solid solubility in Ge (smaller than 1%), growth of high Sn composition single crystal GeSn is difficult. At present, devices prepared with GeSn materials are still in the research and development stage, so they have not been widely used in production.

To the best of our knowledge, there has yet to be a review article that systematically reported on GeSn material growth and counterpart optoelectronic devices using the CVD technique. UHVCVD [101–104], RPCVD [105–110], PECVD [111–113], LPCVD [114–117], and APCVD [118,119] are discussed in this review, with a focus on identifying processes that can be transferred for the commercial production of GeSn. The objective of this comprehensive review article is to provide readers with a full understanding of the recent experimental advancements in GeSn material growth using CVD, as well as their optoelectronic applications. However, due to the large numbers of publications in this area, the authors of this work only selected articles with significant scientific impacts.

2. Research Progress for GeSn CVD Growth and Its Potential Applications

So far, several types of growth techniques, such as MBE, magnetron sputtering, and CVD have been used to grow GeSn materials. CVD is the dominant growth method in the industry, so more easily transferable. Therefore, we decided to review GeSn CVD growth and its potential applications.

2.1. Potential Applications

A literature survey revealed that GeSn materials have numerous potential applications, including Si-based, integrated, high-efficiency light sources [120–122]; high-mobility electronic devices [92–100]; low-cost, Si-based, high-performance shortwave infrared (SWIR) imaging sensors [63–65]; Si-based photovoltaics [123]; optical signal encoding in the mid-infrared range [124,125]; high-performance logic applications [126,127]; Si-based integrated thermoelectrics as wearable devices [128,129]; Si-based spintronics [130,131]; Si-based integrated reconfigurable dipoles [132,133]; and Si-based quantum computing [134,135] (Figure 1). GeSn-related fundamental research and development applications have also been extensively investigated (Figure 2).

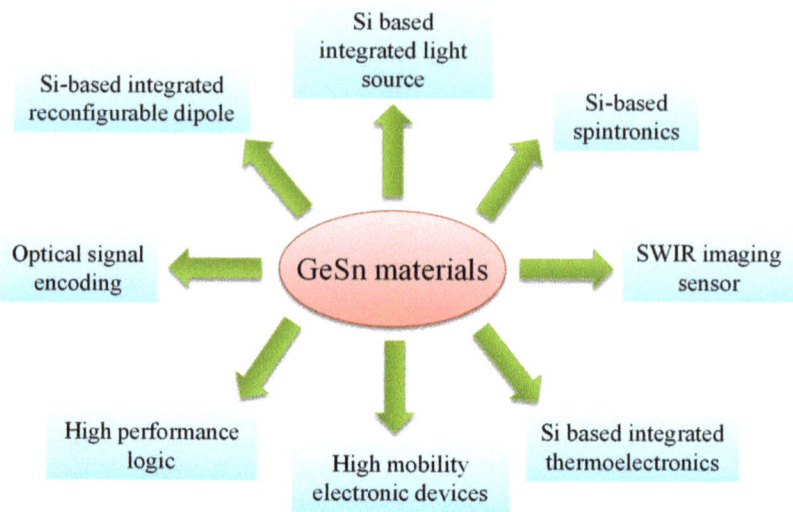

Figure 1. Potential applications of GeSn materials in different research areas.

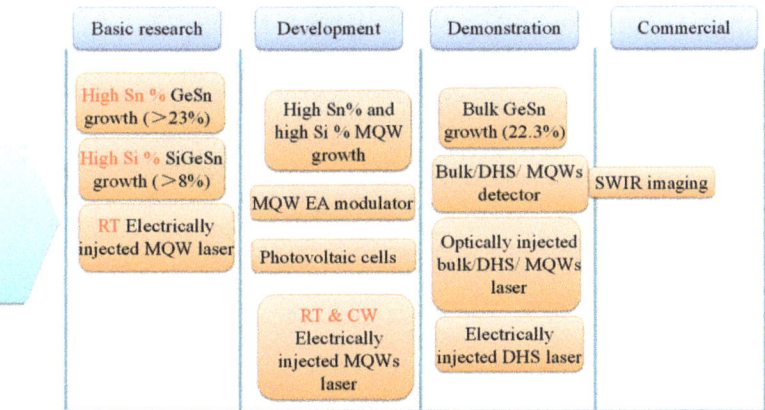

Figure 2. Optoelectronic applications of GeSn as a function of technology readiness level (GeSn transistors, which are still in the technical development stage, are not shown here due to space limitations).

Figure 2 shows the optoelectronic applications of GeSn as a function of technology readiness level. It can be observed that GeSn detectors are getting closer to the low-cost SWIR imaging applications, indicating that GeSn materials have great potential for use in next-generation civilian night-vision and IR cameras [63–65]. However, there are still some technical problems, which are discussed in Section 3. In addition to detectors (which are being rapidly developed), high-quality SiGeSn/GeSn/SiGeSn MQW growth, room-temperature, CW, and electrically injected SiGeSn/GeSn/SiGeSn MQW lasers; MQW electro-absorption (EA) modulators; and photovoltaic cells are in the research and development stage.

2.2. Research Progress for GeSn CVD Growth

In 2001, Kouvetakis's group from Arizona State University (ASU) first reported a GeSn alloy on oxidized and oxidized-free Si using UHVCVD [136]; since then, extensive GeSn CVD growth-related research works have been carried out. In 2003, SnD_4 and SiH_3GeH_3 were used as reaction precursors, and single-phase SiGeSn on a GeSn buffer was first achieved on Si via UHVCVD at 350 °C [137]. To create GeSn materials with higher Sn contents, SnD_4 and Ge_2H_6 were chosen as Sn and Ge precursors, respectively; the experimental results showed that SnD_4 is helpful for low-temperature growth, and its reaction with Ge_2H_6 can create GeSn with an Sn content of up to 25% [138] (Figure 3). The crystallinity, bandgap, lattice constants, optical properties, photoresponses, photocurrents, and Raman scattering results of GeSn materials grown by UHVCVD have been systematically demonstrated [139–144]. In order to grow GeSn at extremely low temperatures, some authors used Ge_3H_8 and Ge_4H_{10} as Ge precursors [145,146]. By using this method, single crystalline GeSn alloys were successfully deposited at temperatures ranging from 300 to 330 °C, the growth rate of the allows was able to meet industrial requirements, and the traditional SK growth mode was avoided. Finally, the authors concluded that Ge_3H_8 is a superior solution to grow GeSn alloys via UHVCVD [145,146]. Compared with previously reported reaction precursors (SnD_4/Ge_2H_6), the growth rate of the SnD_4/Ge_3H_8 combination was found to be improved 3–4 times. For this reason, a 1 μm thick GeSn layer with an Sn content of up to 9% was implemented, and room temperature photoluminescence spectra were observed, indicating that GeSn has great potential to be utilized as a gain medium for a Group IV laser. Later, SiGeSn growth at ultralow temperatures (from 290 to 330 °C) using Ge_4H_{10}, Si_4H_{10}, and SnD_4 were reported [147–149].

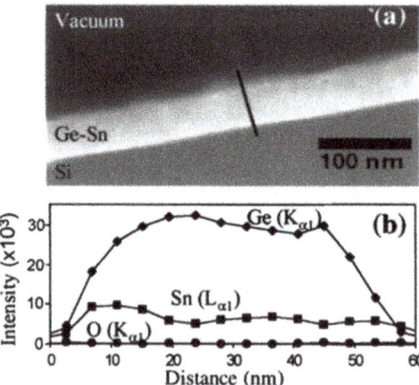

Figure 3. (**a**) Scanning transmission electron microscopy (STEM) image and (**b**) EDX cross-sectional profile of the GeSn with Sn contents of up to 25%. Reproduced with permission from [138], AIP Publishing, 2001.

Although there have been many foundational studies on GeSn growth via CVD investigated, SnD_4 has high costs, is incompatible with the industry, and is unstable at room temperature. For these reasons, other precursors such as $SnCl_4$ have been explored. IMEC and KTH were the first groups to propose GeSn growth using commercially available reaction precursors ($SnCl_4/Ge_2H_6$).

Due to the fact that $SnCl_4$ is liquid at room temperature, these groups evaporated $SnCl_4$ using a bubbler that was connected to an RPCVD chamber. Experimental results showed that defect-free doped and undoped GeSn layers with Sn contents of up to 8% were created using RPCVD at atmosphere conditions. Thermal stability was further determined by annealing at different conditions (400 °C for 10 min, 400 °C for 30 min, 500 °C for 10 min, and 500 °C for 30 min); the (004) omega-2 theta scan of as-grown and annealed $Ge_{0.92}Sn_{0.08}$ samples were compared (Figure 4a). For the sample annealed at 500 °C for 30 min, the diffraction peaks of GeSn and Ge widened and a clear GeSn peak shift was observed, suggesting possible Ge–Sn interdiffusion. To further confirm this assumption, secondary ion mass spectroscopy (SIMS) was conducted. From the SIMS results, the authors concluded that APCVD-grown GeSn with 8% Sn content was stable at the annealing condition of 500 °C for 30 min (Figure 4b). This work paved the way for GeSn growth using both commercially available reaction precursors and CVD production equipment.

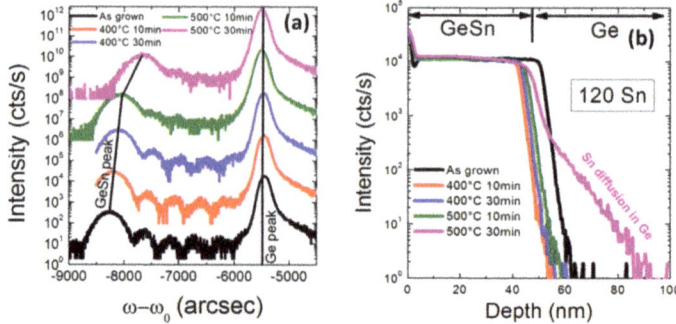

Figure 4. Comparison of (**a**) (004) omega-2 theta scans and (**b**) Sn content profiles of as-grown and annealed $Ge_{0.92}Sn_{0.08}$ samples under different annealing conditions. Reproduced with permission from [150], AIP Publishing, 2011.

Since then, there has been a sharp increase in the scientific knowledge of GeSn CVD growth, as shown by a number of publications (Figure 5a). The number of publications on GeSn CVD growth grew dramatically in 2013 and reaches 19 in 2018 (Figure 5a). The rapid development of GeSn CVD growth techniques has meant that the number of GeSn optoelectronic device publications followed the similar tendency (Figure 5b): (i) following the world's first demonstration of a GeSn detector, GeSn detector-related publications grew from 1 in 2008 to 30 in 2019; (ii) since the world's first demonstration of an optically pumped GeSn laser, publications related to GeSn lasers continually increased from 10 in 2015 to 25 in 2019, and the majority of these laser publications reported experimental results; (iii) there are still few publications regarding GeSn modulators, and a CVD-grown modulator has not been achieved (the majority of the modulator publications have been theoretical investigations).

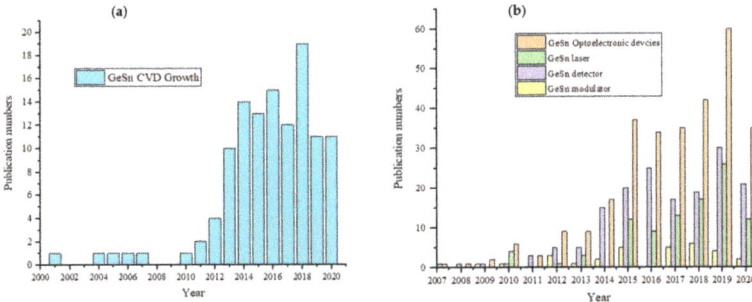

Figure 5. (a) Number of publications/year on GeSn materials grown by the CVD technique; (b) number of the publications/year on GeSn optoelectronic devices (theoretical calculations and conference proceedings are included).

To help readers to understand the research status of CVD growth techniques, Figure 6 summarizes research on GeSn CVD growth since the introduction of CVD in 2001 in terms of the research institution, growth chamber, year of deposition, and corresponding reference. Figure 6 shows several types of growth chambers, such as UHVCVD (baby color dot), RPCVD (dark color dot), APCVD (red color dot), PECVD (orange color dot), LPCVD (coffee color dot), and RTCVD (green color dot), that have been used to grow GeSn materials. Following pioneer works from ASU and IMEC, research groups from KTH Royal Institute of Technology (KTH), Applied Materials Inc (AM), PGI (Peter Grünberg Institute), and UA (University of Arkansas) started researching GeSn growth using CVD technology in 2013. Since then, research groups from ASM, University of Warwick (UW), National Taiwan University (NTWU), and Université de Montréal (EPM) have also researched GeSn CVD growth. Among all CVD growth technologies, RPCVD growth chamber is most widely accepted due to its commercial availability and more easily transferability (six research groups have used RPCVD chambers to grow GeSn). After the successful demonstration of the low-temperature growth of high-quality Ge on Si using PECVD, plasma-enhanced techniques came to be regarded as promising methods to grow GeSn materials. Thus, plasma-enhanced GeSn growth techniques aroused researchers' attentions from UA and ASU.

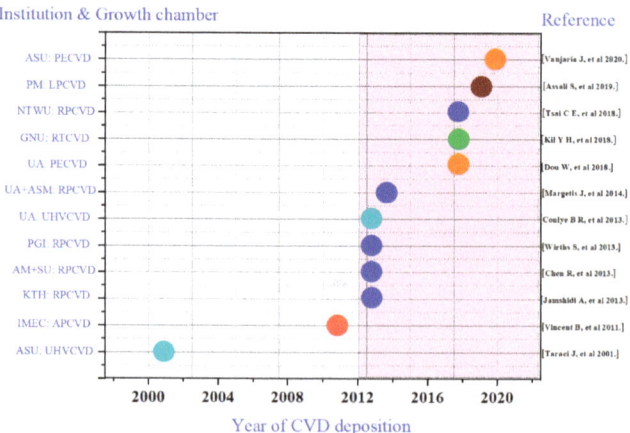

Figure 6. List of GeSn CVD growth papers by different groups.

2.3. GeSn CVD Growth Strategy

To have a full understanding of the GeSn CVD growth strategy, it is necessary to calibrate the Ge growth at low temperatures (below 450 °C). After calibration, the flow rate between Ge precursor and Sn precursor needs to be taken into consideration due to the possible etching effect of the generated Cl* species on the GeSn surface. Therefore, there is a critical flow rate, and the growth rate for GeSn growth has to be high enough to overcome the etching rate. More importantly, the effects of temperature, pressure, carrier gas, and strain relaxation on material growth must be canvassed.

2.3.1. Temperature and Pressure Effect on GeSn Growth

Previous GeSn CVD growth work has demonstrated that Sn content is closely related to growth temperature because the decreasing temperature moves the growth conditions further from equilibrium, thus increasing Sn content. Therefore, we summarize most GeSn CVD growth results in Figure 7. In GeSn growth using the $SnCl_4$/GeH_4 reaction precursor combination, $SnCl_4$ and GeH_4 lose their reactivity at a temperature of 280 °C and growth is totally ceased. Below 285 °C, GeH_4 is not well-adsorbed, which may suggest the generation of GeH_2 and/or 2H. Therefore, the growth temperature for GeSn RPCVD growth with the $SnCl_4$/GeH_4 reaction precursor combination is usually higher than 280 °C. For the Ge_2H_6 and $SnCl_4$ precursor combination, GeSn growth temperature could be as low as 275 °C.

Significantly, UA demonstrated GeSn growth using PECVD with the commercially available GeH_4 and $SnCl_4$; low-temperature growth at 350 °C for GeSn epitaxy on an Si substrate was achieved with an Sn content of up to 6% [113]. By using a 1064 pulsed laser as the light source, a PL signal was also observed at the peak wavelength of 2000 nm, as shown in Figure 8 (Spot III).

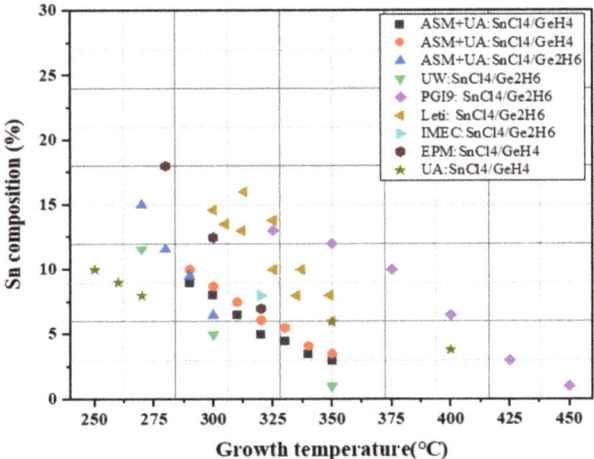

Figure 7. Temperature effect on the Sn content from different research groups (UW, UA, and EPM denote the University of Warwick, University of Arkansas, and Université de Montréal, respectively) [113,115–117,151–156].

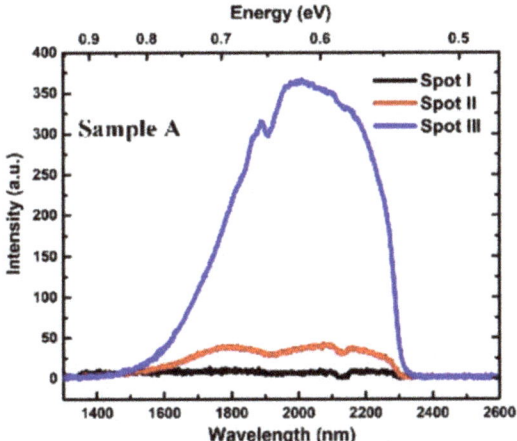

Figure 8. Room-temperature PL spectra for GeSn grown by PECVD technology (the Sn content is 6%). Reproduced from [114], open access by OSA Library, 2018.

Their follow-up work verified that the PECVD system was able to grow a high-Sn-content (>10%, with an PL emission peak at approximately 2100 nm) GeSn layer at ultralow temperatures (250, 260, and 270 °C) [157] (Figure 9). The realization of GeSn PECVD growth at such low temperatures using a $SnCl_4/GeH_4$ precursor combination mainly benefits from plasma-assisted reactivity improvements [157]. With proper growth optimization, the Sn content of the GeSn grown by PECVD should be higher than that of other CVD chambers. Compared to GeH_4, Ge_2H_6 is more reactive and possesses lower growth temperature capabilities, indicating that the reactivity of the Ge-hydride is the only limiting factor for low-temperature GeSn growth. For GeSn RPCVD growth using GeH_4, Sn incorporation was found to drastically decrease at ~285 °C, whereas the growth temperature limit for using Ge_2H_6 was found to be 270 °C [153].

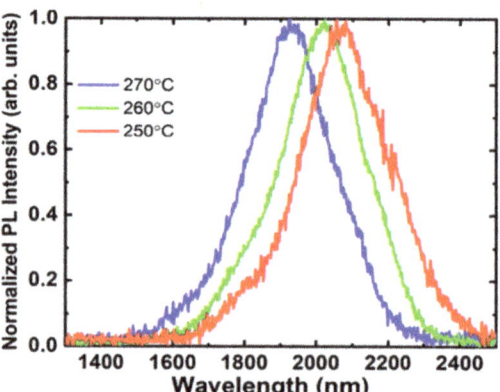

Figure 9. PL spectra for the GeSn grown at temperatures of 250, 260, and 270 °C. Reproduced from [157], open access by ScholarWorks@UARK.

For GeSn growth in a UHVCVD chamber [158–161], growth pressure is usually kept in the range of 1×10^{-4}–2.5×10^{-4} Torr, and Sn content rises with decreasing growth temperatures. Even when different combinations of precursors (SnD_4/Ge_2H_6, SnD_4/Ge_3H_8, and SnD_4/Ge_4H_{10}) are chosen, similar Sn content variation trends are observed (Figure 10). However, growth temperatures with different precursor combinations are varied; the lowest reported growth temperatures for SnD_4/Ge_2H_6, SnD_4/Ge_3H_8, and SnD_4/Ge_4H_{10} are 250, 350, and 150 °C, respectively [145,146,161]. Different from UHVCVD, pressures for GeSn growth in LPCVD and APCVD chambers have been found to range from 10 to 760 Torr [115–117,150]. The surface morphology of a layer GeSn grown by APCVD is shown in Figure 11, where surfaces are milky and pyramidical defects are observed at pressures of 10 and 100 Torr; this issue can solved by further increasing the growth rate (keep the $SnCl_4$ constant and increase the Ge_2H_6 gas flow).

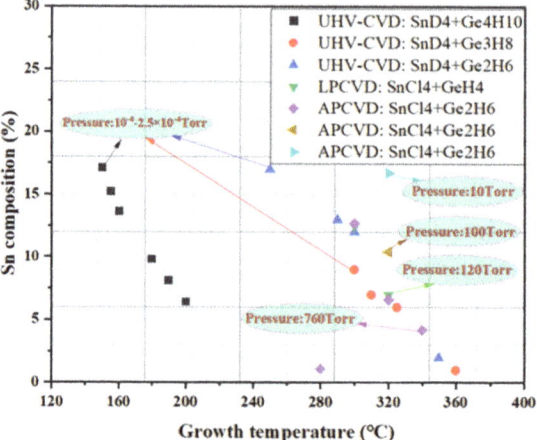

Figure 10. Temperature effect on Sn content from UHVCVD, LPCVD, and APCVD growth.

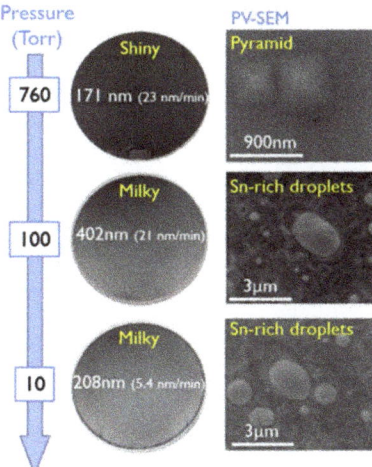

Figure 11. GeSn surface morphology vs. growth pressure (growth temperature: 320 °C; growth pressure: 10, 100, and 760 Torr; precursors: Ge_2H_6 and $SnCl_4$). Reproduced with permission from [119], IOP Publishing, 2018.

For GeSn APCVD growth at a temperature of 320 °C, pressure was found to be a main factor in the growth of high-Sn-content GeSn materials (the achieved Sn contents at 10 and 760 Torr were 16.7% and 6.6%, respectively) [119]. For LPCVD growth at 120 Torr and 320 °C, the Sn content for GeSn was almost the same as that of APCVD.

2.3.2. Carrier Gas Effect on GeSn Growth

The effect of carrier gas on GeSn CVD growth is important and of great significance for the good mixing of precursor gases in a CVD chamber [105,152,153]. In contrast to pure Ge growth, the GeSn CVD growth mechanism has changed due to the introduction of Sn precursors, which have made GeSn CVD growth more complex. In several instances, a thickness reduction or an absence of GeSn occurs when choosing N_2 as the carrier gas; this indicates that the growth rate has already changed and is below the etching rate from HCl. Furthermore, the Sn content of GeSn grown with an N_2 carrier gas is different from that grown with an H_2 carrier gas (Sn% difference is usually approximately 1%; see Figure 12). This Sn content reduction may be mainly attributed to the lower growth rate found when using N_2 as the carrier gas.

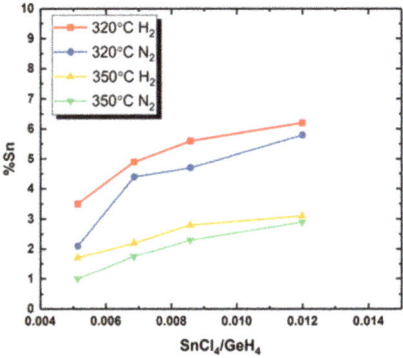

Figure 12. Sn content vs. $SnCl_4/GeH_4$ ratio (growth temperatures: 320 and 350 °C; carrier gases: N_2 and H_2; precursors: GeH_4 and $SnCl_4$). Reproduced from [153], open access by ASU library.

2.3.3. Strain Relaxation Effect on GeSn CVD Growth

F. Gencarelli et al. discovered a composition-dependent strain relaxation mechanism, and they found that high-Sn-content materials show a classical strain relaxation behavior [162]. Their AFM results showed that the island size and density of their low-Sn-content GeSn layers increased with strain relaxation degree (Figure 13) for the following reasons: higher amounts of Sn precursors were needed for high-Sn-content GeSn growth, extra Cl doses were exposed to the surface of GeSn and thus likely avoided Ge–Sn diffusion, Cl atoms could be regarded as the surfactants to mediate the enhancement of island size and density.

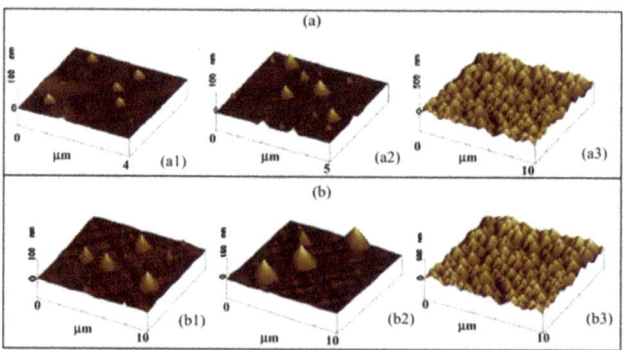

Figure 13. AFM images of (**a**) GeSn with an Sn content of 6.4%; the strain relaxations for (**a1**), (**a2**), and (**a3**) are 8%, 33%, and 75%, respectively. (**b**) GeSn with a strain relaxation of 75%; the Sn contents of (**b1**), (**b2**), and (**b3**) are 12.6%, 8.1%, and 6.4%, respectively. Reproduced with permission from [162], IOP Publishing, 2012.

Later, high-quality GeSn with a world-record high Sn content of 22.3% was crafted after the discovery of strain-relaxation-enhanced (SRE) GeSn CVD growth mechanism [113], thus showing that compressive strain is the primary limiting factor for achieving greater Sn incorporation under an Sn oversaturation condition (Figure 14). In this research, the following growth strategy was proposed: (i) for first GeSn layer growth, they used a growth recipe of 9–12% Sn (the Sn content ranged from 8.8 to 11.9%); (ii) for second GeSn layer growth, they used the same growth recipe, and the $SnCl_4$ flow fraction increased by ~8% compared to the first GeSn layer (the Sn content ranged from 12.5 to 16.5%); and (iii) for third GeSn layer growth, they used the same growth recipe, and the $SnCl_4$ flow fraction increased by ~8% compared to the second GeSn layer. It should be noted that the grading rate of Sn incorporation was well-designed to suppress the growth breakdown. Inspired by the discovery of the SRE GeSn CVD growth mechanism, S. Assali et al. grew a high-quality GeSn layer with 15% Sn using low pressure chemical vapor deposition (LPCVD) in 2018 [115,116].

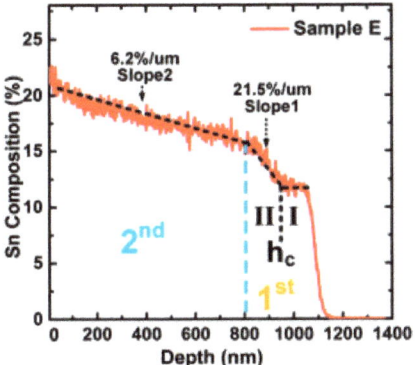

Figure 14. SIMS result for the GeSn sample with an Sn content of up to 22.3% (the maximum Sn contents for regions I, II, and III were 11.9%, 15.5%, and 22.3%, respectively). Reproduced from [113], Springer Nature, open access, 2018.

2.4. Doping for GeSn

Mainstream GeSn doping technologies, such as ion implantation and in situ CVD doping, have been intensively studied for future electronics and photonics applications. Low contact resistivity plays a vital role in the creation of high-performance devices. Table 1 presents a summary of reported B, BF_2^+, and P-doped GeSn via ion implantation in terms of year, institution, Sn content, doping type, doping concentration, activation temperature, and contact metal. Additionally, Tables 2 and 3 present summaries of B-doped GeSn, P-doped GeSn, and As-doped GeSn in terms of year, institution, Sn content, doping type, doping concentration, contact metal, and contact resistivity.

Table 1. Summary of reported B, BF_2^+, and P-doped GeSn via ion implantation in terms of year, institution, Sn content, doping type, doping concentration, activation temperature, and contact metal.

Year	Institution	Sn Content (%)	N-Type	P-Type	Doping Concentration (cm^{-3})	Activation Temperature (°C)	Contact Metal	Ref.
2011	Nagoya University	2–13	—	√	B: 8×10^{19}	350–550	Ni	[163]
2011	CAS-IOS	3	—	√	BF_2^+: ——	400	Al	[164]
2012	NUS and CAS-IOS	2.4	√	—	P: 2.1×10^{19}	400	Al	[165]
2012	NUS and CAS-IOS	4.2	√	√	P: 1×10^{21} BF_2^+: ——	400	Ni	[166]
2012	NUS and CAS-IOS	3–5.3	—	√	BF_2^+: $>1 \times 10^{20}$	300–500	—	[167]
2013	NUS	2.4	√	—	P: 2.1×10^{21}	400	Al	[168]
2013	NUS	4.2	√	√	P: $>1 \times 10^{20}$ BF_2^+: $>1 \times 10^{20}$	400	Ni	[169]
2013	NUS	4.2	√	—	P: ——	450	Ni	[170]
2013	NUS	4.2	√	—	P: ——	400	Ni	[171]

Table 1. Cont.

Year	Institution	Sn Content (%)	N-Type	P-Type	Doping Concentration (cm^{-3})	Activation Temperature (°C)	Contact Metal	Ref.
2013	NUS and CAS-IOS	5.3	—	✓	BF$_2^+$: ——	350	Ni and Ni–Pt	[172]
2013	NUS and CAS-IOS	4.1	—	✓	BF$_2^+$: ——	—	—	[173]
2013	Stanford University	7	✓	✓	P: —— BF$_2^+$: ——	400	Ti/Ni	[174]
2014	NUS and AM	2.4	✓	—	Hot P$^+$: >1 × 10^{20}	450	Ti/Ni	[175]
2014	NUS and AM	2.6	✓	—	P: >1 × 10^{20}	—	—	[176]
2015	CAS-IOS	3.2	✓	—	P: 7.64 × 10^{20}	500	Ni/Al	[177]
2016	Xidian University	4	—	✓	BF$_2^+$: ——	—	Ni	[178]
2016	CAS-IOS	8	—	✓	B: ——	300	Ni/Al	[179]
2017	Xidian University	4	—	✓	BF$_2^+$: ——	—	Ni	[180]
2017	National Taiwan University	8	✓	—	P: ——	300–350	Ni	[181]
2019	CAS-IOS	6	—	✓	BF$_2^+$: ——	450	Ni/Al	[182]
2020	CAS-IOS	9	✓	✓	B: —— P: ——	500	Ni Al/Ti/Au	[183]
2020	National Chiao Tung University	2.8	—	✓	BF$_2^+$: ——	400	Al	[184]

2.4.1. Ion Implantation for GeSn

Ion implantation is a widely used technique for doping semiconductor materials, and its advantages include low-temperature operation, precise dose control, good uniformity, and extremely small lateral diffusion. The research and development of GeSn's ion implantation technology is also of great significance for future device application. So far, researchers have carried out extensive research into GeSn ion implantation technology (although most GeSn has been grown in MBE chambers, which are also significant).

Phosphorus has been widely adopted for ion implantation to achieve efficient N-type doping in GeSn layers because its doping concentrations usually ranges from 2.1×10^{19} to 2.1×10^{21} cm^{-3}. For the P-type doping, there are two options: boron and BF$_2^+$. The highest P-type doping concentration can reach up to 1×10^{20} cm^{-3}.

2.4.2. In Situ GeSn CVD Doping

Optoelectronic devices, such as GeSn LEDs, GeSn lasers, and GeSn detectors, generally need highly doped GeSn for efficient carrier recombination and low contact resistance. Electronic devices, such as GeSn MOSFETs, GeSn TFETs, GeSn FinFETs, and GeSn GAAFETs (gate-all-around), require lower ohmic contacts, higher dopant concentrations, and selective doping. The use of in situ doping technology for GeSn is an attractive route for improving the performance of optoelectronic and electronic devices because it enables the doping of GeSn at low temperatures with a high doping efficiency and selective doping. Indeed, GeSn transitions from an indirect to direct bandgap material with an Sn content as high as 10%, and this property has led to research interest in Si-based, high-efficiency light sources. The first electrically injected GeSn lasers were recently demonstrated with Sn contents of 11% and 15%. It is definitely true that we require better solutions to create direct bandgap, high-quality doped GeSn, and the selection of an appropriate reaction doping gas and the optimization of epitaxial process are vital for this purpose. To this end, the growth of B-doped GeSn, P-doped GeSn, and As-doped GeSn using CVD has been reported by several institutions, as summarized in Table 2. However, there are several key points to consider: (I) Sn loss occurs for B-doped GeSn CVD growth, indicating that there is a competition between Sn and B atoms [150,185]; (II) excess partial pressure for PH_3 contributes to poor material quality due to P segregation; (III) B_2H_6 partial pressure has no degradation effect on material quality, though it increases the activation doping concentration; (IV) more P could be incorporated into Ge and GeSn by using high order precursors; (V) boron δ-doping layers are helpful for highly doped GeSn growth, and the maximum B concentration can reach up to 1×10^{20} cm^{-3}; and (VI) the doping efficiency of As-doped GeSn is better than that of P-doped GeSn [110].

Table 2. Summary of reported B-doped GeSn, P-doped GeSn, and As-doped GeSn in terms of year, institution, Sn content, doping type, doping concentration, and contact metal.

Year	Institution	Sn Content (%)	N-Type	P-Type	Doping Concentration (cm^{-3})	Contact Metal	Ref.
2009	ASU	2	√	——	P: 1×10^{20}	Cr/Au	[57]
2011	IMEC	8	——	√	B: 1.7×10^{19}	——	[150]
2013	KTH Royal Institute of Technology	9.4	√	√	B: 5×10^{18} P: 1×10^{20}	——	[186,187]
2016	PGI 9	8 and 11	√	√	B: 2×10^{19} P: 1×10^{20}	——	[188]
2016	PGI 9	8.5 and 15	√	√	B: 4×10^{18} P: 7.5×10^{19}	——	[189]
2017	ASM	9	√	——	As: $>2 \times 10^{20}$	——	[110]
2017	ASM and IMEC	1.4	——	√	B: 2×10^{20}	——	[190]
2018	National Taiwan University	10	√	√	B: Sn loss P: No Sn loss	——	[191]
2019	National Taiwan University	>12	——	√	B: $>1 \times 10^{21}$	Ti	[192]
2019	Leti	10 and 15	√	——	P: 5×10^{20}	——	[193]
2020	National Taiwan University	2, 4.7, and 13	——	√	B: 2.1×10^{20} for 2% Sn	Ti	[194]
2020	Leti	6.5	√	√	B: 5.2×10^{19} P: 2.2×10^{20}	——	[195]
2020	National Taiwan University	4.7	——	√	B: 1.9×10^{20}	Ti	[196]
2021	National Taiwan University	9	√	——	P: 1.3×10^{20}	Ni	[197]

Table 3. Summary of reported B-doped GeSn, P-doped GeSn, and As-doped GeSn in terms of institution, Sn content, doping type, doping concentration, and contact metal.

Year	Institution	Sn Composition (%)	N-Type	P-Type	Doping Concentration (cm^{-3})	Contact Metal	Contact Resistivity ($\Omega \cdot cm^2$)	Ref.
2014	Institute of Microelectronics, Chinese Academy of Sciences	4	—	—	—	Ni	—	[187]
2018–2020	National Taiwan University	9	√	—	P: 1.3×10^{20}	Ni	1.5×10^{-7}	[191,192,194,196,197]
		2, 4.7, and 13	—	√	B: 2.1×10^{20} for 2% Sn	Ti	4.1×10^{-10} for 2% Sn	
		4.7	—	√	B: 1.9×10^{20}	Ti	1.1×10^{-9}	
		>12	—	√	B: $>1 \times 10^{21}$	Ti	4.1×10^{-10}	
		10	√	—	P: 1.3×10^{20}	Ni	1.1×10^{-7}	
		9	√	√	B: 4×10^{17} P: —	Ni	3.8×10^{-8}	
2020	Leti	6.5	√	√	B: 5.2×10^{19} P: 2.2×10^{20}	—	—	[195]
2020	Université de Montréal	11	√	√	B: $\times 10^{19}$ As: $\times 10^{20}$	—	—	[198]
2019	University College Cork	8	—	—	—	Ti, Ni, and Pt	—	[199]
2013–2019	NUS	5, 7, and 8	—	√	Ga: 3.4×10^{20}	Ti	4.4×10^{-10} for 7% Sn	[165–173,200–203]
		8.5	—	√	Ga: 3.2×10^{20}	—	—	
		5	—	√	Ga: —	Ti	9.3×10^{-10}	
		5	—	√	Ga: —	Ni	2×10^{-10}	
		5	—	√	Ga: 1.6×10^{20}	Ni	1.4×10^{-9}	
		2.4	√	—	P: 2.1×10^{19}	Al	4×10^{-3}	
2012	NUS and CAS-IOS	5.3	—	√	BF$_2^+$: 5.7×10^{20}	Ni	1.6×10^{-5}	[204]
2015–2020	CAS-IOS	7	√	—	Sb: 5×10^{20}	Ni	1.3×10^{-6}	[179,182,183,205,206]
		8	√	—	Sb: 3×10^{19}	Ni/Al	6.2×10^{-5}	
		7	√	—	Sb: 5×10^{19}	Ni	1.3×10^{-6}	
		3.2	√	—	P: 7.64×10^{19}	Ni/Al	2.26×10^{-4}	
		7	—	√	P: 2.44×10^{19}	Ni/Al	1.9×10^{-6}	

2.4.3. GeSn Ohmic Contact

Among the summarized GeSn contact works is that of Henry. H. Radamson et al., who proposed a novel method to improve the thermal stability of the Ni–GeSn contact. It is well-known that carbon stabilize NiSiGe materials, so after GeSn growth, they implanted C into GeSn. In Figure 15, we can see that the NiGeSn film with C was more uniform than the NiGeSn film without C. Characterization results indicated that the presence of C not only led to the improved thermal stability but also tended to change the preferred orientation of NiGeSn [187]. A comparison work with different contact metals (10 nm of Ni, Ti, and Pt) [199] showed that Ni–GeSn was the most promising candidate due to its low sheet resistance and low formation temperature (below 400 °C). Moreover, Pt–GeSn showed better behavior in terms of thermal stability compared to Ni–GeSn and Ti–GeSn. Because Sn loss occurs during B-doped GeSn CVD growth, it is still challenging to create low contact resistivity p-type GeSn contacts with high Sn contents, a challenge that is particularly critical for GeSn lasers and GeSn TFETs [207,208].

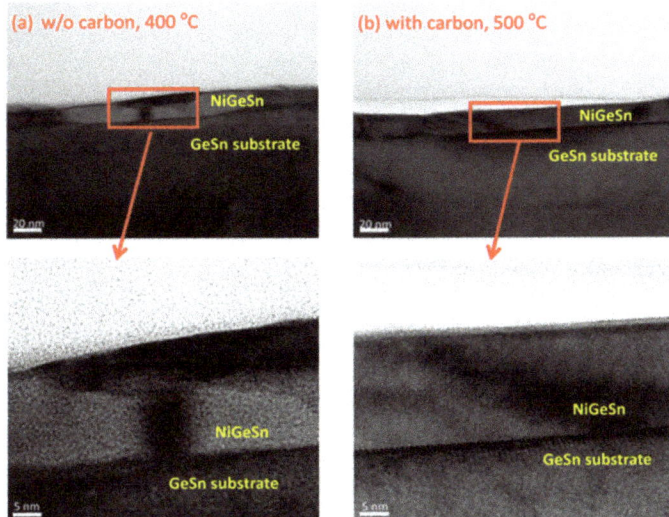

Figure 15. TEM images for Ni–GeSn interface (**a**) annealing at 400 °C without C and (**b**) annealing at 400 °C with C. Reproduced from [187], IOP Publishing, open access, 2015.

3. Research Progress for GeSn Detectors
3.1. GeSn Photoconductive Detector

Photoconductive detector, which can also be defined as metal–semiconductor–metal (MSM) detector, is regarded as the simplest structure to achieve detection. In this type of structure, two Schottky junctions are designed and the total layer structure does not require any doping. Therefore, it can only work at a high bias voltage due to the existence of high contact resistance. However, the capacitance of a photoconductive detector is quite low, which is helpful for high-speed detection. Based on the photoconductive structure, researchers have put great effort into GeSn photoconductive detectors (Figure 16). Table 4 shows the reported performance levels of GeSn photoconductive detectors grown by CVD technology.

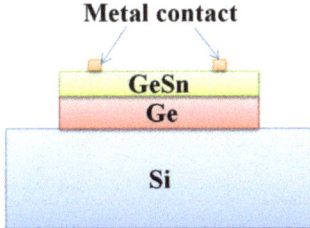

Figure 16. Cross-sectional schematic of a device structure for a GeSn photoconductive detector.

As previously reported, IMEC mastered low-cost and commercially available cutting-edge GeSn growth technology in 2011 (Ge_2H_6/GeH_4 precursor combination) [151]. Subsequently, they further grew a GeSn/Ge MQWs structure, and they also fabricated a photoconductive detector [58]. In 2014, Benjamin, R. Conley et al. reported the temperature-dependent spectral responses and detectivity of GeSn photoconductors with Sn contents ranging from 0.9 to 7% [59]. For a GeSn photoconductor with 7.0% Sn, a maximum wavelength response of 2100 nm was achieved. Experimental results showed that low-temperature responsivity was two orders of magnitude higher than room-temperature responsivity at 1550 nm, and the maximum specific detectivity was 1×10^9 cm·$Hz^{1/2}$/W at 77 K. In the same year, Benjamin, R. Conley et al. further extended the spectral response using a GeSn layer with 10% Sn [60]. The room-and low-temperature (77 K) wavelength cutoffs for the GeSn detector were found to be 2400 and 2200 nm, respectively. Maximum peak responsivity was observed as 1.63 A/W at 77 K due to photoconductive gain. More importantly, the specific detectivity was increased by about five times compared to the previously reported result (a GeSn photoconductor with 7.0% Sn), indicating that the material quality of the GeSn layer with 10% Sn was greatly improved (Figure 17).

Table 4. Summary of reported GeSn photoconductive detectors in terms of Sn content, GeSn thickness, device structure, wavelength cutoff, and responsivity.

Year	Sn Composition	GeSn Thickness	Structure	Cutoff	Responsivity	Ref.
2012	9%	13 or 20 nm	GeSn/Ge 3QWs	2200 nm	0.1 A/W at 5 V	[58]
2014	0.9%	327 nm	Bulk	1800 nm	——	[59]
	3.2%	76 nm		1900 nm	——	
	7.0%	240 nm		2100 nm	0.18 A/W at 10 V	
2014	10%	95 nm	Bulk	2400 nm	1.63 A/W at 50 V	[60]
2015	10%	95 nm	Bulk	2400 nm	0.26 A/W	[61]
2019	12.5%	140 and 660 nm	Bulk	2950 nm	2 A/W	[63]
	15.9%	250 and 670 nm		3200 nm	0.044 A/W	
	15.7%	165, 585, and 254 nm		3400 nm	0.0072 A/W	
	17.9%	310, 550, and 260 nm		3350 nm	0.0038 A/W	
	20%	450 and 950 nm		3650 nm	0.0067 A/W	
	22.3%	380 and 830 nm		3650 nm	0.0032 A/W	

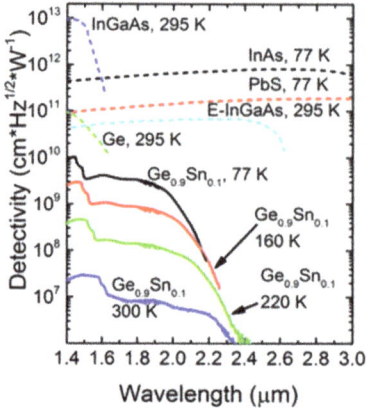

Figure 17. Specific detectivity for a $Ge_{0.9}Sn_{0.1}$ photoconductive detector at temperatures of 77, 160, 220, and 300 K. Reproduced from [60], OSA Publishing, open access, 2014.

In 2019, Huong Tran et al. reported a GeSn photoconductor with high Sn contents (the maximum Sn contents of the top GeSn layer were 12.5%, 15.9%, 15.7%, 17.9%, 20%, and 22.3%) [63]. As the Sn content increased, the cutoff wavelength shifted toward longer wavelength due to the bandgap shrinkage. From 77 to 300 K, the cutoff wavelengths were 3200–3650 nm for the GeSn photoconductor with 22.3% Sn. It is worth noting that this D* value was superior to that of a PbSe detector at the given wavelength range and was comparable to that of a commercial extended-InGaAs detector (4×10^{10} cm·Hz$^{1/2}$·W^{-1}) at the same wavelength range (Figure 18). Even at 300 K, the passivated device showed better results D* than the PbSe detector from 1500 to 2200 nm.

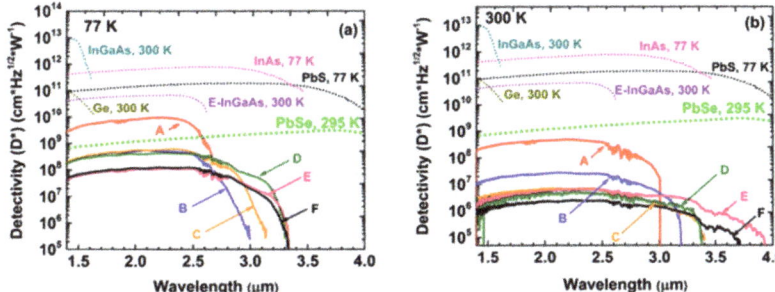

Figure 18. Specific detectivity for a GeSn photoconductive detector at the temperatures of (**a**) 77 K and (**b**) 300 K (the Sn contents for samples A–F were 12.5%, 15.9%, 15.7%, 17.9, 20.0%, and 22.3%, respectively). Reproduced with permission from [63], American Chemical Society, 2019.

To enable a comprehensive overview of the use of GeSn photoconductive materials for infrared detection applications, Figure 19 illustrates the Sn content vs. cutoff wavelength for reported GeSn photoconductive detectors. For GeSn with an Sn incorporation of 0.9–12.5%, the photoconductive detector wavelength coverage was found to range from 1800 to 2950 nm, indicating that GeSn with Sn contents of up to 12.5% or 13% is very promising for SWIR applications. For GeSn with an Sn incorporation of 15.9–22.3%, the photoconductive detector wavelength coverage was found to range from 3200 to 3650 nm, suggesting potential mid wavelength infrared (MWIR) applications. For wavelengths from 3650 to 5000 nm, no detectors have been reported. However, GeSn photoconductive detector performance is limited by current growth technology and Sn distribution uniformity in total layer structures, which causes a low responsivity (the responsivity values are listed in the table above).

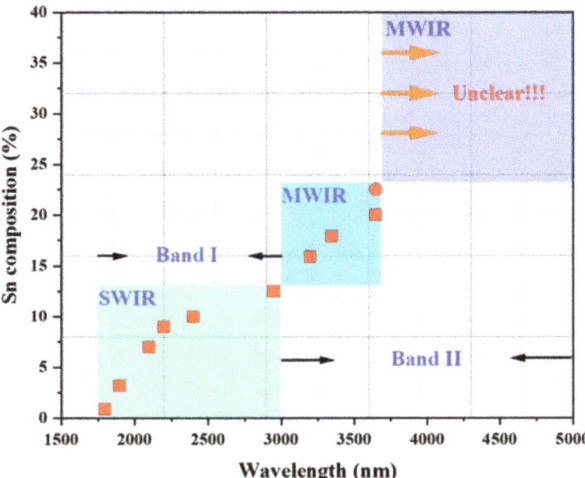

Figure 19. Sn content vs. wavelength of a GeSn photoconductive detector, indicating that GeSn is a promising absorber in SWIR and MWIR detection applications.

3.2. GeSn PIN Detector

The PIN detector is the most common and widely used detector type for Si-based optoelectronics applications. One side of a PIN detector device is for p-type doping, and the other side is for n-type doping; as such, the built-in electric field is able to locate the intrinsic region [18]. A typical cross-sectional schematic diagram of a GeSn PIN detector is shown in Figure 20, and the major device performance values for reported GeSn PIN detectors are summarized in Table 5.

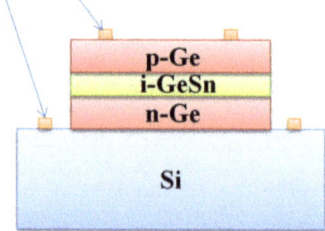

Figure 20. Cross-sectional schematic of typical device structure for a GeSn detector.

Table 5. Summary of reported GeSn PIN detectors in terms of Sn content, GeSn thickness, device structure, wavelength cutoff, and responsivity.

Year	Sn Composition	GeSn Thickness	Structure	Cutoff	Responsivity	Ref
2009	2%	350 nm	n–GeSn/i–GeSn/P-Si	1750 nm	-	[57]
2016	7%	200 nm	p–Ge/i–GeSn/n–Ge	2200 nm	0.15 A/W at 1 V	[62]
	10%	200 nm		2600 nm	0.07 A/W at 1 V	
2018	11%	700 nm	p–Ge/p–GeSn/i–GeSn/n–GeSn/n–Ge	2650 nm	0.32 A/W	[65]
2019	8%	25 nm	p^+–Ge/i–QWs/n^+–Ge	2000 nm	0.2 A/W	[66]

In 2009, Jay Mathews et al. demonstrated the first GeSn photodetector with 2% Sn content; 350 nm $Ge_{0.98}Sn_{0.02}$ was directly grown on a B-doped Si (100) substrate in an UHVCVD system (the carrier concentration in the Si wafer was 4.3×10^{19} cm^{-3}) [57]. Three cycles of post-growth annealing were carried out to decrease the TDDs in $Ge_{0.98}Sn_{0.02}$. Afterwards, n-doped $Ge_{0.98}Sn_{0.02}$ was further deposited, and its carrier concentration was found to be approximately 7.5×10^{19} cm^{-3}. Using the abovementioned layer structure, a circular GeSn photodetector was fabricated. To evaluate the quantum efficiency of the $Ge_{0.98}Sn_{0.02}$ photodetector, the circular mesa was continuously illuminated via a halogen source and 1270, 1300, 1550, and 1620 nm lasers. The $Ge_{0.98}Sn_{0.02}$ detector quantum efficiencies were higher than those in comparable pure Ge device designs processed at low temperatures (Figure 21). Additionally, the wavelength cutoff was extended to at least 1750 nm, which means that a GeSn photodetector with 2% Sn content can cover the entire telecommunication band.

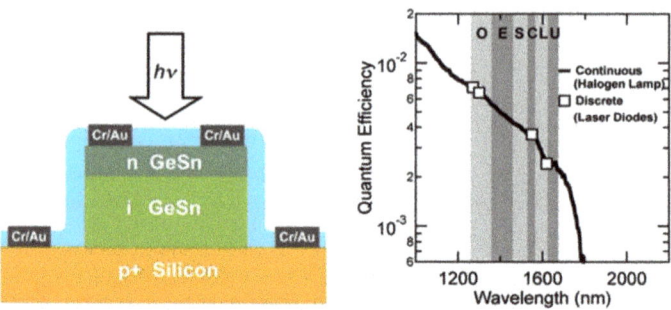

Figure 21. Cross-sectional schematic of a GeSn photodetector and its quantum efficiency as a function of wavelength. Reproduced with permission from [57], AIP Publishing, 2009.

In 2018, Huong Tran et al. fabricated GeSn photodetectors with 700 nm thick GeSn layers using the p–Ge/p–Ge$_{0.91}$Sn$_{0.09}$/i–Ge$_{0.89}$Sn$_{0.11}$/n–Ge$_{0.89}$Sn$_{0.11}$/n–Ge layer structure (all layers were grown by RPCVD) [65]. In order to obtain detailed and accurate external reading of quantum efficiency, all GeSn photodetectors were illuminated with a 2000 nm laser. Room-temperature peak responsivity and external quantum efficiency were measured to be 0.32 A/W at 2000 nm and 20%, respectively. When the GeSn detector was illuminated by a 1550 nm laser, its external quantum efficiency reached up to 22%. Different from the previously reported thin film photoconductor, the thick film photoconductor showed an extended wavelength cutoff (2650 nm) due to the reduced strain relaxation and enhanced light absorption in the thick GeSn film. Nevertheless, the peak specific detectivity for the GeSn detector was compared to other commercial infrared detectors at a wavelength range from 1400 to 3000 nm, which showed that peak specific detectivity of the GeSn detector at 2000 nm was only one order of magnitude lower than that of the extended-InGaAs detector (Figure 22). To improve device performance, Xu S, et al. attempted to create a GeSn/Ge MQW detector [67,68], a GeSnOI detector [69], and a photon-trapping microstructure GeSn/Ge MQW detector [209].

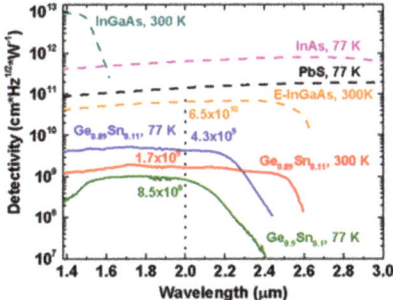

Figure 22. Specific detectivity for a Ge$_{0.89}$Sn$_{0.11}$ photodetector at the temperatures of 77 and 300 K. Reproduced with permission from [65], AIP Publishing, 2018.

Figure 23 summarize the Sn content vs. cut-off wavelength for a reported GeSn PIN detector. For GeSn with an Sn incorporation of 2–11%, the PIN detector wavelength coverage was found to range from 1750 to 2650 nm, indicating that a GeSn PIN detector is very promising for SWIR applications. Due to the limitations of growth technology, PIN detectors at wavelengths from 2650 to 5000 nm have yet to be reported.

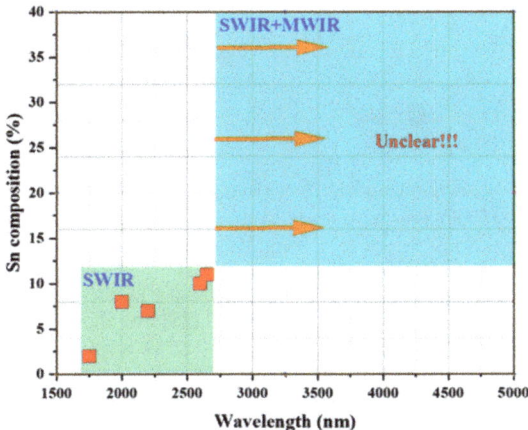

Figure 23. Sn content vs. cut-off wavelength of the GeSn PIN detector.

4. Research Progress for GeSn Lasers

Since Si-based high-efficiency light sources comprise the technical bottleneck for Si-based monolithic optoelectronic integration, researchers have conducted extensive research into Ge and GeSn lasers. Ten years ago, the rapid development of the GeSn CVD growth technique enabled researchers from MIT to demonstrate optically injected and electrically injected Ge lasers at room temperature. The lasing thresholds of these laser devices were very high, which made it difficult to achieve efficient lasing. As a result, more attentions has been paid to the GeSn material due to its direct bandgap property. In this section, we review the latest research on GeSn lasers with different optical cavities, as well as their device performance.

4.1. Optically Injected GeSn Lasers
4.1.1. Optically Injected GeSn Laser with FP Cavity

Based on the GeSn optical gain medium, the world's first optically injected FP cavity GeSn laser was demonstrated at a low temperature [70]. The typical threshold power densities of FP cavity GeSn lasers with cavity lengths of 1 mm, 500 µm, and 250 µm were maintained between 300 and 330 kW/cm^2 (Figure 24a). When the optically injected power density was above its threshold power density, the full width half maximum of the optical emission spectrum was dramatically reduced and the intensity was significantly increased; when the optical injection power density increased to 650 kW/cm^2, the threshold curve tended to be flat (possibly due to a self-heating effect) (Figure 24a). When the optically injected power density increased to 1000 kW/cm^2, the maximum lasing temperature for the GeSn laser with 12% Sn content was 90 K. Figure 24b shows high-resolution laser spectra that indicate the performance of a GeSn laser under multi-mode operation.

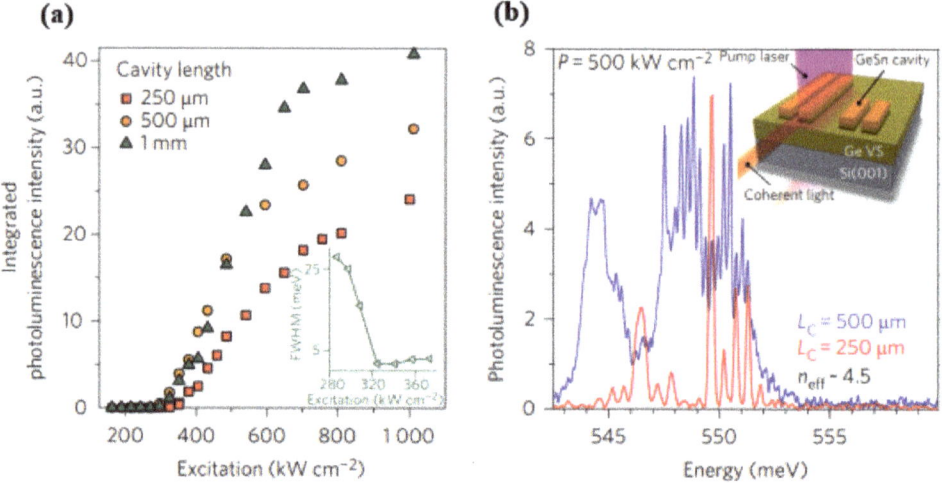

Figure 24. (a) Integrated PL intensity vs. excitation power density for a GeSn FP cavity laser with different cavity lengths; (b) high-resolution laser spectra for a GeSn laser with cavity lengths of 250 and 500 µm. Reproduced with permission from [70], Springer Nature, 2015.

In 2017, Joe Margetis et al. systematically studied the performance of optically injected GeSn lasers with different Sn contents [73]; the Sn contents of samples A–G were 7.3%, 9.9%, 11.4%, 14.4%, 15.9%, 16.6%, and 17.5%, respectively, and the maximum operation temperatures of samples A–G were 77, 110, 140, 160, 77, 140, and 180 K, respectively (Figure 25). Except for sample A (lower Sn content) and sample E (poor material quality), the samples could be lased at 140 K. It is worth noting that the maximum operation temperature of samples D and G were 160 and 180 K, respectively. The results showed that the operating temperature of the optically injected GeSn laser was closely related to the Sn content of GeSn, and the GeSn lasers with higher Sn contents possessed higher operating temperature (except for sample F because of its poor material quality). Therefore, increasing the Sn content in GeSn can effectively increase the operating temperature of the laser device. From the theoretical point of view, the main factors that affect the performance of laser devices are material gain, active layer thickness, device surface roughness, and non-radiative recombination. Therefore, there are differences in the operating temperatures of GeSn laser devices with different Sn contents.

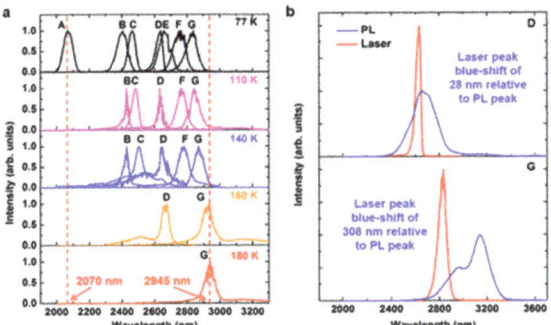

Figure 25. (a) GeSn laser spectra for samples A–G; (b) comparison of the PL and laser spectra of samples D and G. Reproduced with permission from [73], American Chemical Society, 2017.

Thanks to the discovery of the GeSn strain-relaxation-enhanced growth mechanism [88], researchers were able to increase the Sn content of GeSn to 22.3%. In this layer structure, the GeSn buffer layer is grown with a nominal recipe for 11% GeSn. When the thickness of the 11% GeSn layer reaches its critical thickness, internal strain in the GeSn layer gradually relaxes and more Sn atoms can be incorporated into the Ge lattice. Experimental results showed that the strain relaxation growth mechanism could lead to high-Sn-content GeSn alloys (higher than 22.3%). Later, Wei Dou et al. reported an optically injected bulk GeSn laser with an Sn content of up to 22.3% [75]; both 1064 and 1950 nm pulsed lasers were used for optical injection, and the maximum operating temperatures were 150 and 180 K, respectively (Figure 26).

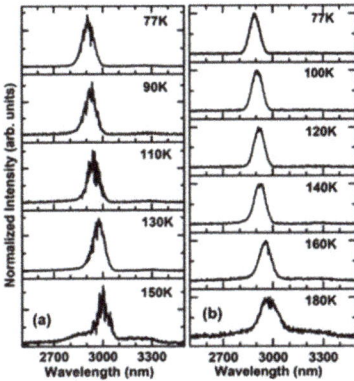

Figure 26. Temperature-dependent lasing spectra for a bulk GeSn laser with an Sn content of up to 22.3%; the optical injection sources were (a) a 1064 nm pulsed laser and (b) a 1950 nm pulsed laser. Reproduced from [75], OSA Publishing, open access, 2018.

In 2019, Yiyin Zhou et al. researched optically injected GeSn lasers (an Sn content of 20%) with different waveguide widths [76]; 1064 and 1950 nm lasers were used for pulsed optical injection characterization (Figure 27). They concluded that the operation temperature for sample A was lower than those of the other samples (the laser operation temperatures under 1064 and 1950 nm pulsed injection were 120 and 140 K, respectively). Moreover, the threshold for sample A was relatively larger than those of the other samples (at 77 K, the thresholds under 1064 and 1950 nm optical pulsed injection were 516 and 132 kW/cm^2, respectively). When the sample width was wider than 20 μm, the operation temperatures of the laser devices could be increased to 260 and 270 K under 1064 and 1950 nm optical pulsed injection, respectively. The reasons for this are as follows: (i) compared with the side wall surface recombination, free carrier absorption loss and non-radiative recombination were the dominant losses at higher temperatures; (ii) the stripe-shaped optical injection light beam had a Gaussian distribution, which may have resulted in absorption occurring in the middle of a wider waveguide (less absorption at the edge of the waveguide); and (iii) the optical confinement factor for sample D was lower, which led to a higher threshold.

1064 nm pumping	A (5 μm)	B (20 μm)	C (100 μm)	D (planar)
threshold @ 77 K (kW/cm^2)	516	384	356	330
peak position @ 77 K (nm)	2641	2802	2965	2961
max operating temp (K)	120	260	260	260
threshold @ max operating temp (kW/cm^2)	903	2990	6055	9587
peak position @ max operating temp (nm)	2970	3432	3444	3334
1950 nm pumping	A (5 μm)	B (20 μm)	C (100 μm)	D (planar)
threshold @ 77 K (kW/cm^2)	132	88	47	74
peak position @ 77 K (nm)	2703	3022	3272	2997
max operating temp (K)	140	270	270	270
threshold @ max operating temp (kW/cm^2)	364	886	796	1105
peak position @ max operating temp (nm)	2780	3414	3462	3354

Figure 27. Summary of laser performance under 1064 and 1950 nm pulsed laser injection (the cavity widths for samples A, B, C, and D were 5 μm, 20 μm, 100 μm, and planar, respectively). Reproduced with permission from [76], ACS Publishing, 2019.

The simplest optical cavity is that of Fabry–Pérot, which consists of two parallel reflecting surfaces that allow coherent light to travel through the whole cavity. Due to the directness difference between GeSn alloys with different contents, we summarize the reported operation temperatures for GeSn with different Sn contents in Figure 28. Operation temperatures were found to increase with more Sn incorporation, indicating that operation temperature is closely related to the directness of GeSn. Different from narrow bulk devices, broad bulk devices (with a cavity width greater than 20 μm) possess higher operation temperatures, possibly due to the following two reasons: (1) they have higher optical gains, and (2) they are wider and thus have higher optical injection efficiencies. However, the operation temperature for a GeSn laser with 22.3% Sn incorporation was found to be the same as that of a GeSn laser with 17.5% Sn incorporation, which means that there were many point defects in the high-Sn-content GeSn layer. For clarification, we also summarize the devices performance for the published FP cavity optically pumped GeSn laser (Table 6).

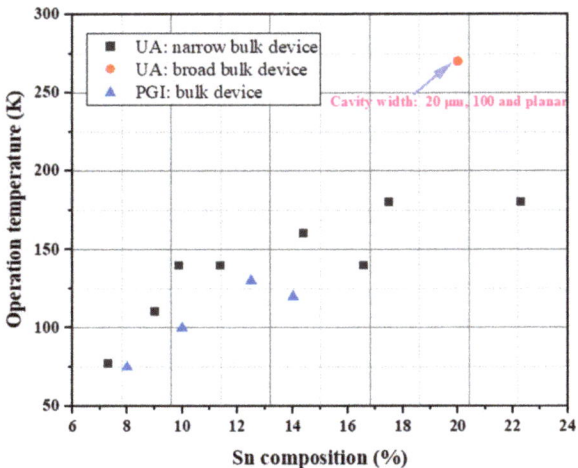

Figure 28. Maximum operation temperature vs. Sn content for optical pumped FP cavity GeSn laser (under pulsed 1064 nm laser) [70,74–76].

Table 6. Summary of the reported optically pumped FP cavity GeSn lasers in terms of structure, Sn content, thickness, cavity width, pumping laser, maximum operation temperature (T_{max}), and threshold.

Year	Structure	Sn (%)	Thickness (nm)	Cavity Width (μm)	Pumping	T_{max} (K)	Threshold (kW/cm^2)	Ref
2015	Bulk	12.6	560	5	Pulsed 1064 nm	90	1000 at 90 K 325 at 20 K	[70]
2016	Hetero	11	260 and 760	5	Pulsed 1064 nm	110	68 at 10 K 166 at 90 K 398 at 110 K	[71]
2017	Bulk	7.3	210 and 680	5	Pulsed 1064 nm	77	300 at 77 K	[73]
		9.9	280 and 850			140	117 at 77 K	
		11.4	180 and 660			140	160 at 77 K	
		14.4	250 and 670			160	138 at 77 K	
		15.9	210 and 450			77	267 at 77 K	
		16.6	160, 680, and 290			140	150 at 77 K	
		17.8	310, 550, and 260			180	171 at 77 K	
2018	Bulk	22.3	380 and 830	5	Pulsed 1064 nm	150	203 at 77 K 609 at 150 K	[75]
					Pulsed 1950 nm	180	137 at 77 K	
2018	QWs	13.8	22 (4×)	—	Pulsed 1950 nm	20	—	[83]
		14.4	31 (4×)	—		90	25 at 10 K 480 at 90 K	
2019	Bulk	20	450 and 970	5	Pulsed 1064 nm	120	516 at 77 K	[76]
				20		260	384 at 77 K	
				100		260	356 at 77 K	
				planar		260	330 at 77 K	
				5	Pulsed 1950 nm	140	132 at 77 K	
				20		270	88 at 77 K	
				100		270	47 at 77 K	
				planar		270	74 at 77 K	

4.1.2. Optically Injected GeSn Laser with WGM Cavity

In 2016, Daniela Stange et al. realized a self-suspending microdisk GeSn laser for the first time [74] (Figure 29). The laser spectrum is shown in Figure 30. It can be seen in the figure that the maximum working temperatures of samples A and B were 80 and 140 K, respectively. Compared with sample B, the lasing spectrum of sample A was blue-shifted due to its higher content. Although the operation temperature for sample A was lower than that of sample B, the threshold for sample A was lower than that of sample B (the thresholds of samples A and B were 125 and 220 kW/cm^2 at 50 K, respectively).

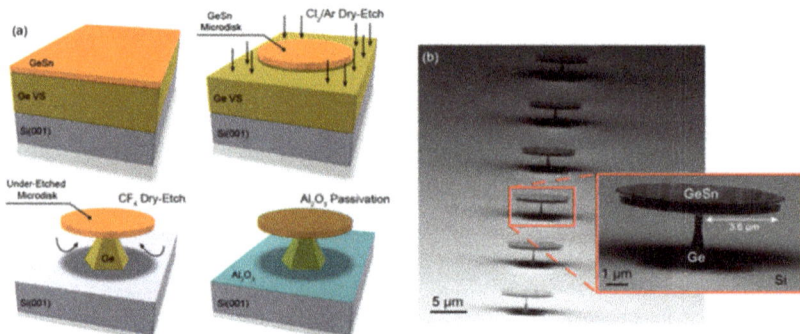

Figure 29. (**a**) Process flow for GeSn microdisk; (**b**) SEM image of GeSn microdisk with an Sn content of 12.5% (diameter was 8 μm). Reproduced with permission from [74], American Chemical Society, 2016.

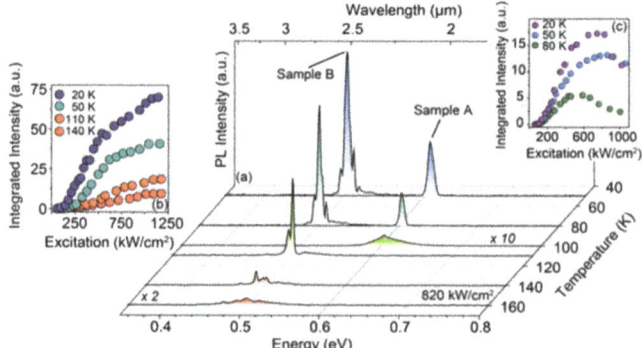

Figure 30. (**a**) Temperature-dependent lasing spectra for samples A and B (the Sn contents for samples A and B were 8.5% and 12.5%, respectively); (**b**,**c**) L–L curves for samples A and B, respectively. Reproduced with permission from [74], American Chemical Society, 2016.

In 2020, Anas Elbaz et al. reported a CW optically injected GeSn microdisk laser with a low Sn content for the first time [80]. Compared with high-Sn-content GeSn, low-Sn-content GeSn has fewer internal point defects and better material quality. After its growth, a low-Sn-content GeSn layer was transferred to an Si substrate with SiN and Al layers. Then, the Si substrate, Ge buffer layer, and defective GeSn layers are removed; only 40 nm, high-quality, low-Sn-content GeSn was left. Finally, the transferred GeSn layer was patterned into independent GeSn/SiN microdisks supported by Al microdisk pillars (Figure 31). The lasing spectrum in Figure 31 shows the continuous wave light injection laser spectrum of a GeSn microdisk with a diameter of 7 μm at 25 K: below the threshold, a light emission spectrum with a wide half-width (red line) was obtained under an optical injection power of 0.5 mW; above the threshold, lasing emission characteristics were obvious under the optical injection power of 6.4 mW. Under the pulsed optical injection and CW light injection, the maximum operating temperatures of the laser device were 90 and 50 K, respectively.

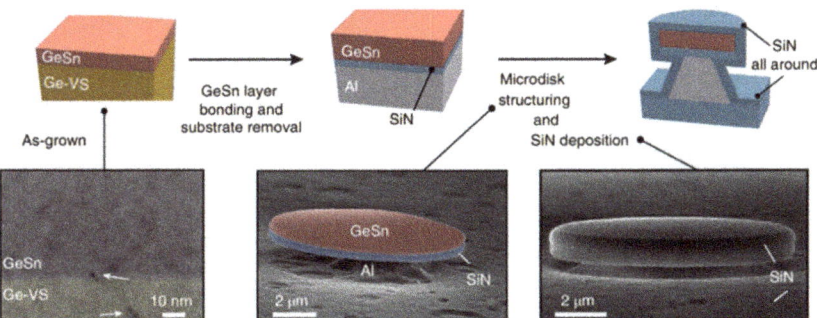

Figure 31. Fabrication process for a GeSn microdisk laser with SiN$_x$ all-around. Reproduced with permission from [80], Springer Nature, 2020.

In 2020, Anas Elbaz et al. created an optically injected GeSn microdisk laser after proper defect management [81,82], indicating that the threshold was greatly reduced compared to that of a GeSn microdisk laser without defect management (the lasing threshold reduction was 1 order of magnitude higher compared to examples in the literature). They also found that the maximum lasing temperature for the optically injected GeSn microdisk laser, with Sn contents ranging from 7% to 10.5%, only weakly depended on Sn content. Apart from the directness of the GeSn active region, the experimental results indicated that nonradiative recombinations and point defects are the main obstacles for high-temperature lasing (Figure 32).

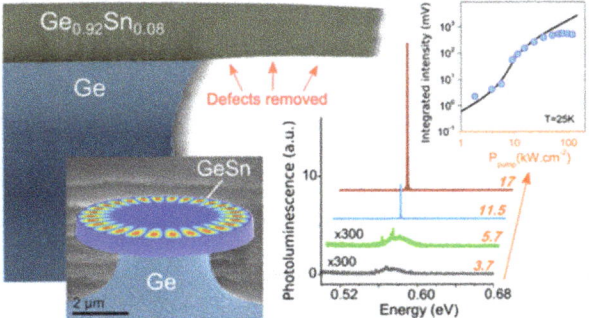

Figure 32. GeSn microdisk laser with removed defects under the disk. Reproduced with permission from [82], American Chemical Society, 2020.

The abovementioned GeSn microdisk laser results show that both pulsed and CW injection have been achieved (Table 7). Especially for CW lasing, this is the most direct evidence to verify that GeSn can withstand a CW injection test. To gain a better understanding of GeSn microdisk lasers, we summarize the operation temperatures for GeSn lasers with different Sn contents in Figure 33. For the pulsed injection, the operation temperature for the GeSn microdisk laser followed a similar trend to that of an FP cavity GeSn laser (the operation temperature increased with Sn content). However, the operation temperature for the heterostructure and quantum well GeSn laser was lower than that of bulk laser, suggesting that there is still room to improve the operation temperatures of heterostructure and quantum well lasers. For CW injection, it seems that operation temperature enhancement is not that sensitive to Sn content, though it brings efficient heat dissipation.

Table 7. Summary of the reported optically pumped WGM cavity GeSn lasers in terms of structure, Sn content, thickness, disk size, pumping laser, maximum operation temperature (T_{max}), and threshold.

Year	Structure	Sn (%)	Thickness (nm)	Disk Size (μm)	Pumping	T_{max} (K)	Threshold (kW/cm²)	Ref
2016	Bulk	8.5	800	8	Pulsed 1064 nm	90	125 at 50 K	[74]
		12.5	560	8		130	220 at 50 K	
2018	Hetero	16	418	20	Pulsed 1064 nm	230	134 at 15 K	[78]
							375 at 135 K	
							640 at 190 K	
							790 at 230 K	
2018	Hetero	14.5	380	8	Pulsed 1064 nm	100	300 ± 25 at 20 K; 250 at 50 K	[84,86]
					Pulsed 1550 nm	120	420 ± 10 at 20 K	
	MQW-A	13.3	22 (10×)	8	Pulsed 1064 nm	100	35 ± 4 at 20 K	
					Pulsed 1550 nm	120	45 ± 3 at 20 K	
	MQW-B	13.5	12 (10×)		Pulsed 1064 nm		No lasing	
					Pulsed 1550 nm	20	—	
2020	Bonded bulk	5.4	40	9	Pulsed 1064 nm	85	0.8 at 25 K	[80]
				12		100		
				9	CW 1550 nm	72	1.1 at 25 K	
2020	Low TDD bulk	7	500	7	CW 1550 nm	80–95	10	[81,82]
		8.1		6			8	
				5			8.9 at 25 K	
		10.5		8			11.6 at 25 K	

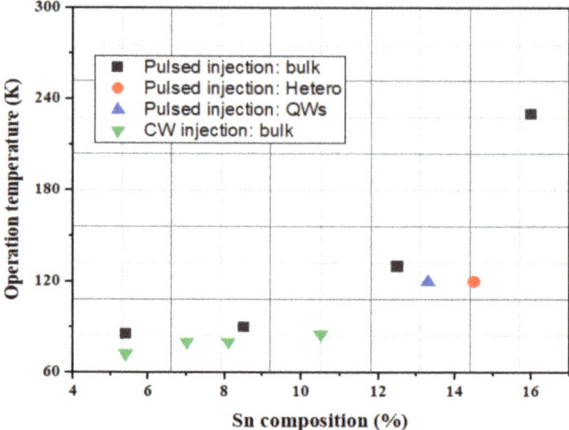

Figure 33. Maximum operation temperature vs. Sn content for an optically pumped WGM cavity GeSn laser (under pulsed 1064 nm laser).

4.1.3. Optically Injected GeSn Laser with Other Microcavities

In addition to those on FP cavity and microdisk cavity GeSn lasers, there have been publications on hexagonal photonic crystal (PC) and micro-bridge GeSn lasers. In 2018, Q.M. Thai et al. reported optically injected GeSn laser with 16% Sn content for the first time [72]. By introducing defects in the photonic crystal defect cavity (such as removing the central hole), the periodic structure around the photonic band gap were able to provide optical feedback to the microcavity. The experimental results showed that the maximum working temperature of the hexagonal photonic crystal GeSn laser was 60 K, and the threshold values at 15 and 60 K were 227 and 340 kW/cm², respectively (Figure 34).

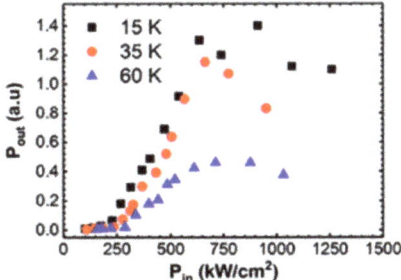

Figure 34. L–L curves for a photonic crystal GeSn laser with an Sn content of up to 16%. Reproduced with permission from [72], AIP Publishing, 2018.

In 2019, Jerémie Chrétien et al. explored a novel approach to create a direct bandgap GeSn material via strain redistribution, thereby controlling band structure and lasing wavelength [77]. Tensile-strained GeSn micro-bridge heterostructures were optically injected using pulsed 1064 and 2650 nm lasers (Figure 35), and the maximum operation temperature for the L = 75 μm micro-bridge structure laser was 273 K, which indicates that the operation temperature was very close to room temperature.

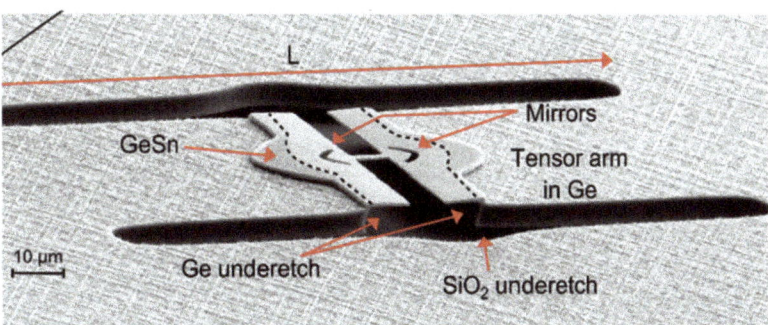

Figure 35. L–L curves for a micro-bridge GeSn laser with an Sn content of up to 16%. Reproduced with permission from [77], American Chemical Society, 2019.

4.2. Electrically Injected GeSn Lasers

Different from optically injected GeSn lasers, electrically injected GeSn lasers are more suitable for practical applications. However, electrically injected GeSn lasers are more challenging to create due to the GeSn active gain medium having to overcome the extra metal absorption loss and more free carrier absorption (FCA) losses. Theoretical predication for the realm of possibility of electrically injected GeSn/SiGeSn lasers can be traced back to ten years ago, when Greg Sun et al. presented modelling and simulation results for an electrically injected SiGeSn/GeSn/SiGeSn double heterostructure laser with an Sn contents ranging from 6 to 12% [210]; they found that this type of laser requires cooling in the temperature range of 100–200 K after taking radiative, nonradiative, and Auger recombinations into consideration. Afterwards, Greg Sun et al. theoretically proposed that the lattice matched that of an $Si_{0.1}Ge_{0.75}Sn_{0.15}/Ge_{0.9}Sn_{0.1}/Si_{0.1}Ge_{0.75}Sn_{0.15}$ MQW laser [211], and they found that modal gain was very sensitive to the QW number in the active region and SiGeSn/GeSn/SiGeSn MQW could operate up to room temperature with a 2300 nm emission wavelength. For the SiGeSn/GeSn/SiGeSn MQW laser with 20 QWs, the optical confinement factor was calculated to be 0.74, and the modal gain was able to exceed 100/cm at a pumping current density of 3 kA/cm^2, which was sufficient to attain room-temperature lasing.

In 2020, Yiyin Zhou et al. reported the first electrically injected FP cavity GeSn/SiGeSn laser on Si with a lasing temperature of up to 100 K; its minimum threshold was approximately 598 A/cm^2 [89,90] (Figure 36). This work was regarded as an essential achievement for Si-based on-chip light source in the development of Si-based OEICs. Later, the effects of cap layer, cap layer thickness, and Sn content in the active region on the operating temperature, threshold, and emission wavelength were further systematically studied [89,90]. Experimental results showed that: (I) an SiGeSn cap had a better optical confinement effect than a GeSn cap; (II) the optical confinement factor was improved via changing the SiGeSn cap layer thickness; and (III) the use of a GeSn laser with an Sn content of up to 15% did not significantly improve device performance.

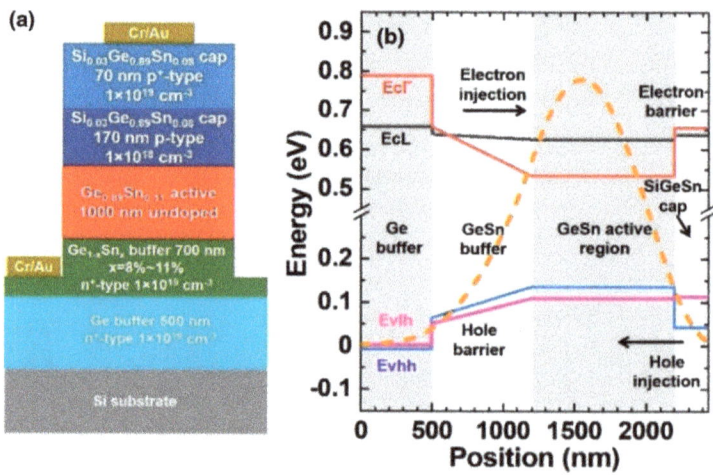

Figure 36. (a) Cross-sectional device structure for the first electrically injected FP cavity GeSn/SiGeSn laser; (b) calculated band structure and fundamental TE mode profile. Reproduced from [89], OSA Publishing, open access, 2020.

5. GeSn Transistors

In addition to the rapid advancement of GeSn detectors and GeSn lasers grown by CVD technology, there have been some achievements in the field of GeSn transistors due to their mobility properties. In the hyper-scaling era, the quest for high-performance and low-power transistors is continuing and intensifying. One of the key technology enablers of these goals is that of channel materials with high carrier mobility and direct band gap structures [212,213]. GeSn films have emerged as the most promising candidate for next generation nano-electronic devices of computing due to their excellent properties, including ultrahigh hole mobility, band structures with direct and low band gaps, Si-based CMOS compatibility, and low thermal budget, all of which are of great importance for ultrahigh density devices and 3D integration in the hyper-scaling era. Anisotropy at the top of the GeSn valence band makes the effective mass of light hole rapidly decrease with increases of Sn content and the transport capacity rapidly increase. GeSn is a very promising channel material for the next generation pMOSFET, and its hole mobility is even higher than that of Ge. The hole mobility of Ge pMOSFET is increased by more than 10 times with respect to Si devices. In addition, compressive strain can improve the mobility of a GeSn channel by decreasing the effective mass of the hole carrier. GeSn is generally grown on Si substrates using Ge as the buffer layer, and GeSn subjects the Ge buffer layer to compressive strain since the Sn lattice constant is greater than that of Ge. As GeSn materials are compatible with Si-CMOS technology, a few research groups have studied GeSn-based transistors (Table 8 lists the reported transistors with CVD-grown GeSn layers).

Table 8. Summary of reported transistors with GeSn layers grown by CVD technology in terms of institution, transistor type, Sn content, subthreshold swing (SS), I_{on}/I_{off} ratio, and V_{DS}.

Year	Institution	Transistor Type	Sn Composition (%)	SS (mV/dec)	I_{on}/I_{off}	V_{DS} (V)	Refs
2017	University of Notre Dame	Ge/GeSn p-type TFETs	11 and 12.5	215	9.2×10^3	−0.5	[99]
2017	NUS	GeSn FinFET on GeSnOI	8	79	$>10^4$	−0.5	[93]
2017	National Taiwan University	Vertically Stacked GeSn Nanowire pGAAFETs	6 and 10	84	-	−1	[214]
2017	National Taiwan University	GeSn N-FinFETs	8	138	10^3	-	[94]
2018	National Taiwan University	GeSn N-Channel MOSFETs	4.5	180	-	-	[215]
2018	National Taiwan University	Vertically Stacked 3-GeSn-Nanosheet pGAAFETs	7	108	5×10^3	−0.5	[91]
2020	PGI 9	Vertical heterojunction GeSn/Ge gate-all-around nanowire pMOSFETs	8	130	3×10^6	−0.5	[216]

Tunnel-field-effect transistors (TFETs) features subthreshold swings (SS) below 60 mV/decade at room temperature, which also enable a decreased power supply without discounting the off-current. Although Si-TFETs have been reported with SS below 60 mV/decade at low current, band-to-band tunneling (BTBT) is limited by its indirect bandgap property and low SS at high current. Therefore, researchers have investigated GeSn with high Sn contents (12% and 15% Sn incorporation; Figure 37) to create high-performance GeSn TFETs [217]. A higher Sn content enhances device performance, but the subthreshold swing is affected by the increased leakage level. For ultrasmall supply voltages, the device structure should be optimized to improve device characteristics. Using Ge/GeSn heterostructure pTFETs led to the improvements of the BTBT rate. Thus, higher on-current and lower off-current were achieved simultaneously. Christian et al. reported the fabrication and characterization of Ge/GeSn pTFETs (Figure 38), and they recorded a low accumulation capacitance of 3 μF/cm² [99]. Moreover, their room-temperature (RT) current–voltage characteristics showed that the Ge/GeSn pTFETs with the 11% Sn content had the highest BTBT current (Figure 39).

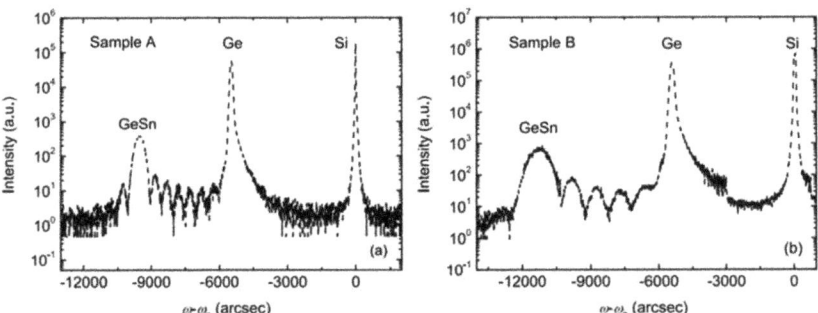

Figure 37. HR–XRD curves for GeSn samples with (**a**) 12% and (**b**) 15% Sn incorporation. Reproduced with permission from [217], IEEE, 2017.

To suppress the short channel effects (SCEs) of multi-gate transistors, Dianlei et al. investigated the p-FinFETs with a CVD-grown GeSn channel [93]. For GeSn p-FinFETs grown on GeSnOI substrates with 8% Sn incorporation (Figure 40), compressive strain and hole mobility were found to be −0.9% and 208 cm²/V·s, respectively. Record low SS of 79 mV/decade for GeSn p-FETs were also achieved.

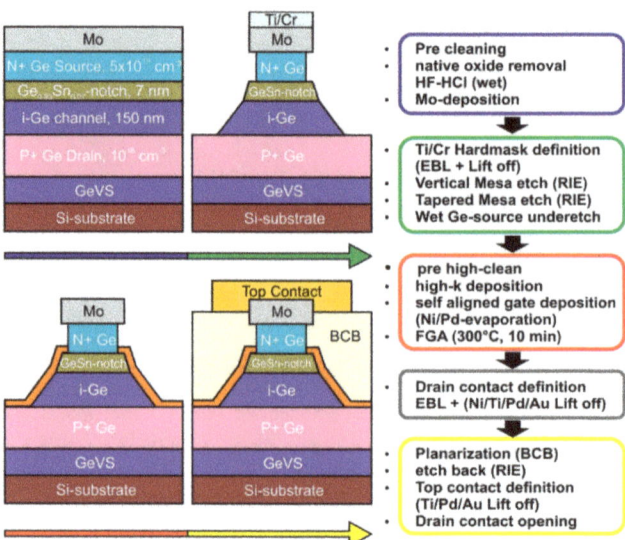

Figure 38. Process flow for Ge/GeSn vertical heterojunction pTFETs. Reproduced with permission from [99], IEEE, 2017.

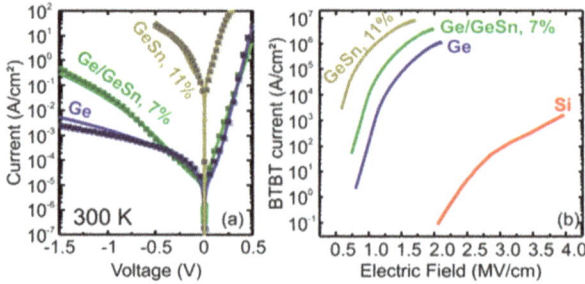

Figure 39. (**a**) RT current–voltage characteristics for GeSn p-i-n diode; (**b**) extracted BTBT current vs. electric field. Reproduced with permission from [99], IEEE, 2017.

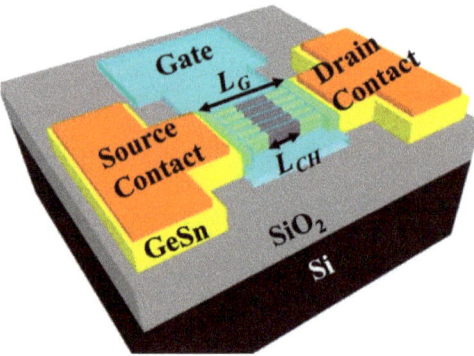

Figure 40. 3D diagram and highlights of the first GeSn FinFETs grown on a GeSnOI substrate. Reproduced with permission from [93], IEEE, 2018.

Compared with FinFETs, gate-all-around (GAA) FETs hold better electrostatic control, which can reduce the SCEs for the gate-length scaling. With down-scaling came the proposition of a vertically stacked Si channel for GAAFETs in order to improve drive current [218,219]. Yu-Shiang Huang et al. systematically investigated the strain response, LF noise, and temperature-dependence properties of vertically stacked GeSn nanowire pGAAFETs [214] (Figure 41). Their experimental results showed that: (I) I_{on} = 1850 µA/µm was improved with higher Sn incorporation; (II) the 6.3% extra enhancement of I_{on} was observed due to the uniaxial compressive strain that occurred when using wafer bending; and (III) the SS for one-nanowire and stacked two-nanowire GAAFETs were 84 and 88 mV/dec, respectively. To further improve the drive current for GAAFETs at a given footprint (Figure 42), vertically stacked 3-GeSn nanosheet pGAAFETs were studied and the I_{on} was increased 1975 µA/µm at $V_{DS} = -1$ V.

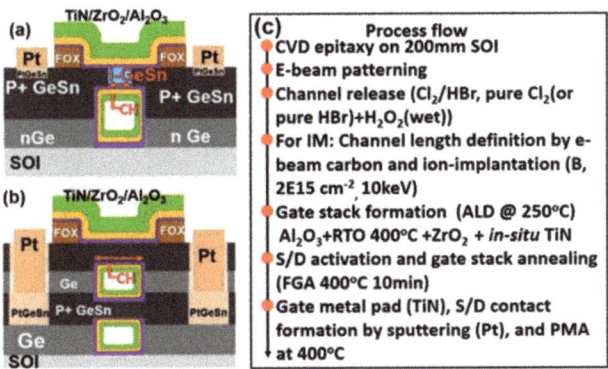

Figure 41. Schematic cross-sectional view of (**a**) single GeSn channel, (**b**) stacked GeSn nanowire pGAAFETs, and (**c**) process flow for devices. Reproduced with permission from [214], IEEE, 2017.

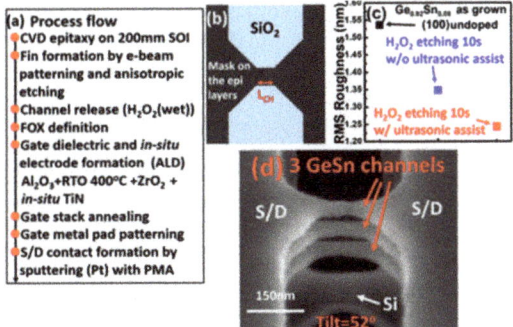

Figure 42. (**a**) Process flow for vertically stacked 3-GeSn nanosheet pGAAFETs; (**b**) top view after the fin formation; (**c**) RMS value for as-grown GeSn; (**d**) SEM image of stacked 3-GeSn nanosheets. Reproduced with permission from [91], IEEE, 2018.

Furthermore, a top–down approach was utilized to fabricate vertical heterojunction GeSn/Ge GAA nanowire pMOSFETs (Figure 43); with proper optimization, a record high I_{on}/I_{off} (3×10^6) was achieved [216].

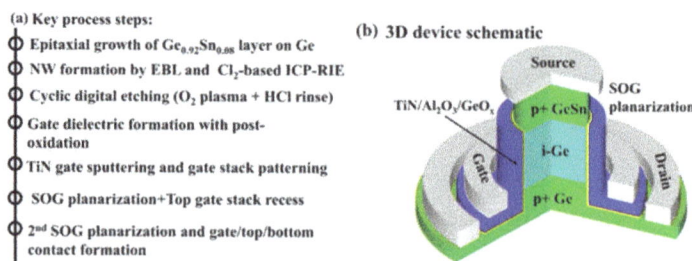

Figure 43. (a) Fabrication process and (b) 3D schematic of single vertical heterojunction GeSn/Ge GAA nanowire pMOSFETs. Reproduced with permission from [220], Elsevier, 2020.

Similar to the n–Ge material, n–GeSn suffers from a large resistance in metal-n–GeSn contacts mainly due to a strong Fermi pinning effect. To improve the performance of GeSn n-FETs, Yen Chuang et al. researched GeSn n-FinFETs and n-Channel MOSFETs: n^+–GeSn contact; in situ doped n^+–GeSn was grown by CVD, and Ni was employed as the contact metal [94]. With the increasing Sn content and n-type doping level, contact resistivity reduced to $3.8 \times 10^{-8}\ \Omega/\text{cm}^2$, which may be attributed to the bandgap shrinkage of GeSn (8% Sn incorporation). With the optimized n^+–GeSn contact, the highest drive current and best SS for GeSn n-FinFETs were 108 A/m and 138 mV/dec, respectively (8% Sn incorporation) [91]. To suppress the dopant diffusion for S/D carrier activation, microwave annealing (MWA) was proposed. For GeSn with 4.5% Sn incorporation, GeSn nMOSFETs were found to possess an electron mobility of 440 $\text{cm}^2/\text{V}\cdot\text{s}$, suggesting that CVD-grown GeSn and MWA technologies are very promising for GeSn CMOS applications. For higher electron mobility, a 0.46% tensile strain was introduced to $Ge_{0.96}Sn_{0.04}$; due to the introducing of tensile strain, the carrier population in the Γ valley was higher. Thus, the electron mobility of GeSn nMOSFETs was further improved to 698 $\text{cm}^2/\text{V}\cdot\text{s}$ [215].

This discussion shows that pTFETs, pFin-FETs, pMOSFETs, nMOSFETs, and vertically stacked nanowire pGAAFETs with CVD-grown GeSn layers have been extensively studied; breaking the bottleneck the n-doped or p-doped GeSn CVD growth technology is one of the main routes forward for high-performance GeSn transistors. Uniformly stacked nanowires or nanosheets with low surface roughness are of great importance for 5 nm CMOS technology nodes and beyond. More importantly, It should be noted that Henry. H. Radamson et al. explored $Ni–(GeSn)_x$ contact formation [220]; the strain dependence, phase formation, and thermal stability of $Ni–(GeSn)_x$ were systematically investigated, and they found that an Sn-rich surface impeded the diffusion of Ni, thus paving the way for the optimization of high-performance nanowire pGAAFETs.

6. Conclusions and Outlooks

In summary, the challenges and progress of GeSn CVD growth technology (including in situ doping technology and ohmic contact formation), GeSn lasers, GeSn detectors, and GeSn transistors were reviewed. Due to growth difficulties, such as the large lattice mismatch between GeSn and Si, the low solubility between Ge and Sn, and phase changes for Sn, more effort must be made in improving the quality of high-Sn-content GeSn materials, GeSn/SiGeSn heterostructures, and GeSn/SiGeSn QWs for high-performance electronic and optoelectronic devices, especially GeSn lasers and GeSn TFETs. Sn distribution uniformity and sharp GeSn/SiGeSn interfaces are the key issues in the development of room temperature, CW electrically pumped GeSn lasers. In addition, research on novel Si-based group IV materials, such as CSiGeSn and CSiGe [221–223], may pave the way for better strain compensation and lattice-mismatched laser structures.

Author Contributions: Conceptualization, Y.M., G.W. and H.H.R.; literature survey, Y.M., G.W., Z.K., X.Z., B.X., X.L., H.L., Y.D., B.L. and J.L.; formal analysis, Y.M., G.W., H.H.R., L.D. and J.Z.; project administration, H.H.R.; supervision, G.W. and H.H.R.; writing—original draft preparation, Y.M.;

writing—review and editing, Y.M. and H.H.R. All authors have read and agreed to the published version of the manuscript.

Funding: This research was supported by the construction of a high-level innovation research institute from the Guangdong Greater Bay Area Institute of Integrated Circuit and System (Grant No. 2019B090909006) and the construction of new research and development institutions (Grant No. 2019B090904015), in part by the National Key Research and Development Program of China (Grant No. 2016YFA0301701), the Youth Innovation Promotion Association of CAS (Grant No. Y2020037), and the National Natural Science Foundation of China (Grant No. 92064002).

Data Availability Statement: The data presented in this study are available on request from the corresponding authors.

Conflicts of Interest: The authors declare no conflict of interest.

References

1. Soref, R. The past, present, and future of silicon photonics. *IEEE J. Sel. Top. Quantum Electron.* **2006**, *12*, 1678–1687. [CrossRef]
2. Soref, R. Silicon photonics: A review of recent literature. *Silicon* **2010**, *2*, 1–6. [CrossRef]
3. Paul, D.J. Silicon photonics: A bright future? *Electron. Lett.* **2009**, *45*, 582–584. [CrossRef]
4. Soref, R.A. Silicon-based optoelectronics. *Proc. IEEE* **1993**, *81*, 1687–1706. [CrossRef]
5. Soref, R.; Buca, D.; Yu, S.Q. Group IV photonics: Driving integrated optoelectronics. *Opt. Photonics News* **2016**, *27*, 32–39. [CrossRef]
6. Radamson, H.; Thylén, L. *Monolithic Nanoscale Photonics-Electronics Integration in Silicon and Other Group IV Elements*; Academic Press: Cambridge, MA, USA, 2014.
7. Geiger, R.; Zabel, T.; Sigg, H. Group IV direct band gap photonics: Methods, challenges, and opportunities. *Front. Mater.* **2015**, *2*, 52. [CrossRef]
8. Zhang, L.; Hong, H.; Li, C.; Chen, S.; Huang, W.; Wang, J.; Wang, H. High-Sn fraction GeSn quantum dots for Si-based light source at 1.55 μm. *Appl. Phys. Express* **2019**, *12*, 055504. [CrossRef]
9. Wang, T.; Wei, W.; Feng, Q.; Wang, Z.; Zhang, J. Telecom InAs quantum-dot FP and microdisk lasers epitaxially grown on (111)-faceted SOI. In Proceedings of the 2020 Conference on Lasers and Electro-Optics (CLEO), Washington, DC, USA, 10–15 May 2020; pp. 1–2.
10. Wei, W.; Feng, Q.; Wang, Z.; Wang, T.; Zhang, J. Perspective: Optically-pumped III–V quantum dot microcavity lasers via CMOS compatible patterned Si (001) substrates. *J. Semicond.* **2019**, *40*, 101303. [CrossRef]
11. Liu, W.K.; Lubyshev, D.; Fastenau, J.M.; Wu, Y.; Bulsara, M.T.; Fitzgerald, E.A.; Urteaga, M.; Ha, W.; Bergman, J.; Brar, B.; et al. Monolithic integration of InP-based transistors on Si substrates using MBE. *J. Cryst. Growth* **2009**, *311*, 1979–1983. [CrossRef]
12. Liow, T.Y.; Ang, K.W.; Fang, Q.; Song, J.F.; Xiong, Y.Z.; Yu, M.B.; Guo, Q.; Kwong, D.L. Silicon modulators and germanium photodetectors on SOI: Monolithic integration, compatibility, and performance optimization. *IEEE J. Sel. Top. Quantum Electron.* **2009**, *16*, 307–315. [CrossRef]
13. Saito, S.; Al-Attili, A.Z.; Oda, K.; Ishikawa, Y. Towards monolithic integration of germanium light sources on silicon chips. *Semicond. Sci. Technol.* **2016**, *31*, 043002. [CrossRef]
14. Yu, H.Y.; Ren, S.; Jung, W.S.; Okyay, A.K.; Miller, D.A.; Saraswat, K.C. High-efficiency pin photodetectors on selective-area-grown Ge for monolithic integration. *IEEE Electron. Device Lett.* **2009**, *30*, 1161–1163.
15. González-Fernández, A.A.; Juvert, J.; Aceves-Mijares, M.; Domínguez, C. Monolithic integration of a silicon-based photonic transceiver in a CMOS process. *IEEE Photonics J.* **2015**, *8*, 1–13. [CrossRef]
16. Li, B.; Li, G.; Liu, E.; Jiang, Z.; Pei, C.; Wang, X. 1.55 μm reflection-type optical waveguide switch based on SiGe/Si plasma dispersion effect. *Appl. Phys. Lett.* **1999**, *75*, 1–3. [CrossRef]
17. Zhao, D.; Shi, B.; Jiang, Z.; Fan, Y.; Wang, X. Silicon-based optical waveguide polarizer using photonic band gap. *Appl. Phys. Lett.* **2002**, *81*, 409–411. [CrossRef]
18. Zhao, X.; Wang, G.; Lin, H.; Du, Y.; Luo, X.; Kong, Z.; Su, J.; Li, J.; Xiong, W.; Miao, Y.; et al. High performance pin photodetectors on Ge-on-insulator platform. *Nanomaterials* **2021**, *11*, 1125. [CrossRef]
19. Michel, J.; Liu, J.; Kimerling, L.C. High-performance Ge-on-Si photodetectors. *Nat. Photonics* **2010**, *4*, 527–534. [CrossRef]
20. Lin, T.Y.; Lin, K.T.; Lin, C.C.; Lee, Y.W.; Shiu, L.T.; Chen, W.Y.; Chen, H.L. Magnetic fields affect hot electrons in silicon-based photodetectors at telecommunication wavelengths. *Mater. Horiz.* **2019**, *6*, 1156–1168. [CrossRef]
21. Marris-Morini, D.; Vivien, L.; Rasigade, G.; Fedeli, J.M.; Cassan, E.; Le Roux, X.; Laval, S. Recent progress in high-speed silicon-based optical modulators. *Proc. IEEE* **2009**, *97*, 1199–1215. [CrossRef]
22. Reed, G.T.; Png, C.E.J. Silicon optical modulators. *Mater. Today* **2005**, *8*, 40–50. [CrossRef]
23. Reed, G.T.; Mashanovich, G.; Gardes, F.Y.; Thomson, D.J. Silicon optical modulators. *Nat. Photonics* **2010**, *4*, 518–526. [CrossRef]
24. Haché, A.; Bourgeois, M. Ultrafast all-optical switching in a silicon-based photonic crystal. *Appl. Phys. Lett.* **2000**, *77*, 4089–4091. [CrossRef]

25. Juan, W.H.; Pang, S.W. High-aspect-ratio Si vertical micromirror arrays for optical switching. *J. Microelectromech. Syst.* **1998**, *7*, 207–213. [CrossRef]
26. Fadaly, E.M.; Dijkstra, A.; Suckert, J.R.; Ziss, D.; van Tilburg, M.A.; Mao, C.; Ren, Y.; Lange, V.; Korzun, K.; Bakkers, E.P.; et al. Direct-bandgap emission from hexagonal Ge and SiGe alloys. *Nature* **2020**, *580*, 205–209. [CrossRef]
27. Sukhdeo, D.S.; Nam, D.; Kang, J.H.; Brongersma, M.L.; Saraswat, K.C. Direct bandgap germanium-on-silicon inferred from 5.7% ⟨100⟩ uniaxial tensile strain. *Photonics Res.* **2014**, *2*, A8–A13. [CrossRef]
28. Sun, X.; Liu, J.; Kimerling, L.C.; Michel, J. Room-temperature direct bandgap electroluminesence from Ge-on-Si light-emitting diodes. *Opt. Lett.* **2009**, *34*, 1198–1200. [CrossRef]
29. Michel, J.; Camacho-Aguilera, R.E.; Cai, Y.; Patel, N.; Bessette, J.T.; Romagnoli, M.; Kimerling, L.C. *An Electrically Pumped Ge-on-Si Laser*; OFC/NFOEC; IEEE: New York, NY, USA, 2012; pp. 1–3.
30. Camacho-Aguilera, R.E.; Cai, Y.; Patel, N.; Bessette, J.T.; Romagnoli, M.; Kimerling, L.C.; Michel, J. An electrically pumped germanium laser. *Opt. Express* **2012**, *20*, 11316–11320. [CrossRef]
31. Liu, J.; Sun, X.; Camacho-Aguilera, R.; Cai, Y.; Kimerling, L.C.; Michel, J. *Optical Gain and Lasing from Band-Engineered Ge-on-Si at Room Temperature*; Institute of Electrical and Electronics Engineers (IEEE): New York, NY, USA, 2010.
32. Koerner, R.; Oehme, M.; Gollhofer, M.; Schmid, M.; Kostecki, K.; Bechler, S.; Widmann, D.; Kasper, E.; Schulze, J. Electrically pumped lasing from Ge Fabry-Perot resonators on Si. *Opt. Express* **2015**, *11*, 14815–14822. [CrossRef]
33. Ghetmiri, S.A.; Du, W.; Margetis, J.; Mosleh, A.; Cousar, L.; Conley, B.R.; Domulevicz, L.; Nazzal, A.; Sun, G.; Soref, R.A.; et al. Direct-bandgap GeSn grown on silicon with 2230 nm photoluminescence. *Appl. Phys. Lett.* **2014**, *105*, 151109. [CrossRef]
34. Grant, P.C.; Margetis, J.; Zhou, Y.; Dou, W.; Abernathy, G.; Kuchuk, A.; Du, W.; Li, B.L.; Tolle, J.; Yu, S.Q.; et al. Direct bandgap type-I GeSn/GeSn quantum well on a GeSn-and Ge-buffered Si substrate. *AIP Adv.* **2018**, *8*, 025104. [CrossRef]
35. Peng, L.; Li, X.; Zheng, J.; Liu, X.; Li, M.; Liu, Z.; Xue, C.; Zuo, Y.; Cheng, B. Room-temperature direct-bandgap electroluminescence from type-I GeSn/SiGeSn multiple quantum wells for 2 μm LEDs. *J. Lumin.* **2020**, *228*, 117539. [CrossRef]
36. Grant, P.C.; Margetis, J.; Zhou, Y.; Dou, W.; Abernathy, G.; Kuchuk, A.; Du, W.; Li, B.; Tolle, J.; Liu, J.; et al. Study of direct bandgap type-I GeSn/GeSn double quantum well with improved carrier confinement. *Nano. Tech.* **2018**, *29*, 465201. [CrossRef] [PubMed]
37. Von Den Driesch, N.; Stange, D.; Wirths, S.; Mussler, G.; Hollander, B.; Ikonic, Z.; Hartmann, J.M.; Stoica, T.; Mantl, S.; Buca, D.; et al. Direct bandgap group IV epitaxy on Si for laser applications. *Chem. Mater.* **2015**, *27*, 4693–4702. [CrossRef]
38. Wang, L.; Zhang, Y.; Wu, Y.; Liu, T.; Miao, Y.; Meng, L.; Jiang, Z.; Hu, H. Effects of Annealing on the Behavior of Sn in GeSn Alloy and GeSn-Based Photodetectors. *IEEE Trans. Electron. Devices.* **2020**, *67*, 3229–3234. [CrossRef]
39. Miao, Y.H.; Hu, H.Y.; Song, J.J.; Xuan, R.X.; Zhang, H.M. Effects of rapid thermal annealing on crystallinity and Sn surface segregation of films on Si (100) and Si (111). *Chin. Phys. B* **2017**, *26*, 127306. [CrossRef]
40. Li, H.; Cui, Y.X.; Wu, K.Y.; Tseng, W.K.; Cheng, H.H.; Chen, H. Strain relaxation and Sn segregation in GeSn epilayers under thermal treatment. *Appl. Phys. Lett.* **2013**, *102*, 251907. [CrossRef]
41. Comrie, C.M.; Mtshali, C.B.; Sechogela, P.T.; Santos, N.M.; van Stiphout, K.; Loo, R.; Wandervorst, W.; Vantomme, A. Interplay between relaxation and Sn segregation during thermal annealing of GeSn strained layers. *J. Appl. Phys.* **2016**, *120*, 145303. [CrossRef]
42. Gurdal, O.; Hasan, M.A.; Sardela, M.R., Jr.; Greene, J.E.; Radamson, H.H.; Sundgren, J.E.; Hansson, G.V. Growth of metastable $Ge_{1-x}Sn_x$/Ge strained layer superlattices on Ge (001) 2×1 by temperature-modulated molecular beam epitaxy. *Appl. Phy. Lett.* **1995**, *67*, 956–958. [CrossRef]
43. Gurdal, O.; Desjardins, P.; Carlsson, J.R.A.; Taylor, N.; Radamson, H.H.; Sundgren, J.E.; Greene, J.E. Low-temperature growth and critical epitaxial thicknesses of fully strained metastable $Ge_{1-x}Sn_x$ ($x \leq 0.26$) alloys on Ge (001) 2 × 1. *J. Appl. Phys.* **1998**, *83*, 162–170. [CrossRef]
44. Ni, W.X.; Ekberg, J.O.; Joelsson, K.B.; Radamson, H.H.; Henry, A.; Shen, G.D.; Hansson, G.V. A silicon molecular beam epitaxy system dedicated to device-oriented material research. *J. Cryst. Growth* **1995**, *157*, 285–294. [CrossRef]
45. Toko, K.; Oya, N.; Saitoh, N.; Yoshizawa, N.; Suemasu, T. 70 °C synthesis of high-Sn content (25%) GeSn on insulator by Sn-induced crystallization of amorphous Ge. *Appl. Phys. Lett.* **2015**, *106*, 082109. [CrossRef]
46. Huo, Y.; Chen, R.; Lin, H.; Kamins, T.I.; Harris, J.S. MBE growth of high Sn-percentage GeSn alloys with a composition-dependent absorption-edge shift. In Proceedings of the 7th IEEE International Conference on Group IV Photonics, Beijing, China, 1–3 September 2010; pp. 344–346.
47. Yang, J.; Hu, H.; Miao, Y.; Dong, L.; Wang, B.; Wang, W.; Su, H.; Xuan, R.; Zhang, H. High-quality GeSn Layer with Sn Composition up to 7% Grown by Low-Temperature Magnetron Sputtering for Optoelectronic Application. *Materials* **2019**, *12*, 2662. [CrossRef]
48. Tran, H.; Pham, T.; Margetis, J.; Zhou, Y.; Dou, W.; Grant, P.C.; Alkabi, S.; Du, W.; Sun, G.; Soref, R.; et al. Study of High Performance GeSn Photodetectors with Cutoff Wavelength Up to 3. In 7 μm for Low-Cost Infrared Imaging. In Proceedings of the 2019 Conference on Lasers and Electro-Optics (CLEO), San Jose, CA, USA; 2019; pp. 1–2.
49. Zheng, J.; Liu, Z.; Zhang, Y.; Zuo, Y.; Li, C.; Xue, C.; Cheng, B.; Wang, Q. Growth of high-Sn content (28%) GeSn alloy films by sputtering epitaxy. *J. Crys. Growth* **2018**, *492*, 29–34. [CrossRef]
50. Bauer, M.; Taraci, J.; Tolle, J.; Chizmeshya, A.V.G.; Zollner, S.; Smith, D.J.; Menendez, J.; Hu, C.; Kouvetakis, J. Ge–Sn semiconductors for band-gap and lattice engineering. *Appl. Phys. Lett.* **2002**, *81*, 2992–2994. [CrossRef]

51. Al-Kabi, S.; Ghetmiri, S.A.; Margetis, J.; Du, W.; Mosleh, A.; Dou, W.; Sun, G.; Soref, R.; Tolle, J.; Yu, S.Q.; et al. Study of High-Quality GeSn Alloys Grown by Chemical Vapor Deposition towards Mid-Infrared Applications. *J. Electron. Mater.* **2016**, *45*, 6251–6257. [CrossRef]
52. Al-Kabi, S.; Ghetmiri, S.A.; Margetis, J.; Du, W.; Mosleh, A.; Alher, M.; Dou, W.; Grant, J.; Sun, G.; Yu, S.Q.; et al. Optical characterization of Si-based $Ge_{1-x}Sn_x$ alloys with Sn compositions up to 12%. *J. Electron. Mater.* **2016**, *45*, 2133–2141. [CrossRef]
53. Dou, W.; Ghetmiri, S.A.; Al-Kabi, S.; Mosleh, A.; Zhou, Y.; Alharthi, B.; Alharthi, B.; Du, W.; Margetis, J.; Yu, S.Q.; et al. Structural and optical characteristics of GeSn quantum wells for silicon-based mid-infrared optoelectronic applications. *J. Electron. Mater.* **2016**, *45*, 6265–6272. [CrossRef]
54. Du, W.; Ghetmiri, S.A.; Margetis, J.; Al-Kabi, S.; Zhou, Y.; Liu, J.; Sun, G.; Soref, R.; Tolle, J.; Li, B.; et al. Investigation of optical transitions in a SiGeSn/GeSn/SiGeSn single quantum well structure. *J. Appl. Phy.* **2017**, *122*, 123102. [CrossRef]
55. Zhou, Y.; Margetis, J.; Abernathy, G.; Dou, W.; Grant, P.C.; Alharthi, B.; Du, W.; Wadsworth, A.; Guo, Q.; Tran, H.; et al. Investigation of SiGeSn/GeSn/SiGeSn quantum well structures and optically pumped lasers on Si. In Proceedings of the 2019 Conference on Lasers and Electro-Optics, San Jose, CA, USA, 5–10 May 2019; p. STu3N-3.
56. Ghetmiri, S.A.; Zhou, Y.; Margetis, J.; Al-Kabi, S.; Dou, W.; Mosleh, A.; Du, W.; Kuchuk, A.; Liu, J.; Yu, S.Q.; et al. Study of a SiGeSn/GeSn/SiGeSn structure toward direct bandgap type-I quantum well for all group-IV optoelectronics. *Opt. Lett.* **2017**, *42*, 387–390. [CrossRef]
57. Mathews, J.; Roucka, R.; Xie, J.; Yu, S.Q.; Menéndez, J.; Kouvetakis, J. Extended performance GeSn/Si (100) p-i-n photodetectors for full spectral range telecommunication applications. *Appl. Phys. Lett.* **2009**, *95*, 133506. [CrossRef]
58. Gassenq, A.; Gencarelli, F.; Van Campenhout, J.; Shimura, Y.; Loo, R.; Narcy, G.; Vincent, B.; Roelkens, G. GeSn/Ge heterostructure short-wave infrared photodetectors on silicon. *Opt. Express* **2012**, *20*, 27297–27303. [CrossRef] [PubMed]
59. Conley, B.R.; Margetis, J.; Du, W.; Tran, H.; Mosleh, A.; Ghetmiri, S.A.; Tolle, J.; Sun, G.; Soref, R.; Li, B.; et al. Si based GeSn photoconductors with a 1.63 A/W peak responsivity and a 2.4 µm long-wavelength cutoff. *Appl. Phys. Lett.* **2014**, *105*, 221117. [CrossRef]
60. Conley, B.R.; Mosleh, A.; Ghetmiri, S.A.; Du, W.; Soref, R.A.; Sun, G.; Margetis, J.; Tolle, J.; Nassem, H.; Yu, S.Q. Temperature dependent spectral response and detectivity of GeSn photoconductors on silicon for short wave infrared detection. *Opt. Express* **2014**, *22*, 15639–15652. [CrossRef] [PubMed]
61. Pham, T.N.; Du, W.; Conley, B.R.; Margetis, J.; Sun, G.; Soref, R.A.; Tolle, J.; Li, B.; Yu, S.Q. Si-based $Ge_{0.9}Sn_{0.1}$ photodetector with peak responsivity of 2.85 A/W and longwave cutoff at 2.4 µm. *Electron. Lett.* **2015**, *51*, 854–856. [CrossRef]
62. Pham, T.; Du, W.; Tran, H.; Margetis, J.; Tolle, J.; Sun, G.; Yu, S.Q. Systematic study of Si-based GeSn photodiodes with 2.6 µm detector cutoff for short-wave infrared detection. *Opt. Express* **2016**, *24*, 4519–4531. [CrossRef] [PubMed]
63. Tran, H.; Pham, T.; Margetis, J.; Zhou, Y.; Dou, W.; Grant, P.C.; Grant, J.; Sun, G.; Tolle, J.; Du, W.; et al. Si-based GeSn photodetectors toward mid-infrared imaging applications. *ACS Photonics* **2019**, *6*, 2807–2815. [CrossRef]
64. Tran, H.; Littlejohns, C.G.; Thomson, D.J.; Pham, T.; Ghetmiri, A.; Mosleh, A.; Margetis, J.; Tolle, J.; Du, W.; Yu, S.Q.; et al. Study of GeSn mid-infrared photodetectors for high frequency applications. *Front. Mater.* **2019**, *6*, 278. [CrossRef]
65. Tran, H.; Pham, T.; Du, W.; Zhang, Y.; Grant, P.C.; Grant, J.M.; Sun, G.; Soref, R.; Margetis, J.; Yu, S.Q.; et al. High performance $Ge_{0.89}Sn_{0.11}$ photodiodes for low-cost shortwave infrared imaging. *J. Appl. Phys.* **2018**, *124*, 013101. [CrossRef]
66. Xu, S.; Wang, W.; Huang, Y.C.; Dong, Y.; Masudy-Panah, S.; Wang, H.; Xiao, G.; Yeo, Y.C. High-speed photo detection at two-micron-wavelength: Technology enablement by GeSn/Ge multiple-quantum-well photodiode on 300 mm Si substrate. *Opt. Express* **2019**, *27*, 5798–5813. [CrossRef]
67. Zhou, H.; Xu, S.; Wu, S.; Huang, Y.C.; Zhao, P.; Tong, J.; Son, B.; Guo, X.; Zhang, D.; Gong, X.; et al. Photo detection and modulation from 1550 to 2000 nm realized by a GeSn/Ge multiple-quantum-well photodiode on a 300-mm Si substrate. *Opt. Express* **2020**, *28*, 34772–34786. [CrossRef]
68. Wu, S.; Xu, S.; Zhou, H.; Jin, Y.; Chen, Q.; Huang, Y.C.; Zhang, L.; Gong, X.; Tan, C.S. High-performance back-illuminated $Ge_{0.92}Sn_{0.08}$/Ge multiple-quantum-well photodetector on Si platform for SWIR Detection. *IEEE J. Sel. Top. Quantum Electron.* **2021**, *28*, 1–9. [CrossRef]
69. Wang, W.; Lei, D.; Huang, Y.C.; Lee, K.H.; Loke, W.K.; Dong, Y.; Xu, S.; Tan, C.; Wang, H.; Yoon, S.; et al. High-performance GeSn photodetector and fin field-effect transistor (FinFET) on an advanced GeSn-on-insulator platform. *Opt. Express* **2018**, *26*, 10305–10314. [CrossRef] [PubMed]
70. Wirths, S.; Geiger, R.; Von Den Driesch, N.; Mussler, G.; Stoica, T.; Mantl, S.; Ikonic, Z.; Luysberg, M.; Buca, D.; Grützmacher, D.; et al. Lasing in direct-bandgap GeSn alloy grown on Si. *Nat. Photonics* **2015**, *9*, 88–92. [CrossRef]
71. Al-Kabi, S.; Ghetmiri, S.A.; Margetis, J.; Pham, T.; Zhou, Y.; Dou, W.; Collier, B.; Quinde, R.; Du, W.; Yu, S.Q.; et al. An optically pumped 2.5 µm GeSn laser on Si operating at 110 K. *Appl. Phys. Lett.* **2016**, *109*, 171105. [CrossRef]
72. Thai, Q.M.; Pauc, N.; Aubin, J.; Bertrand, M.; Chrétien, J.; Chelnokov, A.; Martmann, J.; Reboud, V.; Calvo, V. 2D hexagonal photonic crystal GeSn laser with 16% Sn content. *Appl. Phys. Lett.* **2018**, *113*, 051104. [CrossRef]
73. Margetis, J.; Al-Kabi, S.; Du, W.; Dou, W.; Zhou, Y.; Pham, T.; Grant, P.; Ghetmiri, S.; Mosleh, A.; Li, B.; et al. Si-based GeSn lasers with wavelength coverage of 2–3 µm and operating temperatures up to 180 K. *ACS Photonics* **2017**, *5*, 827–833. [CrossRef]
74. Stange, D.; Wirths, S.; Geiger, R.; Schulte-Braucks, C.; Marzban, B.; von den Driesch, N.; Mussler, G.; Zabel, T.; Stoica, T.; Buca, D.; et al. Optically pumped GeSn microdisk lasers on Si. *ACS Photonics* **2016**, *3*, 1279–1285. [CrossRef]

75. Dou, W.; Zhou, Y.; Margetis, J.; Ghetmiri, S.A.; Al-Kabi, S.; Du, W.; Liu, J.; Sun, G.; Soref, R.; Yu, S.Q.; et al. Optically pumped lasing at 3 μm from compositionally graded GeSn with tin up to 22.3%. *Opt. Lett.* **2018**, *43*, 4558–4561. [CrossRef]
76. Zhou, Y.; Dou, W.; Du, W.; Ojo, S.; Tran, H.; Ghetmiri, S.A.; Liu, J.; Sun, G.; Soref, R.; Yu, S.Q.; et al. Optically pumped GeSn lasers operating at 270 K with broad waveguide structures on Si. *ACS Photonics* **2019**, *6*, 1434–1441. [CrossRef]
77. Chrétien, J.; Pauc, N.; Armand Pilon, F.; Bertrand, M.; Thai, Q.M.; Casiez, L.; Bernier, N.; Dansas, H.; Gergaud, P.; Hartmann, J.; et al. GeSn lasers covering a wide wavelength range thanks to uniaxial tensile strain. *ACS Photonics* **2019**, *6*, 2462–2469. [CrossRef]
78. Reboud, V.; Gassenq, A.; Pauc, N.; Aubin, J.; Milord, L.; Thai, Q.M.; Bertrand, M.; Guilloy, K.; Rouchon, D.; Calvo, V.; et al. Optically pumped GeSn micro-disks with 16% Sn lasing at 3.1 μm up to 180 K. *Appl. Phys. Lett.* **2017**, *111*, 092101. [CrossRef]
79. Du, W.; Thai, Q.M.; Chrétien, J.; Bertrand, M.; Casiez, L.; Zhou, Y.; Margetis, J.; Pauc, N.; Reboud, V.; Yu, S.Q.; et al. Study of Si-based GeSn optically pumped lasers with micro-disk and ridge waveguide structures. *Front. Phys.* **2019**, *7*, 147. [CrossRef]
80. Elbaz, A.; Buca, D.; von den Driesch, N.; Pantzas, K.; Patriarche, G.; Zerounian, N.; Herth, E.; Checory, X.; Sauvage, S.; El Kurdi, M.; et al. Ultra-low-threshold continuous-wave and pulsed lasing in tensile-strained GeSn alloys. *Nat. Photonics* **2020**, *14*, 375–382. [CrossRef]
81. Kurdi, M.E.; Elbaz, A.; Wang, B.; Sakat, E.; Herth, E.; Patriarche, G.; Pantzas, K.; Sagnes, I.; Sauvage, S.; Buca, D. Tensile Strain Engineering and Defects Management in GeSn Laser Cavities. *ECS Trans.* **2020**, *98*, 61. [CrossRef]
82. Elbaz, A.; Arefin, R.; Sakat, E.; Wang, B.; Herth, E.; Patriarche, G.; Foti, A.; Ossikovski, R.; Sauvage, S.; Checoury, X.; et al. Reduced lasing thresholds in GeSn microdisk cavities with defect management of the optically active region. *ACS Photonics* **2020**, *7*, 2713–2722. [CrossRef]
83. Margetis, J.; Zhou, Y.; Dou, W.; Grant, P.C.; Alharthi, B.; Du, W.; Tran, H.; Ojo, S.; Liu, J.; Yu, S.Q.; et al. All group-IV SiGeSn/GeSn/SiGeSn QW laser on Si operating up to 90 K. *Appl. Phys. Lett.* **2018**, *113*, 221104. [CrossRef]
84. Stange, D.; von den Driesch, N.; Zabel, T.; Armand-Pilon, F.; Rainko, D.; Marzban, B.; Zaumseil, P.; Hartmann, J.M.; Ikonic, Z.; Buca, D.; et al. GeSn/SiGeSn heterostructure and multi quantum well lasers. *ACS Photonics* **2018**, *5*, 4628–4636. [CrossRef]
85. Thai, Q.M.; Pauc, N.; Aubin, J.; Bertrand, M.; Chrétien, J.; Delaye, V.; Chelnokov, A.; Hartmann, J.; Reboud, V.; Calvo, V. GeSn heterostructure micro-disk laser operating at 230 K. *Opt. Express* **2018**, *26*, 32500–32508. [CrossRef]
86. Von den Driesch, N.; Stange, D.; Rainko, D.; Povstugar, I.; Zaumseil, P.; Capellini, G.; Denneulin, T.; Ikonic, Z.; Hartmann, J.; Buca, D.; et al. Advanced GeSn/SiGeSn group IV heterostructure lasers. *Adv. Sci.* **2018**, *5*, 1700955. [CrossRef]
87. Fujisawa, T.; Arai, M.; Saitoh, K. Microscopic gain analysis of modulation-doped GeSn/SiGeSn quantum wells: Epitaxial design toward high-temperature lasing. *Opt. Express* **2019**, *27*, 2457–2464. [CrossRef]
88. Dou, W.; Benamara, M.; Mosleh, A.; Margetis, J.; Grant, P.; Zhou, Y.; Al-Kabi, S.; Du, W.; Tolle, J.; Yu, S.Q.; et al. Investigation of GeSn strain relaxation and spontaneous composition gradient for low-defect and high-Sn alloy growth. *Sci. Rep.* **2018**, *8*, 1–11. [CrossRef] [PubMed]
89. Zhou, Y.; Miao, Y.; Ojo, S.; Tran, H.; Abernathy, G.; Grant, J.M.; Amoah, S.; Salamo, G.; Du, W.; Yu, S.Q.; et al. Electrically injected GeSn lasers on Si operating up to 100 K. *Optica* **2020**, *7*, 924–928. [CrossRef]
90. Zhou, Y.; Ojo, S.; Miao, Y.; Tran, H.; Grant, J.M.; Abernathy, G.; Amoah, S.; Bass, J.; Salamo, G.; Yu, S.Q.; et al. Electrically injected GeSn lasers with peak wavelength up to 2.7 micrometer at 90 K. *arXiv Preprint* **2020**, arXiv:2009.12229.
91. Huang, Y.S.; Lu, F.L.; Tsou, Y.J.; Ye, H.Y.; Lin, S.Y.; Huang, W.H.; Liu, C.W. Vertically stacked strained 3-GeSn-nanosheet pGAAFETs on Si using GeSn/Ge CVD epitaxial growth and the optimum selective channel release process. *IEEE Electron Device Lett.* **2018**, *39*, 1274–1277. [CrossRef]
92. Lei, D.; Lee, K.H.; Bao, S.; Wang, W.; Masudy-Panah, S.; Yadav, S.; Kumar, A.; Dong, Y.; Kang, Y.; Xu, S.; et al. The first GeSn FinFET on a novel GeSnOI substrate achieving lowest S of 79 mV/decade and record high Gm, int of 807 μS/μm for GeSn P-FETs. In Proceedings of the 2017 Symposium on VLSI Technology, Tyoto, Japan, 5–8 June 2017; pp. T198–T199.
93. Lei, D.; Lee, K.H.; Huang, Y.C.; Wang, W.; Masudy-Panah, S.; Yadav, S.; Kumar, A.; Dong, Y.; Kang, Y.; Yeo, Y.C.; et al. Germanium-tin (GeSn) P-channel fin field-effect transistor fabricated on a novel GeSn-on-insulator substrate. *IEEE Trans. Electron Devices* **2018**, *65*, 3754–3761. [CrossRef]
94. Gupta, S.; Vincent, B.; Yang, B.; Lin, D.; Gencarelli, F.; Lin, J.Y.; Chen, R.; Richard, O.; Bender, H.; Saraswat, K.C.; et al. Towards high mobility GeSn channel nMOSFETs: Improved surface passivation using novel ozone oxidation method. In Proceedings of the 2012 International Electron Devices Meeting, San Francisco, CA, USA, 10–13 December 2012; pp. 16.2.1–16.2.4.
95. Fang, Y.C.; Chen, K.Y.; Hsieh, C.H.; Su, C.C.; Wu, Y.H. N-MOSFETs formed on solid phase epitaxially grown GeSn film with passivation by oxygen plasma featuring high mobility. *ACS Appl. Mater. Inter.* **2015**, *7*, 26374–26380. [CrossRef]
96. Kang, Y.; Han, K.; Kong, E.Y.J.; Lei, D.; Xu, S.; Wu, Y.; Huang, Y.; Gong, X. The first GeSn gate-all-around nanowire P-FET on the GeSnOI substrate with channel length of 20 nm and subthreshold swing of 74 mV/decade. In Proceedings of the 2019 International Symposium on VLSI Technology, Systems and Application (VLSI-TSA), Hsinchu, Taiwan, 22–25 April 2019; pp. 1–2.
97. Chuang, Y.; Huang, H.C.; Li, J.Y. GeSn N-FinFETs and NiGeSn contact formation by phosphorus implant. In Proceedings of the 2017 Silicon Nanoelectronics Workshop (SNW), IEEE Conference, Kyoto, Japan, 4–5 June 2017; pp. 97–98.
98. Pandey, R.; Schulte-Braucks, C.; Sajjad, R.N.; Barth, M.; Ghosh, R.K.; Grisafe, B.; Sharma, P.; Driesch, N.; Vohra, A.; Datta, S.; et al. Performance benchmarking of p-type $In_{0.65}Ga_{0.35}As/GaAs_{0.4}Sb_{0.6}$ and $Ge/G_{0.93}Sn_{0.07}$ hetero-junction tunnel FETs. In Proceedings of the 2016 IEEE International Electron Devices Meeting (IEDM), San Francisco, CA, USA, 3–7 December 2016; pp. 1–4.

99. Schulte-Braucks, C.; Pandey, R.; Sajjad, R.N.; Barth, M.; Ghosh, R.K.; Grisafe, B.; Loo, R.; Mantl, S.; Buca, D.; Datta, S.; et al. Fabrication, characterization, and analysis of Ge/GeSn heterojunction p-type tunnel transistors. *IEEE Trans. Electron Devices* **2017**, *64*, 4354–4362. [CrossRef]
100. Wang, H.; Han, G.; Jiang, X.; Liu, Y.; Zhang, J.; Hao, Y. Improved performance in GeSn/SiGeSn TFET by hetero-line architecture with staggered tunneling junction. *IEEE Trans. Electron Devices* **2019**, *66*, 1985–1989. [CrossRef]
101. Mosleh, A.; Alher, M.; Cousar, L.C.; Du, W.; Ghetmiri, S.A.; Al-Kabi, S.; Dou, W.; Sun, G.; Soref, R.; Yu, S.Q.; et al. Buffer-free GeSn and SiGeSn growth on Si substrate using in situ SnD_4 gas mixing. *J. Electron. Mater.* **2016**, *45*, 2051–2058. [CrossRef]
102. Grant, P.C.; Dou, W.; Alharthi, B.; Grant, J.M.; Tran, H.; Abernathy, G.; Mosleh, A.; Du, W.; Li, B.; Yu, S.Q.; et al. UHV-CVD growth of high quality GeSn using $SnCl_4$: From material growth development to prototype devices. *Opt. Mater. Express* **2019**, *9*, 3277–3291. [CrossRef]
103. Cook, C.S.; Zollner, S.; Bauer, M.R.; Aella, P.; Kouvetakis, J.; Menendez, J. Optical constants and interband transitions of $Ge_{1-x}Sn_x$ alloys (x < 0.2) grown on Si by UHV-CVD. *Thin Solid Film.* **2004**, *455*, 217–221.
104. Xu, C.; Gallagher, J.; Senaratne, C.; Brown, C.; Fernando, N.; Zollner, S.; Kouvetakis, J.; Menendez, J. Doping and strain dependence of the electronic band structure in Ge and GeSn alloys. In *APS March Meeting Abstracts*; American Physical Society: College Park, MD, USA, 2015; L14-011.
105. Margetis, J.; Mosleh, A.; Ghetmiri, S.A.; Al-Kabi, S.; Dou, W.; Du, W.; Bhargava, N.; Yu, S.; Profijt, H.; Tolle, J.; et al. Fundamentals of $Ge_{1-x}Sn_x$ and $Si_yGe_{1-x-y}Sn_x$ RPCVD epitaxy. *Mater. Sci. Semicond. Process.* **2017**, *70*, 38–43. [CrossRef]
106. Margetis, J.; Ghetmiri, S.A.; Du, W.; Conley, B.R.; Mosleh, A.; Soref, R.; Yu, S.; Tolle, J. Growth and characterization of epitaxial $Ge_{1-X}Sn_x$ alloys and heterostructures using a commercial CVD system. *ECS Tran.* **2014**, *64*, 711. [CrossRef]
107. Chen, R.; Huang, Y.C.; Gupta, S.; Lin, A.C.; Sanchez, E.; Kim, Y.; Saraswat, K.; Kamins, T.; Harris, J.S. Material characterization of high Sn-content, compressively-strained GeSn epitaxial films after rapid thermal processing. *J. Crys. Growth* **2013**, *365*, 29–34. [CrossRef]
108. Wirths, S.; Buca, D.; Mussler, G.; Tiedemann, A.T.; Holländer, B.; Bernardy, P.; Stoica, T.; Grutzmacher, D.; Mantl, S. Reduced pressure CVD growth of Ge and $Ge_{1-x}Sn_x$ alloys. *ECS J. Solid State Sci. Tech.* **2013**, *2*, N99. [CrossRef]
109. Zhang, L.; Chen, Q.; Wu, S.; Son, B.; Lee, K.H.; Chong, G.Y.; Tan, C.S. Growth and Characterizations of GeSn Films with High Sn Composition by Chemical Vapor Deposition (CVD) Using Ge_2H_6 and $SnCl_4$ for Mid-IR Applications. *ECS Trans.* **2020**, *98*, 91. [CrossRef]
110. Kohen, D.; Vohra, A.; Loo, R.; Vandervorst, W.; Bhargava, N.; Margetis, J.; Tolle, J. Enhanced B doping in CVD-grown GeSn: B using B δ-doping layers. *J. Crys. Growth* **2018**, *483*, 285–290. [CrossRef]
111. Vanjaria, J.; Arjunan, A.C.; Salagaj, T.; Tompa, G.S.; Yu, H. PECVD Growth of Composition Graded SiGeSn Thin Films as Novel Approach to Limit Tin Segregation. *ECS J. Solid State Sci. Tech.* **2020**, *9*, 034009. [CrossRef]
112. Vanjaria, J.V. Growth and Characterization of Si-Ge-Sn Semiconductor Thin Films using a Simplified PECVD Reactor. Ph.D. Thesis, Arizona State University, Tempe, AZ, USA, 2020.
113. Dou, W.; Alharthi, B.; Grant, P.C.; Grant, J.M.; Mosleh, A.; Tran, H.; Du, W.; Li, B.; Naseem, H.; Yu, S.Q.; et al. Crystalline GeSn growth by plasma enhanced chemical vapor deposition. *Opt. Mater. Express* **2018**, *8*, 3220–3229. [CrossRef]
114. Li, Z. Room Temperature Lasing in GeSn Alloys. Ph.D. Thesis, University of Dayton, Dayton, OH, USA, 2015.
115. Assali, S.; Nicolas, J.; Mukherjee, S.; Dijkstra, A.; Moutanabbir, O. Atomically uniform Sn-rich GeSn semiconductors with 3.0–3.5 μm room-temperature optical emission. *Appl. Phys. Lett.* **2018**, *112*, 251903. [CrossRef]
116. Assali, S.; Nicolas, J.; Moutanabbir, O. Enhanced Sn incorporation in GeSn epitaxial semiconductors via strain relaxation. *J. Appl. Phys.* **2019**, *125*, 025304. [CrossRef]
117. Assali, S.; Attiaoui, A.; Del Vecchio, P.; Mukherjee, S.; Kumar, A.; Moutanabbir, O. Epitaxial growth of atomically-sharp GeSn/Ge/GeSn tensile strained (≥1.5%) quantum well on Si. *Bull. Am. Phys. Soc.* **2020**, *65*, 6.
118. Gupta, S.; Chen, R.; Vincent, B.; Lin, D.; Magyari-Kope, B.; Caymax, M.; Dekoster, J.; Harris, J.; Nishi, Y.; Saraswat, K.C. GeSn channel n and p MOSFETs. *ECS Trans.* **2013**, *50*, 937. [CrossRef]
119. Loo, R.; Shimura, Y.; Ike, S.; Vohra, A.; Stoica, T.; Stange, D.; Buca, D.; Kohen, D.; Margetis, J.; Tolle, J. Epitaxial GeSn: Impact of process conditions on material quality. *Semicond. Sci. Technol.* **2018**, *33*, 114010. [CrossRef]
120. Sun, G.; Cheng, H.H.; Menendez, J.; Khurgin, J.B.; Soref, R.A. Strain-free Ge/GeSiSn quantum cascade lasers based on L-valley intersubband transitions. *Appl. Phys. Lett.* **2007**, *90*, 251105. [CrossRef]
121. Cong, H.; Yang, F.; Xue, C.; Yu, K.; Zhou, L.; Wang, N.; Cheng, B.; Wang, Q. Multilayer graphene–GeSn quantum well heterostructure SWIR light source. *Small* **2018**, *14*, 1704414. [CrossRef]
122. Sun, G.; Yu, S.Q. The SiGeSn approach towards Si-based lasers. *Solid State Electron.* **2013**, *83*, 76–81. [CrossRef]
123. Conley, B.R.; Naseem, H.; Sun, G.; Sharps, P.; Yu, S.Q. High efficiency MJ solar cells and TPV using SiGeSn materials. In Proceedings of the 2012 38th IEEE Photovoltaic Specialists Conference, Austin, TX, USA, 3–8 June 2012; pp. 001189–001192.
124. Li, J.; Yu, P.; Cheng, H.; Liu, W.; Li, Z.; Xie, B.; Chen, S.; Tian, J. Optical polarization encoding using graphene-loaded plasmonic metasurfaces. *Adv. Opt. Mater.* **2016**, *4*, 91–98. [CrossRef]
125. Israelsen, N.M.; Petersen, C.R.; Barh, A.; Jain, D.; Jensen, M.; Hannesschläger, G.; Tidemand-Lichtenberg, P.; Pedersen, C.; Podoleanu, A.; Bang, O. Real-time high-resolution mid-infrared optical coherence tomography. *Light Sci. Appl.* **2019**, *8*, 1–13. [CrossRef]

126. Kim, D.H.; Kim, T.W.; Baek, R.H.; Kirsch, P.D.; Maszara, W.; Del Alamo, J.A.; Antoniadis, D.; Urteaga, M.; Brar, B.; Seo, K.S.; et al. High-performance III–V devices for future logic applications. In Proceedings of the 2014 IEEE International Electron Devices Meeting, San Francisco, CA, USA, 15–17 December 2014; pp. 1–4.
127. Rachmady, W.; Agrawal, A.; Sung, S.H.; Dewey, G.; Chouksey, S.; Chu-Kung, B.; Elbaz, G.; Fischer, P.; Huang, C.; Kavalieros, J.; et al. 300 mm heterogeneous 3D integration of record performance layer transfer germanium PMOS with silicon NMOS for low power high performance logic applications. In Proceedings of the 2019 IEEE International Electron Devices Meeting (IEDM), San Francisco, CA, USA, 7–11 December 2019; pp. 1–4.
128. Yamasaka, S.; Watanabe, K.; Sakane, S.; Takeuchi, S.; Sakai, A.; Sawano, K.; Nakamura, Y. Independent control of electrical and heat conduction by nanostructure designing for Si-based thermoelectric materials. *Sci. Rep.* **2016**, *6*, 1–8. [CrossRef]
129. Noroozi, M.; Hamawandi, B.; Toprak, M.S.; Radamson, H.H. Fabrication and thermoelectric characterization of GeSn nanowires. In Proceedings of the 2014 15th International Conference on Ultimate Integration on Silicon (ULIS), Stockholm, Sweden, 7–9 April 2014; pp. 125–128.
130. Pearton, S. Silicon-based spintronics. *Nat. Mater.* **2004**, *3*, 203–204. [CrossRef]
131. Hortamani, M.; Sandratskii, L.; Kratzer, P.; Mertig, I. Searching for Si-based spintronics by first principles calculations. *New J. Phys.* **2009**, *11*, 125009. [CrossRef]
132. Su, H.; Hu, H.; Mousavi, P.; Zhang, H.; Wang, B.; Miao, Y. Silicon-based high-integration reconfigurable dipole with SPiN. *Solid-State Electron.* **2019**, *154*, 20–23. [CrossRef]
133. Su, H.; Hu, H.; Zhang, H.; Miao, Y. Investigation of a Silicon-Based High Integration Reconfigurable Dipole. *Prog. Electrom. Res. Lett.* **2018**, *79*, 135–141. [CrossRef]
134. Tai, C.T.; Chiu, P.Y.; Liu, C.Y.; Kao, H.S.; Harris, C.T.; Lu, T.M.; Hsieh, C.; Chang, S.; Li, J.Y. Strain Effects on Rashba Spin-Orbit Coupling of 2D Hole Gases in GeSn/Ge Heterostructures. *Adv. Mater.* **2021**, 2007862. [CrossRef]
135. Marchionni, A.; Zucchetti, C.; Ciccacci, F.; Finazzi, M.; Funk, H.S.; Schwarz, D.; Oehme, M.; Schulze, J.; Bottegoni, F. Inverse spin-Hall effect in GeSn. *Appl. Phys. Lett.* **2021**, *118*, 212402. [CrossRef]
136. Tolle, J.; Roucka, R.; D'Costa, V.; Menendez, J.; Chizmeshya, A.; Kouvetakis, J. Sn-based Group-IV Semiconductors on Si: New Infrared Materials and New Templates for Mismatched Epitaxy. *MRS Online Proc. Lib.* **2005**, *891*, 1–6. [CrossRef]
137. Bauer, M.; Ritter, C.; Crozier, P.A.; Ren, J.; Menendez, J.; Wolf, G.; Kouvetakis, J. Synthesis of ternary SiGeSn semiconductors on Si (100) via Sn_xGe_{1-x} buffer layers. *Appl. Phys. Lett.* **2003**, *83*, 2163–2165. [CrossRef]
138. Taraci, J.; Tolle, J.; Kouvetakis, J.; McCartney, M.R.; Smith, D.J.; Menendez, J.; Santana, M.A. Simple chemical routes to diamond-cubic germanium–tin alloys. *Appl. Phys. Lett.* **2001**, *78*, 3607–3609. [CrossRef]
139. Aella, P.; Cook, C.; Tolle, J.; Zollner, S.; Chizmeshya, A.V.G.; Kouvetakis, J. Optical and structural properties of $Si_xSn_yGe_{1-x-y}$ alloys. *Appl. Phys. Lett.* **2004**, *84*, 888–890. [CrossRef]
140. Bauer, M.R.; Cook, C.S.; Aella, P.; Tolle, J.; Kouvetakis, J.; Crozier, P.A.; Chizmeshya, A.; Zollner, S. SnGe superstructure materials for Si-based infrared optoelectronics. *Appl. Phys. Lett.* **2003**, *83*, 3489–3491. [CrossRef]
141. Li, S.F.; Bauer, M.R.; Menéndez, J.; Kouvetakis, J. Scaling law for the compositional dependence of Raman frequencies in SnGe and GeSi alloys. *Appl. Phys. Lett.* **2004**, *84*, 867–869. [CrossRef]
142. Roucka, R.; Yu, S.Q.; Tolle, J.; Fang, Y.Y.; Wu, S.N.; Menendez, J.; Kouvetakis, J. Photoresponse at 1. In 55 µm in GeSn epitaxial films grown on Si. In Proceedings of the LEOS 2007-IEEE Lasers and Electro-Optics Society Annual Meeting Conference Proceedings, Lake Buena Vista, FL, USA, 21–25 October 2007; pp. 178–179.
143. Mathews, J.; Roucka, R.; Yu, S.Q.; Tolle, J.; Kouvetakis, J.; Menendez, J. Photocurrent Measurements on Novel Group IV Semiconductor Alloys. In *APS Four Corners Section Meeting Abstracts 2007*; American Physical Society: College Park, MD, USA, 2007; E1-015.
144. D'costa, V.R.; Tolle, J.; Roucka, R.; Poweleit, C.D.; Kouvetakis, J.; Menendez, J. Raman scattering in $Ge_{1-y}Sn_y$ alloys. *Solid State Commun.* **2007**, *144*, 240–244. [CrossRef]
145. Grzybowski, G.; Beeler, R.T.; Jiang, L.; Smith, D.J.; Kouvetakis, J.; Menendez, J. Next generation of $Ge_{1-y}Sn_y$ (y = 0.01–0.09) alloys grown on Si(100) via Ge_3H_8 and SnD_4: Reaction kinetics and tunable emission. *Appl. Phys. Lett.* **2012**, *101*, 072105. [CrossRef]
146. Grzybowski, G.; Jiang, L.; Beeler, R.T.; Watkins, T.; Chizmeshya, A.V.; Xu, C.; Menendez, J.; Kouvetakis, J. Ultra-low-temperature Epitaxy of Ge-based semiconductors and Optoelectronic structures on Si(100): Introducing higher order Germanes (Ge_3H_8, Ge_4H_{10}). *Chem. Mater.* **2012**, *24*, 1619–1628. [CrossRef]
147. Beeler, R.T.; Xu, C.; Smith, D.J.; Grzybowski, G.; Menendez, J.; Kouvetakis, J. Compositional dependence of the absorption edge and dark currents in $Ge_{1-x-y}Si_xSn_y/Ge(100)$ photodetectors grown via ultra-low-temperature epitaxy of Ge_4H_{10}, Si_4H_{10}, and SnD_4. *Appl. Phys. Lett.* **2012**, *101*, 221111. [CrossRef]
148. Kouvetakis, J.; Gallagher, J.; Menéndez, J. Direct gap Group IV semiconductors for next generation Si-based IR photonics. *MRS Online Proc. Libr.* **2014**, *1666*, 24–35. [CrossRef]
149. Mircovich, M.A.; Xu, C.; Ringwala, D.A.; Poweleit, C.D.; Menéndez, J.; Kouvetakis, J. Extended Compositional Range for the Synthesis of SWIR and LWIR $Ge_{1-y}Sn_y$ Alloys and Device Structures via CVD of SnH_4 and Ge_3H_8. *ACS Appl. Electron. Mater.* **2021**, *3*, 3451–3460. [CrossRef]
150. Vincent, B.; Gencarelli, F.; Bender, H.; Merckling, C.; Douhard, B.; Petersen, D.H.; Hansen, O.; Henrichsen, H.; Meersschaut, J.; Caymax, M.; et al. Undoped and in-situ B doped GeSn epitaxial growth on Ge by atmospheric pressure-chemical vapor deposition. *Appl. Phys. Lett.* **2011**, *99*, 152103. [CrossRef]

151. Margetis, J.; Yu, S.Q.; Li, B.; Tolle, J. Chemistry and kinetics governing hydride/chloride chemical vapor deposition of epitaxial Ge$_{1-x}$Sn$_x$. *J. Vacu. Sci. Technolo. A Vacu. Surf. Film.* **2019**, *37*, 021508. [CrossRef]
152. Patchett, D. Germanium-tin-silicon Epitaxial Structures Grown on Silicon by Reduced Pressure Chemical Vapour Deposition. Ph.D. Thesis, University of Warwick, Coventry, UK, 2016.
153. Margetis, J. RPCVD Growth of Epitaxial Si-Ge-Sn Alloys for Optoelectronics Applications. Ph.D. Thesis, Arizona State University, Tempe, AZ, USA, 2018.
154. Wirths, S. Group IV Epitaxy for Advanced Nano-and Optoelectronic Applications. Ph.D. Thesis, Halbleiter-Nanoelektronik, Jülich, Germany, 2016.
155. Aubin, J.; Hartmann, J.M.; Gassenq, A.; Rouviere, J.L.; Robin, E.; Delaye, V.; Cooper, D.; Mollard, N.; Reboud, V.; Calvo, V. Growth and structural properties of step-graded, high Sn content GeSn layers on Ge. *Semicond. Sci. Technolo.* **2017**, *32*, 094006. [CrossRef]
156. Aubin, J. Low Temperature Epitaxy of Si, Ge, and Sn Based Alloys. Ph.D. Thesis, Université Grenoble Alpes, Grenoble, France, 2017.
157. Grant, J.M. Investigation of Critical Technologies of Chemical Vapor Deposition for Advanced (Si) GeSn Materials. Ph.D. Thesis, University of Arkansas, Fayetteville, AR, USA, 2019.
158. Xu, C.; Wallace, P.M.; Ringwala, D.A.; Chang, S.L.; Poweleit, C.D.; Kouvetakis, J.; Menéndez, J. Mid-infrared (3–8 μm) Ge$_{1-y}$Sn$_y$ alloys (0.15 < y < 0.30): Synthesis, structural, and optical properties. *Appl. Phys. Lett.* **2019**, *114*, 212104.
159. Xu, C.; Ringwala, D.; Wang, D.; Liu, L.; Poweleit, C.D.; Chang, S.L.; Zhuang, H.; Menendez, J.; Kouvetakis, J. Synthesis and fundamental studies of Si-compatible (Si) GeSn and GeSn mid-IR systems with ultrahigh Sn contents. *Chem. Mater.* **2019**, *31*, 9831–9842. [CrossRef]
160. Hu, T. Synthesis and Properties of Sn-Based Group IV Alloys. Ph.D. Thesis, Arizona State University, Tempe, AZ, USA, 2019.
161. Wallace, P.M. Expanding the Optical Capabilities of Germanium in the Infrared Range through Group IV and III-V-IV Alloy Systems. Ph.D. Thesis, Arizona State University, Tempe, AZ, USA, 2018.
162. Gencarelli, F.; Vincent, B.; Demeulemeester, J.; Vantomme, A.; Moussa, A.; Franquet, A.; Kumar, A.; Bender, H.; Meersschaut, J.; Heyns, M.; et al. Crystalline properties and strain relaxation mechanism of CVD grown GeSn. *ECS J. Solid State Sci. Technol.* **2013**, *2*, P134. [CrossRef]
163. Takeuchi, S.; Shimura, Y.; Nishimura, T.; Vincent, B.; Eneman, G.; Clarysse, T.; Demeulemeester, J.; Vantomme, A.; Loo, R.; Zaima, S.; et al. Ge$_{1-x}$Sn$_x$ stressors for strained-Ge CMOS. *Solid-State Electron.* **2011**, *60*, 53–57. [CrossRef]
164. Su, S.; Cheng, B.; Xue, C.; Wang, W.; Cao, Q.; Xue, H.; Hu, W.; Zhang, G.; Zuo, Y.; Wang, Q. GeSn pin photodetector for all telecommunication bands detection. *Opt. Express* **2011**, *19*, 6400–6405. [CrossRef] [PubMed]
165. Han, G.; Su, S.; Wang, L.; Wang, W.; Gong, X.; Yang, Y.; Ivana; Guo, P.; Guo, C.; Yeo, Y.C.; et al. Strained germanium-tin (GeSn) N-channel MOSFETs featuring low temperature N+/P junction formation and GeSnO$_2$ interfacial layer. In Proceedings of the Symposium on VLSI Technology (VLSIT), IEEE Conference, Honolulu, HI, USA, 12–14 June 2012; pp. 97–98.
166. Yang, Y.; Su, S.; Guo, P.; Wang, W.; Gong, X.; Wang, L.; Low, K.; Zhang, G.; Xue, C.; Cheng, B.; et al. Towards direct band-to-band tunneling in p-channel tunneling field effect transistor (TFET): Technology enablement by germanium-tin (GeSn). In Proceedings of the 2012 International Electron Devices Meeting, IEEE Conference, San Francisco, CA, USA, 10–13 December 2012; pp. 1–4.
167. Han, G.; Su, S.; Zhou, Q.; Wang, L.; Wang, W.; Zhang, G.; Xue, C.; Cheng, B.; Yeo, Y.C. BF^{2+} ion implantation and dopant activation in strained Germanium-tin (Ge$_{1-x}$Sn$_x$) epitaxial layer. In Proceedings of the 2012 12th International Workshop on Junction Technology, Shanghai, China, 14–15 May 2012; pp. 106–108.
168. Guo, P.; Zhan, C.; Yang, Y.; Gong, X.; Liu, B.; Cheng, R.; Wang, W.; Pan, J.; Zhang, Z.; Yeo, Y.C.; et al. Germanium-Tin (GeSn) N-channel MOSFETs with low temperature silicon surface passivation. In Proceedings of the 2013 International Symposium on VLSI Technology, Systems and Application (VLSI-TSA), Hsinchu, Taiwan, 22–24 April 2013; pp. 1–2.
169. Yang, Y.; Han, G.; Guo, P.; Wang, W.; Gong, X.; Wang, L.; Low, K.; Yeo, Y.C. Germanium–tin p-channel tunneling field-effect transistor: Device design and technology demonstration. *IEEE Trans. Electron Devices* **2013**, *60*, 4048–4056. [CrossRef]
170. Han, G.; Su, S.; Yang, Y.; Guo, P.; Gong, X.; Wang, L.; Wang, W.; Guo, C.; Zhang, G.; Yeo, Y.C.; et al. High Hole Mobility in Strained Germanium-Tin (GeSn) Channel pMOSFET Fabricated on (111) Substrate. *ECS Trans.* **2013**, *50*, 943. [CrossRef]
171. Tong, Y.; Han, G.; Liu, B.; Yang, Y.; Wang, L.; Wang, W.; Yeo, Y.C. Ni(Ge$_{1-x}$Sn$_x$) Ohmic Contact Formation on N-Type Ge$_{1-x}$Sn$_x$ Using Selenium or Sulfur Implant and Segregation. *IEEE Trans. Electron Devices* **2013**, *60*, 746–752. [CrossRef]
172. Yeo, Y.C.; Han, G.; Gong, X.; Wang, L.; Wang, W.; Yang, Y.; Guo, P.; Liu, B.; Su, S.; Cheng, B.; et al. Tin-Incorporated Source/Drain and Channel Materials for Field-Effect Transistors. *ECS Trans.* **2013**, *50*, 931. [CrossRef]
173. Gong, X.; Han, G.; Su, S.; Cheng, R.; Guo, P.; Bai, F.; Yang, Y.; Zhou, Q.; Liu, B.; Yeo, Y.C.; et al. Uniaxially strained germanium-tin (GeSn) gate-all-around nanowire PFETs enabled by a novel top-down nanowire formation technology. In Proceedings of the 2013 Symposium on VLSI Technology, Kyoto, Japan, 11–13 June 2013; pp. T34–T35.
174. Gupta, S.; Huang, Y.C.; Kim, Y.; Sanchez, E.; Saraswat, K.C. Hole Mobility Enhancement in Compressively Strained Ge$_{0.93}$Sn$_{0.07}$ pMOSFETs. *IEEE Electron Device Lett.* **2013**, *34*, 831–833. [CrossRef]
175. Wang, L.; Liu, B.; Gong, X.; Guo, P.; Zhou, Q.; Chua, L.H.; Zou, W.; Hatem, C.; Henry, T.; Yeo, Y.C. Self-crystallization and reduced contact resistivity by hot phosphorus ion implant in germanium-tin alloy. In Proceedings of the Technical Program-2014 International Symposium on VLSI Technology, Systems and Application (VLSI-TSA), Hsinchu, Taiwan, 28–30 April 2014; pp. 1–2.
176. D'Costa, V.R.; Wang, L.; Wang, W.; Lim, S.L.; Chan, T.K.; Chua, L.H.; Henry, T.; Zou, W.; Hatem, C.; Yeo, Y.C.; et al. Towards simultaneous achievement of carrier activation and crystallinity in Ge and GeSn with heated phosphorus ion implantation: An optical study. *Appl. Phys. Lett.* **2014**, *105*, 122108. [CrossRef]

177. Zhang, X.; Zhang, D.; Zheng, J.; Liu, Z.; He, C.; Xue, C.; Zhang, G.; Li, C.; Cheng, B.; Wang, Q. Formation and characterization of Ni/Al Ohmic contact on n^+-type GeSn. *Solid-State Electron.* **2015**, *114*, 178–181. [CrossRef]
178. Zhou, J.; Han, G.; Li, Q.; Peng, Y.; Lu, X.; Zhang, C.; Sun, Q.; Zhang, D.; Hao, Y. Ferroelectric HfZrO$_x$Ge and GeSn PMOSFETs with Sub-60 mV/decade subthreshold swing, negligible hysteresis, and improved Ids. In Proceedings of the 2016 IEEE International Electron Devices Meeting (IEDM), San Francisco, CA, USA, 3–7 December 2016; pp. 1–4.
179. Cong, H.; Xue, C.; Zheng, J.; Yang, F.; Yu, K.; Liu, Z.; Zhang, X.; Cheng, B.; Wang, Q. Silicon based GeSn pin photodetector for SWIR detection. *IEEE Photon. J.* **2016**, *8*, 1–6. [CrossRef]
180. Zhou, J.; Han, G.; Peng, Y.; Liu, Y.; Zhang, J.; Sun, Q.Q.; Zhang, D.; Hao, Y. Ferroelectric negative capacitance GeSn PFETs with sub-20 mV/decade subthreshold swing. *IEEE Electron Device Lett.* **2017**, *38*, 1157–1160. [CrossRef]
181. Liu, Y.C.; Huang, Y.S.; Lu, F.L.; Ye, H.Y.; Tu, C.T.; Liu, C.W. Novel vertically stacked Ge0.85Si0.15 nGAAFETs above a Si channel with low SS of 76 mV/dec by underneath Si channel and enhanced Ion (1.7 X at VOV = VDS = 0.5 V) by Ge$_{0.85}$Si$_{0.15}$ channels. *Semicond. Sci. Technol.* **2020**, *35*, 055010. [CrossRef]
182. Zhao, Y.; Wang, N.; Yu, K.; Zhang, X.; Li, X.; Zheng, J.; Xue, C.; Cheng, B.; Li, C. High performance silicon-based GeSn p–i–n photodetectors for short-wave infrared application. *Chi. Phys. B* **2019**, *28*, 128501. [CrossRef]
183. Peng, L.; Li, X.; Liu, Z.; Liu, X.; Zheng, J.; Xue, C.; Zuo, Y.; Cheng, B. Horizontal GeSn/Ge multi-quantum-well ridge waveguide LEDs on silicon substrates. *Photonics Res.* **2020**, *8*, 899–903. [CrossRef]
184. Tsui, B.Y.; Liao, H.H.; Chen, Y.J. Degradation Mechanism of Ge N^+-P Shallow Junction with Thin GeSn Surface Layer. *IEEE Trans. Electron Devices* **2020**, *67*, 1120–1125. [CrossRef]
185. Tsai, C.E.; Lu, F.L.; Chen, P.S.; Liu, C.W. Boron-doping induced Sn loss in GeSn alloys grown by chemical vapor deposition. *Thin Solid Film.* **2018**, *660*, 263–266. [CrossRef]
186. Radamson, H.H.; Noroozi, M.; Jamshidi, A.; Thompson, P.E.; Östling, M. Strain engineering in GeSnSi materials. *ECS Trans.* **2013**, *50*, 527. [CrossRef]
187. Liu, Q.; Geilei, W.; Guo, Y.; Ke, X.; Radamson, H.; Liu, H.; Zhao, C.; Luo, J. Improvement of the thermal stability of nickel stanogermanide by carbon pre-stanogermanidation implant into GeSn substrate. *ECS J. Solid State Sci. Technolo.* **2014**, *4*, P67. [CrossRef]
188. Von den Driesch, N.; Stange, D.; Wirths, S.; Rainko, D.; Mussler, G.; Stoica, T.; Ikonic, Z.; Hartmann, J.; Mantl, S.; Buca, D.; et al. Direct bandgap GeSn light emitting diodes for short-wave infrared applications grown on Si. In Proceedings of the Silicon Photonics XI, International Society for Optics and Photonics, San Francisco, CA, USA, 13–18 February 2016; Volume 9752, p. 97520C.
189. Stange, D.; Von Den Driesch, N.; Rainko, D.; Schulte-Braucks, C.; Wirths, S.; Mussler, G.; Tiedemann, A.; Stoica, T.; Hartmann, J.; Buca, D.; et al. Study of GeSn based heterostructures: Towards optimized group IV MQW LEDs. *Opt. Express* **2016**, *24*, 1358–1367. [CrossRef]
190. Bhargava, N.; Margetis, J.; Tolle, J. As doping of Si–Ge–Sn epitaxial semiconductor materials on a commercial CVD reactor. *Semicond. Sci. Technol.* **2017**, *32*, 094003. [CrossRef]
191. Lu, F.L.; Tsai, C.E.; Huang, C.H.; Ye, H.Y.; Lin, S.Y.; Liu, C.W. Record Low Contact Resistivity (4.4×10^{-10} Ω-cm^2) to Ge Using In-situ B and Sn Incorporation by CVD With Low Thermal Budget ($\leq$400 °C) and without Ga. In Proceedings of the 2019 Symposium on VLSI Technology, Kyoto, Japan, 9–14 June 2019; pp. T178–T179.
192. Tsai, C.E.; Lu, F.L.; Liu, Y.C.; Ye, H.Y.; Liu, C.W. Low Contact Resistivity to Ge Using In-Situ B and Sn Incorporation by Chemical Vapor Deposition. *IEEE Trans. Electron Devices* **2020**, *67*, 5053–5058. [CrossRef]
193. Frauenrath, M.; Hartmann, J.M.; Nolot, E. Boron and Phosphorous Doping of GeSn for Photodetectors and Light Emitting Diodes. *ECS Trans.* **2020**, *98*, 325. [CrossRef]
194. Lu, F.L.; Liu, Y.C.; Tsai, C.E.; Ye, H.Y.; Liu, C.W. Record Low Contact Resistivity to Ge: B (8.1×10^{-10} Ω-cm^2) and GeSn: B (4.1×10^{-10} Ω-cm^2) with Optimized [B] and [Sn] by In-situ CVD Doping. In Proceedings of the 2020 IEEE Symposium on VLSI Technology, Honolulu, HI, USA, 16–19 June 2020; pp. 1–2.
195. Baert, B. Impact of Electron Trap States on the Transport Properties of GeSn Semiconducting Heterostructures Assessed by Electrical Characterizations. Ph.D. Thesis, Université de Liège, Liège, Belgium, 2016.
196. Li, H.; Cheng, H.H.; Lee, L.C.; Lee, C.P.; Su, L.H.; Suen, Y.W. Electrical characteristics of Ni Ohmic contact on n-type GeSn. *Appl. Phys. Lett.* **2014**, *104*, 241904. [CrossRef]
197. Chuang, Y.; Liu, C.Y.; Kao, H.S.; Tien, K.Y.; Luo, G.L.; Li, J.Y. Schottky Barrier Height Modulation of Metal/n-GeSn Contacts Featuring Low Contact Resistivity by in Situ Chemical Vapor Deposition Doping and NiGeSn Alloy Formation. *ACS Appl. Electron. Mater.* **2021**, *3*, 1334–1340. [CrossRef]
198. Abdi, S.; Atalla, M.; Assali, S.; Kumar, A.; Groell, L.; Koelling, S.; Moutanabbir, O. Towards Ultra-Low Specific Contact Resistance on P-Type and N-Type Narrow Bandgap GeSn Semiconductors. In *ECS Meeting Abstracts*; IOP Publishing: Montreal, QC, Canada, 2020; Volume 22, p. 1322.
199. Galluccio, E.; Petkov, N.; Mirabelli, G.; Doherty, J.; Lin, S.Y.; Lu, F.L.; Liu, C.; Holmes, J.; Duffy, R. Formation and characterization of Ni, Pt, and Ti stanogermanide contacts on Ge$_{0.92}$Sn$_{0.08}$. *Thin Solid Film.* **2019**, *690*, 137568. [CrossRef]
200. Wu, Y.; Xu, H.; Han, K.; Gong, X. Thermal Stability and Sn Segregation of Low-Resistance Ti/p^+-Ge$_{0.95}$Sn$_{0.05}$ Contact. *IEEE Electron Device Lett.* **2019**, *40*, 1575–1578. [CrossRef]

201. Wu, Y.; Xu, H.; Chua, L.H.; Han, K.; Zou, W.; Henry, T.; Gong, X. A Novel Fast-Turn-Around Ladder TLM Methodology with Parasitic Metal Resistance Elimination, and 2×10^{-10} $\Omega \cdot cm^2$ Resolution: Theoretical Design and Experimental Demonstration. In Proceedings of the 2019 Symposium on VLSI Technology, Kyoto, Japan, 9–14 June 2019; pp. T150–T151.
202. Wu, Y.; Wang, W.; Masudy-Panah, S.; Li, Y.; Han, K.; He, L.; Zhang, Z.; Lei, D.; Xue, S.; Gong, X.; et al. Sub-10^{-9} $\Omega \cdot cm^2$ Specific Contact Resistivity (Down to 4.4×10^{-10} $\Omega \cdot cm^2$) for Metal Contact on Ga and Sn Surface-Segregated GeSn Film. *IEEE Trans. Electron Devices* **2018**, *65*, 5275–5281. [CrossRef]
203. Wu, Y.; Luo, S.; Wang, W.; Masudy-Panah, S.; Lei, D.; Liang, G.; Gong, X.; Yeo, Y.C. Ultra-low specific contact resistivity (1.4×10^{-9} $\Omega\ cm^2$) for metal contacts on in-situ Ga-doped $Ge_{0.95}Sn_{0.05}$ film. *J. Appl. Phys.* **2017**, *122*, 224503. [CrossRef]
204. Han, G.; Su, S.; Zhou, Q.; Guo, P.; Yang, Y.; Zhan, C.; Wang, L.; Wang, W.; Wang, Q.; Yeo, Y.C.; et al. Dopant Segregation and Nickel Stanogermanide Contact Formation on $p^+Ge_{0.947}Sn_{0.053}$ Source/Drain. *IEEE Electron Device Lett.* **2012**, *33*, 634–636. [CrossRef]
205. Zheng, J.; Zhang, Y.; Liu, Z.; Zuo, Y.; Li, C.; Xue, C.; Cheng, B.; Wang, Q. Fabrication of low-resistance Ni ohmic contacts on n^+-$Ge_{1-x}Sn_x$. *IEEE Trans. Electron Devices* **2018**, *65*, 4971–4974. [CrossRef]
206. Zheng, J.; Wang, S.; Zhang, X.; Liu, Z.; Xue, C.; Li, C.; Zuo, Y.; Cheng, B.; Wang, Q. Ni ($Ge_{1-x-y}Si_xSn_y$) Ohmic Contact Formation on p-type $Ge_{0.86}Si_{0.07}Sn_{0.07}$. *IEEE Electron Device Lett.* **2015**, *36*, 878–880. [CrossRef]
207. Schulte-Braucks, C.; Hofmann, E.; Glass, S.; von den Driesch, N.; Mussler, G.; Breuer, U.; Hartmann, J.; Zaumseil, P.; Zhao, Q.; Buca, D.; et al. Schottky barrier tuning via dopant segregation in NiGeSn-GeSn contacts. *J. Appl. Phys.* **2017**, *121*, 205705. [CrossRef]
208. Farokhnejad, A.; Schwarz, M.; Horst, F.; Iñiguez, B.; Lime, F.; Kloes, A. Analytical modeling of capacitances in tunnel-FETs including the effect of Schottky barrier contacts. *Solid-State Electron.* **2019**, *159*, 191–196. [CrossRef]
209. Zhou, H.; Xu, S.; Lin, Y.; Huang, Y.C.; Son, B.; Chen, Q.; Guo, X.; Lee, K.; Goh, S.; Gong, X.; et al. High-efficiency GeSn/Ge multiple-quantum-well photodetectors with photon-trapping microstructures operating at 2 µm. *Opt. Express* **2020**, *28*, 10280–10293. [CrossRef]
210. Sun, G.; Soref, R.A.; Cheng, H.H. Design of an electrically pumped SiGeSn/GeSn/SiGeSn double-heterostructure midinfrared laser. *J. Appl. Phys.* **2010**, *108*, 033107. [CrossRef]
211. Sun, G.; Soref, R.A.; Cheng, H.H. Design of a Si-based lattice-matched room-temperature GeSn/GeSiSn multi-quantum-well mid-infrared laser diode. *Opt. Express* **2010**, *18*, 19957–19965. [CrossRef]
212. Radamson, H.H.; Luo, J.; Simoen, E.; Zhao, C. *CMOS Past, Present and Future*; Woodhead Publishing: Cambridge, UK, 2018. ISBN 9780081021392.
213. Takagi, S.; Iisawa, T.; Tezuka, T.; Numata, T.; Nakaharai, S.; Hirashita, N.; Moriyama, Y.; Usuda, K.; Toyoda, E.; Sugiyama, N.; et al. Carrier-transport-enhanced channel CMOS for improved power consumption and performance. *IEEE Trans. Electron Devices* **2007**, *55*, 21–39. [CrossRef]
214. Huang, Y.S.; Lu, F.L.; Tsou, Y.J.; Tsai, C.E.; Lin, C.Y.; Huang, C.H.; Liu, C.W. First vertically stacked GeSn nanowire pGAAFETs with I on = 1850 µA/µm (Vov = Vds = −1 V) on Si by GeSn/Ge CVD epitaxial growth and optimum selective etching. In Proceedings of the 2017 IEEE International Electron Devices Meeting (IEDM), San Francisco, CA, USA, 2–6 December 2017; pp. 1–4.
215. Chuang, Y.; Liu, C.Y.; Luo, G.L.; Li, J.Y. Electron Mobility Enhancement in GeSn n-Channel MOSFETs by Tensile Strain. *IEEE Electron. Device Lett.* **2020**, *42*, 10–13. [CrossRef]
216. Liu, M.; Mertens, K.; von den Driesch, N.; Schlykow, V.; Grap, T.; Lentz, F.; Trellenkamp, S.; Hartmann, J.; Knoch, J.; Buca, D.; et al. Vertical heterojunction $Ge_{0.92}Sn_{0.08}$/Ge gate-all-around nanowire pMOSFETs with NiGeSn contact. *Solid-State Electron.* **2020**, *168*, 107716. [CrossRef]
217. Liu, L.; Liang, R.; Wang, G.; Radamson, H.H.; Wang, J.; Xu, J. Investigation on direct-gap GeSn alloys for high-performance tunneling field-effect transistor applications. In Proceedings of the 2017 IEEE Electron Devices Technology and Manufacturing Conference (EDTM), Toyama, Japan, 28 February–2 March 2017; pp. 180–181.
218. Radamson, H.H.; Zhu, H.L.; Wu, Z.H.; He, X.B.; Lin, H.X.; Liu, J.B.; Xiang, J.J.; Kong, Z.Z.; Wang, G.L. State of the Art and Future Perspectives in Advanced CMOS Technology. *Nanomaterials* **2020**, *10*, 1555. [CrossRef] [PubMed]
219. Yin, X.; Zhang, Y.; Zhu, H.; Wang, G.L.; Li, J.J.; Du, A.Y.; Li, C.; Zhao, L.H.; Huang, W.X. Vertical sandwich gate-all-around field-effect transistors with self-aligned high-k metal gates and small effective-gate-length variation. *IEEE Electron. Device Lett.* **2019**, *42*, 8–11.
220. Noroozi, M.; Moeen, M.; Abedin, A.; Toprak, M.S.; Radamson, H.H. Effect of strain on Ni-$(GeSn)_x$ contact formation to GeSn nanowires. *MRS Online Proc. Libr.* **2014**, *1707*, 7–12. [CrossRef]
221. Jamshidi, A.; Noroozi, M.; Moeen, M.; Hallén, A.; Hamawandi, B.; Lu, J.; Hultman, L.; Östling, M.; Radamson, H. Growth of GeSnSiC layers for photonic applications. *Surf. Coat. Technol.* **2013**, *230*, 106–110. [CrossRef]
222. Noroozi, M.; Abedin, A.; Moeen, M.; Östling, M.; Radamson, H.H. CVD growth of GeSnSiC alloys using disilane, digermane, Tin Tetrachloride and methylsilane. *ECS Trans.* **2014**, *64*, 703. [CrossRef]
223. Hållstedt, J.; Blomqvist, M.; Persson, P.Å.; Hultman, L.; Radamson, H.H. The effect of carbon and germanium on phase transformation of nickel on $Si_{1-x-y}Ge_xC_y$ epitaxial layers. *J. Appl. Phys.* **2004**, *95*, 2397–2402. [CrossRef]

Article

High Performance p-i-n Photodetectors on Ge-on-Insulator Platform

Xuewei Zhao [1,2,3], Guilei Wang [2,3,4,*], Hongxiao Lin [2,3,4], Yong Du [2,3], Xue Luo [4], Zhenzhen Kong [2,3], Jiale Su [2], Junjie Li [2,3], Wenjuan Xiong [2,3], Yuanhao Miao [2,4,*], Haiou Li [1], Guoping Guo [1] and Henry H. Radamson [2,3,4,5,*]

1. CAS Key Laboratory of Quantum Information, University of Science and Technology of China, Hefei 230026, China; zhaoxuewei@ime.ac.cn (X.Z.); haiouli@ustc.edu.cn (H.L.); gpguo@ustc.edu.cn (G.G.)
2. Key Laboratory of Microelectronics Devices & Integrated Technology, Institute of Microelectronics, Chinese Academy of Sciences, Beijing 100029, China; linhongxiao@ime.ac.cn (H.L.); duyong@ime.ac.cn (Y.D.); kongzhenzhen@ime.ac.cn (Z.K.); sujiale@ime.ac.cn (J.S.); lijunjie@ime.ac.cn (J.L.); xiongwenjuan@ime.ac.cn (W.X.)
3. Institute of Microelectronics, University of Chinese Academy of Sciences, Beijing 100049, China
4. Research and Development Center of Optoelectronic Hybrid IC, Guangdong Greater Bay Area Institute of Integrated Circuit and System, Guangzhou 510535, China; luoxue@giics.com.cn
5. Department of Electronics Design, Mid Sweden University, Holmgatan 10, 85170 Sundsvall, Sweden
* Correspondence: wangguilei@ime.ac.cn (G.W.); miaoyuanhao@ime.ac.cn (Y.M.); rad@ime.ac.cn (H.H.R.)

Abstract: In this article, we demonstrated novel methods to improve the performance of p-i-n photodetectors (PDs) on a germanium-on-insulator (GOI). For GOI photodetectors with a mesa diameter of 10 μm, the dark current at −1 V is 2.5 nA, which is 2.6-fold lower than that of the Ge PD processed on Si substrates. This improvement in dark current is due to the careful removal of the defected Ge layer, which is formed with the initial growth of Ge on Si. The bulk leakage current density and surface leakage density of the GOI detector at −1 V are as low as 1.79 mA/cm^2 and 0.34 μA/cm, respectively. GOI photodetectors with responsivity of 0.5 and 0.9 A/W at 1550 and 1310 nm wavelength are demonstrated. The optical performance of the GOI photodetector could be remarkably improved by integrating a tetraethylorthosilicate (TEOS) layer on the oxide side due to the better optical confinement and resonant cavity effect. These PDs with high performances and full compatibility with Si CMOS processes are attractive for applications in both telecommunications and monolithic optoelectronics integration on the same chip.

Keywords: GOI; photodetectors; dark current; responsivity

1. Introduction

Silicon complementary metal oxide semiconductor (CMOS) technology could be used to integrate optical data communication and electrical data computing to achieve electron–photon synergy. Combining the advantages of photons and electrons can overcome some of the bottlenecks which microelectronics technology encounters with the development of Moore's Law, and open more directions for the continuation of Moore's Law [1,2]. Photodetectors (PDs) are the core devices of high-performance electro-optic conversion. Si photodetectors (PDs) are not attractive in optical communication because the cut-off wavelength of their absorption spectrum is 1100 nm. Ge has emerged as a leading contender for optoelectronic devices due to its pseudo-direct bandgap behavior, large absorption coefficient in the near-infrared region, lower cost, and compatibility with CMOS processing compared to III–V materials. This makes Ge PDs very promising in optoelectronic applications.

There are still some practical difficulties to overcome in the research of Ge PDs, especially for the Ge epitaxy on Si. People first attempted to epitaxially grow Ge on Si with a low dislocation density, mainly because the mobility of electrons and holes in Ge is higher than that in Si [3]. For high-quality Ge epitaxy on Si, the greatest challenge is the 4.2% lattice mismatch ratio between these two elements. This mismatch will result in high surface roughness after growth and high threading dislocation densities in the Ge epitaxial layer, which will hinder the integration of Si and Ge devices and affect device epitaxy on Si [4–6]. SiGe buffer layers were used to reduce the number of threading dislocations in the Ge layer. By carefully adjusting the composition of the SiGe layers, many threading dislocations can be bent or terminated at the heterojunction interfaces, which greatly reduces the dislocation density in the Ge layer [7,8]. A two-step growth method has also been proposed, which can effectively prevent island-like growth, and subsequent annealing can greatly reduce the linear dislocation density [9–11]. Many efforts have been taken to improve the quality of the Ge layer on the Si substrate in order to improve the performance of Ge normal-incident PDs [12], waveguide PDs (WGPDs) [13,14], metal–semiconductor–metal PDs (MSM PDs) [15], etc. However, the fundamental problem of high defective regions at the epitaxial Ge/Si interface has not yet been solved. Various methods have been widely reported to optimize the process of Ge. Compared to the Ge-on-Si structure, the quality of the Ge layer on Ge-on-insulator (GOI) platforms can be significantly improved. During the preparation of GOI substrates, the low-temperature Ge layer with high defect density was removed, resulting in fewer generation/recombination centers in the Ge crystal [16]. In ref [17], the leakage current of Ge p-i-n photodiodes on a GOI substrate with threading dislocation density (TDD) of ~3.2×10^6 cm^{-2} was reduced by 53-fold from one with a TDD of ~5.2×10^8 cm^{-2}. In addition, the introduction of an insulator layer between the Si and Ge can provide better optical confinement for the Ge active layer, enhancing the optical responses of the devices [18]. The direct absorption edge of Ge at 1550 nm limits the application of Ge PD in the C-band (1530–1560 nm) and L-band (1560–1625 nm). In order to achieve high photodetection efficiency in the Ge layer, the optical absorption coefficient must be enhanced by narrowing the direct bandgap energy. An effective approach to narrow the bandgap is to apply enhanced tensile lattice strain. Tremendous efforts have been made to enhance tensile strain in Ge, including the introduction of GeSn [19,20], microbridge structures [21,22], external stressors of SiN$_x$ [23,24], etc. To date, only a few Ge PD photoelectric detection ranges have been extended due to tensile strain on GOI substrates [25–29], and to the best of our knowledge, there is no report on the performance comparison of p-i-n detectors prepared on GOI with different oxide thickness.

In this work, we propose novel methods to improve the p-i-n photodetectors on the GOI substrate. Compared to the PD with low-temperature Ge on the Si substrate, the responsivity of the GOI detector is remarkably improved from 0.32 to 0.5 A/W under 1550 nm wavelength, and from 0.54 to 0.9 A/W under 1310 nm wavelength, increases of 56% and 67%, respectively. The measurement results show that the GOI PDs have lowest dark current of 2.5 nA because of better crystal quality. We also found that the optical performance of the GOI PD could be remarkably improved by integrating a tetraethylorthosilicate (TEOS) layer on the oxide side due to the better optical confinement and resonant cavity effect. We have realized the fabrication and experimental verification of ultra-low dark current and high responsivity GOI photodetectors, for which the detection range can be extended to 1630 nm. This method provides a good foundation for the realization of single-chip optoelectronic integration on large wafers in the future.

2. Materials and Methods

In this study, the Ge layers were deposited on p-type Si (001) 200 mm wafers in a reduced pressure chemical vapor deposition (RPCVD) reactor (ASM Epsilon 2000, Almere, The Netherlands). Ge epitaxy (epi-Ge) was performed in two steps, at 400 °C (low temperature or LT) and 650 °C (high temperature or HT), followed by an annealing of 820 °C for 10 min. The TDD was estimated to be 2.79×10^7 cm^{-2} in the Ge layer. The top 700 nm Ge layer was boron-doped to form P$^+$-Ge. The growth parameters were carefully optimized to avoid dopant precipitates in the Ge layer [30]. At the same time, a thermal oxide layer of 523 nm was formed on the handle silicon wafer. Later, a 10 nm Al$_2$O$_3$ layer was deposited on the epi-Ge wafer to increase the adhesion for bonding. At this stage, two groups of samples were prepared for bonding; one group was directly bonded to the oxide wafer, but the other one with the TEOS layer was thinned to ~300 nm by chemical mechanical polishing (CMP). Then, the Si of the bonded wafer was removed through a combination of mechanical grinding and wet etching in tetramethylammonium hydroxide (TMAH) solution to form the GOI wafer. Afterwards, a CMP process was applied to remove the defected LT-Ge layer on the top. In these GOI wafers, the final Ge thickness was ~2 µm but the oxide thickness for the sample with no TEOS was 523 nm, whereas for the other one it was ~800 nm. In our experiments, the grown Ge layer on Si was considered as the reference sample.

Finally, a 100 nm thick top n-type Ge layer was formed by ion implantation twice at a dose of 1×10^{15} cm^{-2} and an energy of 18 keV. After each implantation, an annealing treatment at 500 °C in hydrogen for 60 s was performed. The doping level of the n-type layer was estimated as ~2×10^{20} cm^{-3}. The dopant profile of the PIN structure was evaluated by secondary ion mass spectroscopy (Figure 1a).

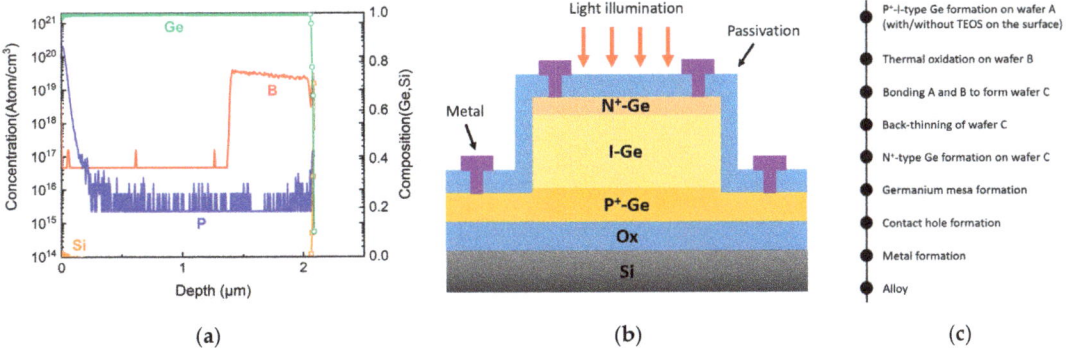

Figure 1. (a) SIMS data, (b) cross-section schematic, and (c) main process flow of GOI p-i-n photodetectors.

Pixels with diameter sizes of 10, 20, 40, 60, 80, and 100 µm were defined and etched down to p-type Ge (bottom contact), followed by a 300 nm thick SiO$_2$ deposition as a passivation layer by plasma-enhanced chemical vapor deposition (PECVD). After contact, electrode holes were formed by dry etching and wet etching, a 10 nm Ni layer was deposited and rapid thermal annealing (RTA)-treated at 450 °C to form NiGe. The NiGe reduced the contact resistivity to 1.3×10^{-5} Ω·cm^2, which was one of the important factors for low dark current in our detectors. Ni is a better choice compared to Co and Pt, which require higher annealing temperatures which cause dopant diffusion in the detector structure [31–33]. Later, a Ti/TiN/AlCu stack with thickness of 50 nm/10 nm/400 nm was deposited and etched to form metal electrodes. A schematic of the processed detector structure is shown in Figure 1b, and the main process flow of the GOI detector is shown in Figure 1c.

Figure 2a displays the cross-sectional transmission electron microscopy (TEM) (Thermo Fisher Talos, Brno, Czech Republic) image of the entire GOI photodetector with the diameter of 100 µm. An enlargement highlights the lattice arrangement of the Ge layer in Figure 2b. The crystalline quality of Ge layer was high and no obvious threading dislocations were observed, because in our process the defected LT-Ge layer was totally removed, and only the high-quality HT-Ge layer remained. The selected area diffraction (SAD) result of the Ge intrinsic layer is shown in the inset of Figure 2b, which has a very regular arrangement of spots, indicating excellent crystal quality. Figure 2c is an enlarged TEM image at the interface, showing a flat interface with no dislocations between the Ge/Al$_2$O$_3$/SiO$_2$ layers. The materials and processes used for device fabrication can be implemented in a standard CMOS process flow.

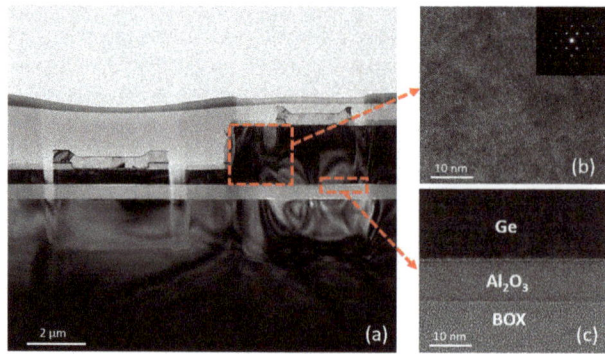

Figure 2. (**a**) Cross-sectional transmission electron microscopy (TEM) image of GOI p-i-n photodetectors. (**b**) High-resolution TEM image with its selected area diffraction (SAD) image (the inset) in the Ge layer, and (**c**) the Ge/Al$_2$O$_3$/SiO$_2$ interface.

3. Results and Discussions

3.1. Dark Current

The p-i-n PDs were characterized with respect to their electrical properties. The dark current of the photodiode is not only an indication of material quality, but also determines optical receiver sensitivity [10]. The dark current–voltage characteristics of the devices on GOI with various mesa radii were measured with a Keithley 4200-SCS semiconductor parameter analyzer (Cleveland, OH, USA) at room temperature, as shown in Figure 3a. The dark current curves exhibited a remarkable rectifying behavior with a high on/off current ratio near 10^8 between 1 and −1 V. Figure 3b displays the dark current comparison of GOI PDs and Ge-on-Si PDs with diameters of 10 μm and 100 μm. For Ge-on-Si PD with 10 μm diameter, the dark currents are 6.4, 15.6 and 30.1 nA at −1, −2 and −3 V. The detector on GOI with TEOS with diameter of 10 μm exhibited dark currents as low as 2.7, 4.7, and 6.6 nA at −1, −2, and −3 V. The detector on GOI without TEOS with a diameter of 10 μm exhibited dark currents as low as 2.5, 3.8 and 5 nA at −1, −2 and −3 V. The dark currents of GOI PDs without TEOS were 7.8, 27 and 100 nA for diameters of 20, 40, and 80 μm at −1 V, respectively. The defects/dislocations in the Ge layer increased the trap-assisted tunneling (TAT) leakage current, and carrier tunnels through the center of the Shockley–Read–Hall (SRH) under a relatively high reverse bias [34]. The GOI PDs exhibited ultra-low dark current because of the absence of a defective region compared to Ge-on-Si PDs.

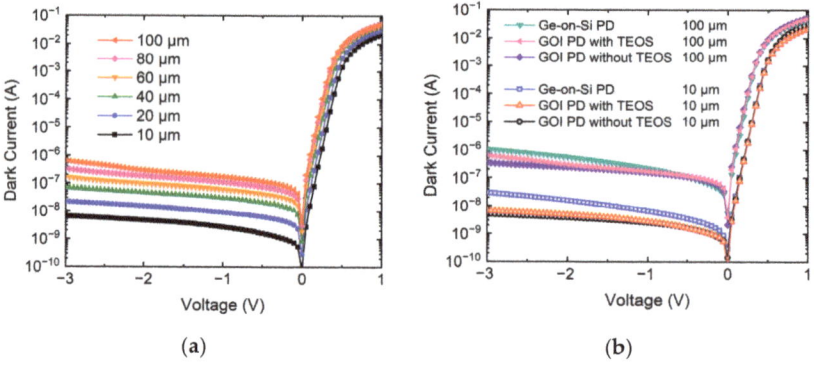

Figure 3. Dark currents of (**a**) GOI detectors with diameters of 10, 20, 40, 60, 80, and 100 μm, and (**b**) detectors on three types of substrates with diameters of 10 and 80 μm.

The dark current density (J_{total}) can be divided into the bulk leakage current density (J_{bulk}) and the surface leakage density (J_{surf}) using the following equation [6]:

$$J_{total} = J_{bulk} + 4\,J_{surf}/D \qquad (1)$$

Figure 4 displays the total dark current densities (J_{total}) of different photodetectors at −1 V versus 1/D, where D is the mesa diameter of the device. Table 1 shows the dark current density comparison of Ge-on-Si PDs and GOI PDs. Compared with Ge-on-Si PDs, both surface leakage current and bulk leakage current of GOI PDs have been significantly reduced. The J_{bulk} and J_{surf} values of the GOI detector without TEOS extracted from Figure 4 were as low as 1.79 mA/cm^2 and 0.34 µA/cm, respectively. This low bulk dark current density of 1.79 mA/cm^2 is one of the lowest reported dark current density values among the Ge p-i-n photodetectors [35,36], which confirms the excellent Ge crystal quality. The J_{surf} of 0.37 µA/cm indicates excellent surface passivation, resulting in lower surface leakage current.

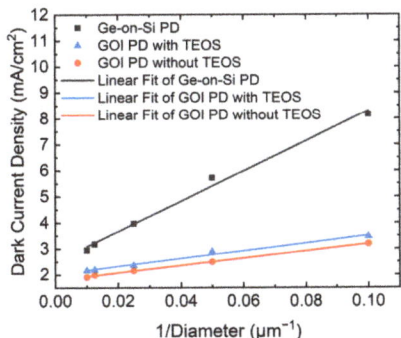

Figure 4. Dark current density (J_{total}) versus 1/D of PDs at −1 V reverse bias.

Table 1. Dark current density comparison of PDs on three types of substrates.

Dark Current Density	Ge-on-Si PD	GOI PD with TEOS	GOI PD without TEOS
J_{bulk} (mA/cm^2)	2.50	2.02	1.79
J_{surf} (µA/cm)	1.45	0.37	0.34

3.2. Responsivity

The photo currents of GOI photodetectors were measured at room temperature using an Agilent B1500A semiconductor parameter analyzer, a probe station, a laser with a wavelength of 1310 nm, and a tunable laser (1500–1630 nm). The photocurrents of the three types of detectors, which are Ge-on-Si PDs, GOI PDs without TEOS, and GOI PDs with TEOS, were measured, respectively. The thickness of the intrinsic Ge layer of all detectors was almost the same (~1.2 µm). The laser output was measured by a calibrated commercial reference detector. The incident light was coupled into the detectors through a single-mode fiber perpendicular to the surface. The spot-size of the fiber was about 3–5 µm. The light power was verified to be 1 mW by a calibrated commercial reference detector. The responsivity parameter is defined as follows:

$$R = I_{ph}/P_o = \eta q \lambda / hc \qquad (2)$$

where I_{ph} is the photocurrent, P_o is the optical power incident on the PD, η is the quantum efficiency, q is the electrical charge, λ is incident light wavelength, h is Planck's constant, and c is the speed of light. The responsivity characteristics of different photodetectors under wavelength of 1550 and 1310 nm are shown in Figure 5a. The optical responsivity of

GOI PD without TEOS at 1550 and 1310 nm was 0.43 and 0.7 A/W at −1 V, corresponding to the external quantum efficiencies of 34.4 and 66.3%, respectively. For GOI PD with TEOS, the responsivity at 1550 and 1310 nm was 0.5 and 0.9 A/W at −1 V, corresponding to the external quantum efficiencies of 40 and 85.2%, respectively. The saturation of the optical responsivity values at 0 V bias revealed that the photodetector configuration allowed a complete photogenerated carrier collection without bias. The responsivities of the GOI PDs with TEOS at 1550 nm were 0.50, 0.47, and 0.45 A/W at −1 V, −2 V, and −3V, respectively. The responsivity at λ = 1310 nm was almost constant throughout the reverse bias region, while the responsivity around the band edge (λ = 1550 nm) slightly decreased with the increasing reverse bias because of the Franz–Keldysh effect (FKE) [37].

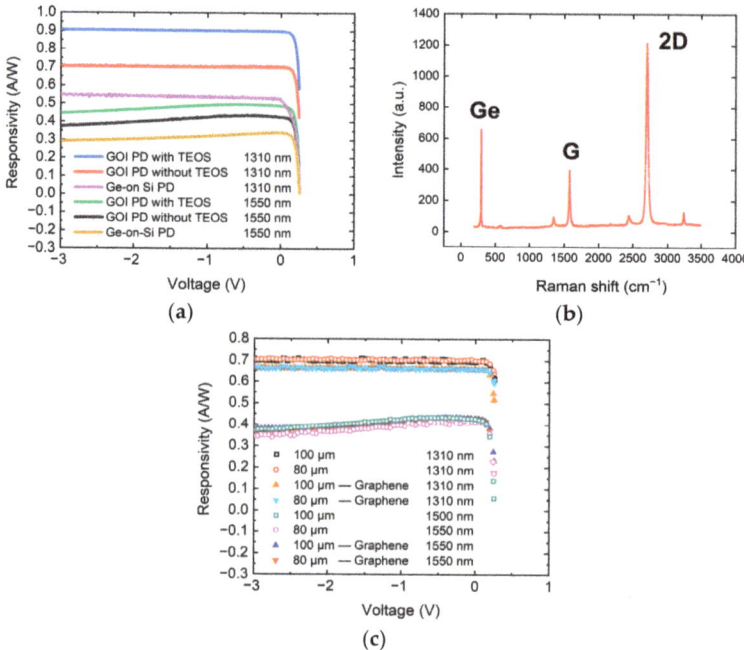

Figure 5. (**a**) Responsivity–voltage characteristics for illuminated PDs with a diameter of 80 μm on three types of detectors. (**b**) Raman spectra of transferred graphene on the detector. (**c**) Responsivity–voltage characteristics for illuminated GOI PDs (without TEOS) with or without graphene on the surface.

In order to improve the performance of Ge PD, an absorption graphene layer was placed on the detector's chip. Raman analysis was performed to confirm the quality of graphene after the transfer process, as shown in Figure 5b. The Ge (≈800 cm^{-1}), G (≈1600 cm^{-1}) and 2D peaks (≈2700 cm^{-1}) with an almost-invisible D peak (≈1300 cm^{-1}) on the SiO$_2$ indicate that the monolayer graphene was transferred successfully. Figure 5c shows the responsivity characteristics of GOI PDs with or without graphene at 1550 nm and 1310 nm, where 80 and 100 μm are the diameters of the detectors. Although the initial idea behind using graphene was to absorb infrared (IR) radiation, in these measurements, the detector with a graphene layer showed no significant improvements in the responsivity values.

In order to study the high-power characteristics of photodetectors, we used a semiconductor parameter analyzer, a probe station, a 1550 nm laser, and an erbium-doped fiber amplifier (EDFA) to measure the photocurrent under different optical powers at room temperature. The laser output was amplified by the EDFA and introduced on the top surface of the photodetectors. Figure 6a shows the saturated photocurrent curves of the GOI PDs without TEOS with 80 μm diameter at a bias voltage from 0 to −2 V. The photocurrent gradually increased until it was saturated as the optical power increased. The saturated photocurrent increased with the increase in the bias voltage, because the intensity of electric field became stronger with the increase in the bias voltage and more photogenerated carriers were brought to the electrode, which eventually led to the saturation of the photocurrent [12]. The photocurrents of GOI PDs without TEOS under 100 mW incident light power were 28, 19.3, and 7.6 mA at bias voltages of −2, −1 and 0 V, respectively.

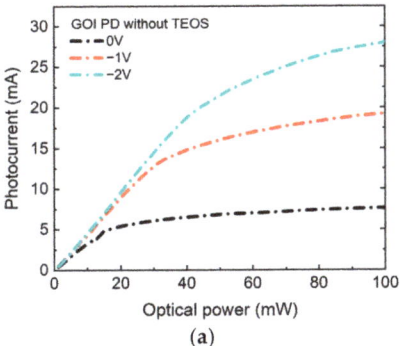
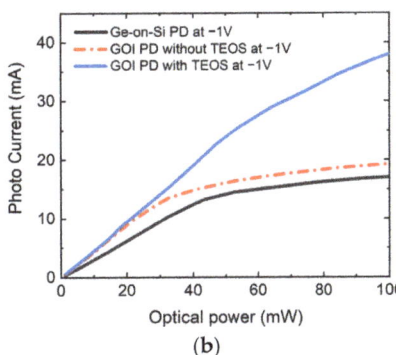

Figure 6. (a) Saturated photocurrent curves of the GOI PD without TEOS with 80 μm diameter at bias voltages of 0, −1, and −2 V. (b) Saturated photocurrent curves of the normal Ge-on-Si photodetector, GOI PD without TEOS, and GOI PD with TEOS at −1 V.

The high-power characteristics of photodetectors on different types of substrates were investigated, as shown in Figure 6b. The black, red, and blue lines represent the saturated photocurrents curves of the Ge-on-Si PD, GOI PD without TEOS, and GOI PD with TEOS with the same diameter of 80 μm, respectively. Compared to the Ge-on-Si PD, the saturated photocurrent of the GOI PD without TEOS was improved from 17 to 19 mA at −1 V. The saturated photocurrent of the GOI PD with TEOS seemed to be higher than 40 mA when the incident light power exceeded 100 mW, which is twice that of the GOI PD without TEOS at −1 V.

Due to the high refractive contrast between Ge(n~4.2), SiO$_2$(n~1.45), and Si(n~3.42), the light propagating in the Ge active layer can experience strong reflection at the Ge/insulator/Si interfaces, achieving better optical confinement in the GOI structure [18]. Due to the resonant cavity effect, the light intensity in the GOI active layer is higher than that of Ge on Si under the same light power irradiation, so the photocurrent of GOI PD is higher than that of Ge-on-Si PD. The schematic illustration of the principle of the dielectric mirror with high and low refractive index layers (Ge/Oxide/Si) is shown in Figure 7a. The thicknesses of Ge and SiO$_2$ are d_1 and d_2. To enhance reflected light inside the Ge layer, A wave and B wave should interfere constructively; this requires the phase difference to be 2π. Thus, d_2 needs to satisfy the following formula [38]:

$$d_2 = m(\lambda/4n) \qquad (3)$$

in which $m = 1, 3, 5 \ldots$ is an odd integer. λ is the free-space wavelength. n is the refractive index of SiO$_2$. When $m = 2, 4, 6 \ldots$ is an even-integer, the A wave and B wave will interfere destructively. For wavelengths of 1550 nm, the calculated d_2 for constructive interference could be 801 nm. The thickness of the SiO$_2$ layer of GOI with TEOS is ~800 nm, which is very close to the calculated result of d_2; therefore, it is more beneficial to form constructive interference in the Ge layer. Furthermore, the GOI with TEOS contained two types of oxide layers, TEOS and thermal oxide. The two-layer oxide structure with slightly different refractive indexes also provided a stronger optical resonant cavity effect for the detector, as shown in Figure 7b. That is, the GOI PD with TEOS had better optical confinement and stronger resonant cavity effect, resulting in its higher photo current than that of GOI PD with only a thermal oxide thickness of 523 nm. Table 2 shows the effect of different SiO_2 thicknesses of GOI PDs on the light waves in the Ge layer. Thus, to obtain higher responsivity under 1550 nm and 1310 nm, the thickness of the oxide layer of GOI PDs should be accurately formed. In addition, the multi-layer structure can be used to realize the enhancement of the resonant cavity effect.

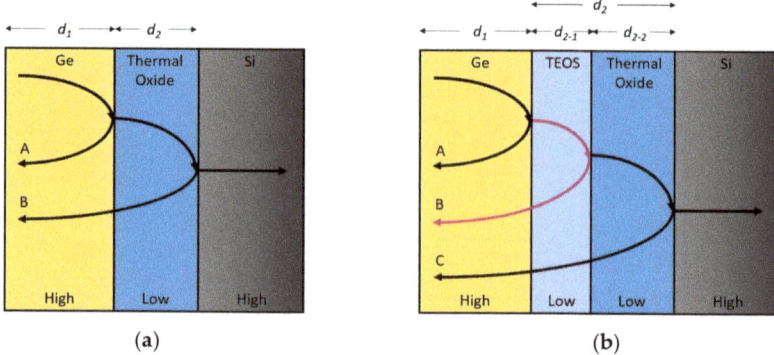

Figure 7. The schematic illustration of the principle of the dielectric mirror with high and low refractive index layers of (**a**) Ge/thermal-oxide/Si, and (**b**) Ge/TEOS/thermal-oxide/Si. Reflected waves A and B interfere constructively if the layer thicknesses d_2 is one-quarter of a wavelength within the layer.

Table 2. The effect of different SiO_2 thicknesses of GOI PDs on the light waves in the Ge layer.

GOI	Oxide Thickness		
m value	1	2	3
effect	Constructive	Destructive	Constructive
1550 nm	267 nm	534 nm	801 nm
1310 nm	226 nm	452 nm	678 nm

3.3. Spectral Response

In order to investigate the broad spectral responsivity of the GOI detector, the spectral response of the photodetector was measured using a Nicolet 8700 Fourier-transform infrared spectrometer (FTIR) (Thermo Scientific, Waltham, MA, USA) with a KBr beam splitter and glow-bar source at room temperature. A commercial InGaAs photodetector was used to calibrate spectrum responsivity. Figure 8a shows the spectral response of the GOI PD without TEOS under zero-bias. The optical responsivities of 0.7, 0.43 and 0.028 A/W at −1 V obtained by laser under 1310 nm, 1550 and 1630 nm are also shown for comparison. The trend of spectral response is consistent with that measured by laser. Compared to the other Ge PDs reported previously [18,26,36], this GOI detector achieved high responsivity in a wide spectral range of 1200~1650 nm. The responsivity spectrum of GOI PD showed strong oscillation structures, indicating the effectiveness of the resonant cavity structure to enhance the responsivity. We strongly believe that the responsivity can be further improved: (i) by engineering the cavity length (the thickness of the Ge active layer and oxide layer); and (ii) by optimizing the device process. Moreover, the time-resolved photocurrent response of the GOI PD with a diameter of 80 μm at −1 V under different incident light powers at a wavelength of 1550 nm is shown in Figure 8b. The consistent and repeatable photocurrent responses are observed without noticeable degradation while the incident light is switched with 5 s period. This indicates that our Ge PDs have low defect density and good performance.

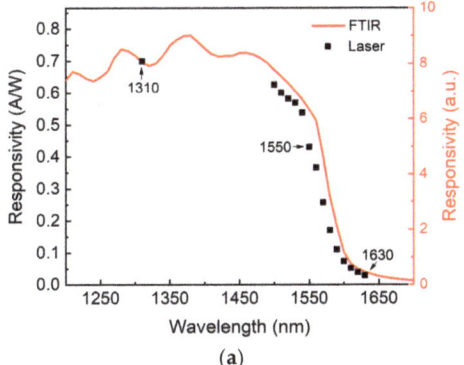
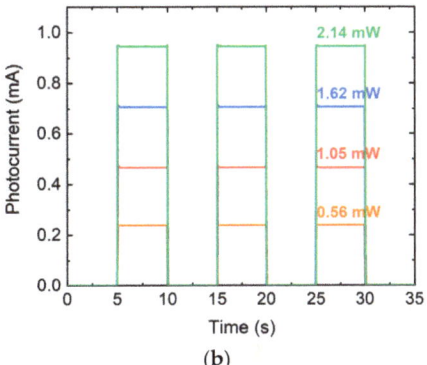

Figure 8. (a) The spectral response of GOI photodetectors between 1200 and 1700 nm, and (b) time-resolved photocurrent of the GOI photodetector at a wavelength of 1550 nm.

Therefore, we conclude that the GOI PDs are promising candidates for telecommunication applications with their extended photodetection range, enhanced optical responsivity, and complementary metal oxide semiconductor compatibility.

For comparison, an overview of the obtained results of the reported Ge-on-Si p-i-n photodetectors are listed in Table 3. To the best of our knowledge, our GOI p-i-n photodetectors proposed in this study exhibit the best comprehensive performance among the reported results.

Table 3. Performance comparison for the top illuminated Ge p-i-n photodetectors.

Year	Active Area (μm^2)	Dark Current at -1 V (nA)	Dark Current Density at -1 V (mA/cm^2)	Responsivity at 1550 nm (A/W)	Responsivity at 1310 nm (A/W)	Quantum Efficiency (1550 nm/1310 nm)	Ref
2005	$\pi \times 2500$	1728 at -2 V *	22 at -2 V	0.56 at -2 V	0.87 at -2 V	45%/82% *	[39]
2008	20×20	6.8 *	1.7	0.2 at -1 V	0.3 at -1 V	16%/28% *	[35]
2009	$\pi \times 25$	102 *	130	0.05 at -2 V	0.2 at -2 V	4%/19% *	[40]
2010	50×50	990	40	0.13 at -1 V	——	10.4%/—	[10]
2013	$\pi \times 100$	120 *	38.3	0.30 at -1 V	——	24%/—	[11]
2017	$\pi \times 36$	5.2	4.6	0.27 at 0 V	0.59 at 0 V	21.6%/55.8%	[36]
2017	$\pi \times 900$	1329 *	47	0.39 at -2 V	——	31.2%/— *	[27] GOI
2019	$\pi \times 25$	4	5	0.12 at -1 V	——	10%/— *	[41]
2020	$\pi \times 56.25$	450	255 *	0.31 at -1 V	0.52 at -1 V	24.8%/49.2%	[12]
2020	$\pi \times 15625$	280 *	0.57	0.28 at -1 V	——	22.4%/— *	[26] GOI
This work	$\pi \times 25$	6.4	8.1	0.32 at -1 V	0.54 at -1 V	25.6%/51.1%	Ge on Si
This work	$\pi \times 25$	2.5	3.2	0.43 at -1 V	0.70 at -1 V	34.4%/66.3%	GOI WO-TEOS
This work	$\pi \times 25$	2.7	3.4	0.50 at -1 V	0.90 at -1 V	40%/85.2%	GOI W-TEOS

The data have been extracted from the references as indicated. * Data calculated using the referenced material.

4. Conclusions

We have demonstrated a GOI substrate with a high-quality strained Ge layer. Optical devices (PDs) were prepared using a standard CMOS process on these substrates. The optical responsivity of the fabricated GOI p-i-n photodetectors with TEOS at 1550 nm and 1310 nm were 0.50 and 0.90 A/W at -1 V, corresponding to the external quantum efficiencies of 40% and 85.2%, respectively. The GOI p-i-n photodetector with both a thermal oxide and tetraethylorthosilicate (TEOS) layer showed the best optical performance due to its better optical confinement and resonant cavity effect. The GOI PDs without TEOS with a mesa diameter of 10 μm exhibited dark currents as low as 2.5 and 3.8 nA at -1 V and -2 V, and exhibited remarkable rectifying behavior with a high on/off current ratio near 10^8 between 1 and -1 V. These high-performance GOI PDs with extended detection range to 1630 nm indicate that the GOI substrates and devices are ideal for telecommunications and Si-based monolithically integrated optoelectronics compatible with the CMOS process.

5. Patents

The patent of resonant cavity substrate structure for improving the responsivity of photodetector and its method has been submitted.

Author Contributions: Conceptualization, X.Z., G.W. and H.H.R.; Data curation, X.Z., G.W., Y.M., H.L. (Haiou Li) and G.G.; Formal analysis, X.Z., G.W., H.L. (Haiou Li), G.G. and H.H.R.; Funding acquisition, G.W., H.L. (Haiou Li), G.G. and H.H.R.; Investigation, X.Z. and Y.M.; Methodology, X.Z., H.L. (Hongxiao Lin), X.L., Z.K., J.S., J.L., W.X., Y.M., H.L. (Haiou Li) and G.G.; Project administration, G.W., H.L. (Haiou Li), G.G. and H.H.R.; Resources, X.Z., H.L. (Hongxiao Lin), Y.D., X.L., Z.K., J.S., J.L. and W.X.; Software, X.Z., Y.D. and Z.K.; Supervision, G.W., H.L. (Haiou Li), G.G. and H.H.R.; Validation, X.Z., H.L. (Hongxiao Lin), X.L. and J.S.; Visualization, X.Z.; Writing—original draft preparation, X.Z.; Writing—review and editing, Y.M. and H.H.R.. All authors have read and agreed to the published version of the manuscript.

Funding: This research was supported by the construction of a high-level innovation research institute from the Guangdong Greater Bay Area Institute of Integrated Circuit and System (Grant No. 2019B090909006) and the projects of the construction of new research and development institutions (Grant No. 2019B090904015), in part by the National Key Research and Development Program of China (Grant No. 2016YFA0301701), the Youth Innovation Promotion Association of CAS (Grant No. Y2020037), and the National Natural Science Foundation of China (Grant No. 92064002).

Data Availability Statement: The data presented in this study are available on request from the corresponding authors.

Conflicts of Interest: The authors declare no conflict of interest.

References

1. Radamson, H.H.; Zhu, H.; Wu, Z.; He, X.; Lin, H.; Liu, J.; Xiang, J.; Kong, Z.; Xiong, W.; Li, J.; et al. State of the Art and Future Perspectives in Advanced CMOS Technology. *Nanomaterials* **2020**, *10*, 1555. [CrossRef]
2. Radamson, H.H.; Simeon, E.; Luo, J.; Wang, G. 2-Scaling and evolution of device architecture. In *CMOS Past, Present and Future*; Radamson, H.H., Luo, J., Simoen, E., Zhao, C., Eds.; Woodhead Publishing: Shaston, UK, 2018. [CrossRef]
3. Michel, J.; Liu, J.; Kimerling, L.C. High-performance Ge-on-Si photodetectors. *Nat. Photonics* **2010**, *4*, 527–534. [CrossRef]
4. Luryi, S.; Kastalsky, A.; Bean, J.C. New infrared detector on a silicon chip. *IEEE Electron. Device Lett.* **1984**, *31*, 1135–1139. [CrossRef]
5. Colace, L.; Gianlorenzo, M.; Galluzzi, F.; Assanto, G.; Capellini, G.; Di Gaspare, L.; Evangelisti, F. Ge/Si (001) Photodetector for Near Infrared Light. *Solid State Phenom.* **1997**, *54*, 55–58. [CrossRef]
6. Yu, H.; Ren, S.; Jung, W.S.; Okyay, A.K.; Miller, D.A.B.; Saraswat, K.C. High-Efficiency p-i-n Photodetectors on Selective-Area-Grown Ge for Monolithic Integration. *IEEE Electron. Device Lett.* **2009**, *30*, 1161–1163. [CrossRef]
7. Huang, Z.; Kong, N.; Guo, X.; Liu, M.; Duan, N.; Beck, A.L.; Banerjee, S.K.; Campbell, J.C. 21-GHz-Bandwidth Germanium-on-Silicon Photodiode Using Thin SiGe Buffer Layers. *IEEE J. Sel. Top. Quantum Electron.* **2006**, *12*, 1450–1454. [CrossRef]
8. Radamson, H.; Thylén, L. Chapter 3—Silicon and Group IV Photonics. In *Monolithic Nanoscale Photonics–Electronics Integration in Silicon and Other Group IV Elements*; Radamson, H., Thylén, L., Eds.; Academic Press: Oxford, UK, 2015; pp. 87–119. [CrossRef]
9. Ni, W.X.; Ekberg, J.O.; Joelsson, K.B.; Radamson, H.; Henry, A.; Shen, G.D.; Hansson, G. A silicon molecular beam epitaxy system dedicated to device-oriented material research. *J. Cryst. Growth* **1995**, *157*, 285–294. [CrossRef]
10. Zhou, Z.; He, J.; Wang, R.; Li, C.; Yu, J. Normal incidence p–i–n Ge heterojunction photodiodes on Si substrate grown by ultrahigh vacuum chemical vapor deposition. *Opt. Commun.* **2010**, *283*, 3404–3407. [CrossRef]
11. Li, C.; Xue, C.; Liu, Z.; Cheng, B.; Li, C.; Wang, Q. High-Bandwidth and High-Responsivity Top-Illuminated Germanium Photodiodes for Optical Interconnection. *IEEE Electron. Device Lett.* **2013**, *60*, 1183–1187. [CrossRef]
12. Li, X.; Peng, L.; Liu, Z.; Liu, X.; Zheng, J.; Zuo, Y.; Xue, C.; Cheng, B. High-power back-to-back dual-absorption germanium photodetector. *Opt Lett.* **2020**, *45*, 1358–1361. [CrossRef]
13. Fang, Q.; Jia, L.; Song, J.; Lim, A.E.; Tu, X.; Luo, X.; Yu, M.; Lo, G. Demonstration of a vertical pin Ge-on-Si photo-detector on a wet-etched Si recess. *Opt Express* **2013**, *21*, 23325–23330. [CrossRef]
14. Chen, H.; Verheyen, P.; De Heyn, P.; Lepage, G.; De Coster, J.; Balakrishnan, S.; Absil, P.; Yao, W.; Shen, L.; Roelkens, G.; et al. −1 V bias 67 GHz bandwidth Si-contacted germanium waveguide p-i-n photodetector for optical links at 56 Gbps and beyond. *Opt. Express.* **2016**, *24*, 4622–4631. [CrossRef]
15. Okyay, A.K.; Nayfeh, A.M.; Saraswat, K.C.; Yonehara, T.; Marshall, A.; McIntyre, P.C. High-efficiency metal-semiconductor-metal photodetectors on heteroepitaxially grown Ge on Si. *Opt. Lett.* **2006**, *31*, 2565–2567. [CrossRef]
16. Lee, K.H.; Bao, S.; Chong, G.Y.; Tan, Y.H.; Fitzgerald, E.A.; Tan, C.S. Defects reduction of Ge epitaxial film in a germanium-on-insulator wafer by annealing in oxygen ambient. *Appl. Mater.* **2015**, *3*. [CrossRef]

17. Son, B.; Lin, Y.; Lee, K.H.; Chen, Q.; Tan, C.S. Dark current analysis of germanium-on-insulator vertical p-i-n photodetectors with varying threading dislocation density. *Int. J. Appl. Phys.* **2020**, *127*. [CrossRef]
18. Ghosh, S.; Lin, K.C.; Tsai, C.H.; Lee, K.H.; Chen, Q.; Son, B.; Mukhopadhyay, B.; Tan, C.S.; Chang, G.E. Resonant-cavity-enhanced responsivity in germanium-on-insulator photodetectors. *Opt Express* **2020**, *28*, 23739–23747. [CrossRef] [PubMed]
19. Radamson, H.H.; Noroozi, M.; Jamshidi, A.; Thompson, P.E.; Ostling, M. Strain Engineering in GeSnSi Materials. *Ecs Trans.* **2013**, *50*, 527–531. [CrossRef]
20. Fang, Y.; Tolle, J.; Roucka, R.; Chizmeshya, A.; Kouvetakis, J.; D'Costa, V.; Menendez, J. Perfectly tetragonal, tensile-strained Ge on Ge1−ySny buffered Si(100). *Appl. Phys. Lett.* **2007**, *90*, 061915. [CrossRef]
21. Sanchez-Perez, J.R.; Boztug, C.; Chen, F.; Sudradjat, F.F.; Paskiewicz, D.M.; Jacobson, R.; Lagally, M.G.; Paiella, R. Direct-bandgap light-emitting germanium in tensilely strained nanomembranes. *Proc. Natl. Acad. Sci. USA* **2011**, *108*, 18893–18898. [CrossRef] [PubMed]
22. Bao, S.; Kim, D.; Onwukaeme, C.; Gupta, S.; Saraswat, K.; Lee, K.H.; Kim, Y.; Min, D.; Jung, Y.; Qiu, H.; et al. Low-threshold optically pumped lasing in highly strained germanium nanowires. *Nat. Commun.* **2017**, *8*, 1845. [CrossRef] [PubMed]
23. Ghrib, A.; El Kurdi, M.; Prost, M.; Sauvage, M.; Checoury, X.; Beaudoin, G.; Chaigneau, M.; Ossikovski, R.; Sagnes, I.; Boucaud, P. All-Around SiN Stressor for High and Homogeneous Tensile Strain in Germanium Microdisk Cavities. *Adv. Opt. Mater.* **2015**, *3*, 353–358. [CrossRef]
24. Jain, J.R.; Hryciw, A.; Baer, T.M.; Miller, D.A.B.; Brongersma, M.L.; Howe, R.T. A micromachining-based technology for enhancing germanium light emission via tensile strain. *Nat. Photonics* **2012**, *6*, 398–405. [CrossRef]
25. Nam, J.H.; Afshinmanesh, F.; Nam, D.; Jung, W.S.; Kamins, T.I.; Brongersma, M.L.; Saraswat, K.C. Monolithic integration of germanium-on-insulator p-i-n photodetector on silicon. *Opt Express* **2015**, *23*, 15816–15823. [CrossRef] [PubMed]
26. Son, B.; Lin, Y.; Lee, K.H.; Wang, Y.; Wu, S.; Tan, C.S. High speed and ultra-low dark current Ge vertical p-i-n photodetectors on an oxygen-annealed Ge-on-Insulator platform with GeOx surface passivation. *Opt. Express* **2020**, *28*, 23978–23990. [CrossRef] [PubMed]
27. Lin, Y.; Lee, K.H.; Bao, S.; Guo, X.; Wang, H.; Michel, J.; Tan, C.S. High-efficiency normal-incidence vertical p-i-n photodetectors on a germanium-on-insulator platform. *Photonics Res.* **2017**, *5*. [CrossRef]
28. Cheng, C.Y.; Tsai, C.H.; Yeh, P.L.; Hung, S.F.; Bao, S.; Lee, K.H.; Tan, C.S.; Chang, G.E. Ge-on-insulator lateral p-i-n waveguide photodetectors for optical communication. *Opt. Lett.* **2020**, *45*, 6683–6686. [CrossRef]
29. Wang, X.; Xiang, J.; Han, K.; Wang, S.; Luo, J.; Zhao, C.; Ye, T.; Radamson, H.H.; Simoen, E.; Wang, W. Physically Based Evaluation of Effect of Buried Oxide on Surface Roughness Scattering Limited Hole Mobility in Ultrathin GeOI MOSFETs. *IEEE Electron. Device Lett.* **2017**, *64*, 2611–2616. [CrossRef]
30. Radamson, H.; Joelsson, K.; Ni, W.X.; Hultman, L.; Hansson, G. Characterization of highly boron-doped Si, Si 1 − x Ge x and Ge layers by high-resolution transmission electron microscopy. *J. Cryst. Growth* **1995**, *157*, 80–84. [CrossRef]
31. Hållstedt, J.; Blomqvist, M.; Persson, P.; Hultman, L.; Radamson, H. The effect of carbon and germanium on phase transformation of nickel on Si1-x-yGexCy epitaxial layers. *Int. J. Appl. Phys.* **2004**, *95*, 2397–2402. [CrossRef]
32. Nur, O.; Willander, M.; Radamson, H.H.; Sardela, M.R.; Hansson, G.V.; Petersson, C.S.; Maex, K. Strain characterization of CoSi2/n-Si0.9Ge0.1/p-Si heterostructures. *Int. J. Appl. Phys.* **1994**, *64*, 440–442. [CrossRef]
33. Nur, O.; Willander, M.; Hultman, L.; Radamson, H.; Hansson, G.; Sardela, M.; Greene, J. CoSi2/Si1-xGex/Si(001) heterostructures formed through different reaction routes: Silicidation-induced strain relaxation, defect formation, and interlayer diffusion*Int. J. Appl. Phys.* **1996**, *78*, 7063–7069. [CrossRef]
34. Eneman, G.; Gonzalez, M.; Hellings, G.; Jaeger, B.; Wang, G.; Mitard, J.; DeMeyer, K.; Claeys, C.; Meuris, M.; Heyns, M.; et al. Trap-Assisted Tunneling in Deep-Submicron Ge pFET Junctions. *ECS Trans.* **2010**, *28*, 143–152. [CrossRef]
35. Colace, L.; Assanto, G.; Fulgoni, D.; Nash, L. Near-Infrared p-i-n Ge-on-Si Photodiodes for Silicon Integrated Receivers. *J. Light. Technol.* **2008**, *26*, 2954–2959. [CrossRef]
36. Liu, Z.; Yang, F.; Wu, W.; Cong, H.; Zheng, J.; Li, C.; Xue, C.; Cheng, B.; Wang, Q. 48 GHz High-Performance Ge-on-SOI Photodetector With Zero-Bias 40 Gbps Grown by Selective Epitaxial Growth. *J. Light. Technol.* **2017**, *35*, 5306–5310. [CrossRef]
37. Schmid, M.; Kaschel, M.; Gollhofer, M.; Oehme, M.; Werner, J.; Kasper, E.; Schulze, J. Franz–Keldysh effect of germanium-on-silicon p–i–n diodes within a wide temperature range. *Thin Solid Films* **2012**, *525*, 110–114. [CrossRef]
38. Kasap, S.; Sinha, R. *Optoelectronics and Photonics: Principles and Practice*, 2nd ed.; University of Saskatchewan: Saskatoon, SK, Canada, 2013.
39. Liu, J.; Michel, J.; Giziewicz, W.; Pan, D.; Wada, K.; Cannon, D.; Jongthammanurak, S.; Danielson, D.; Kimerling, L.; Chen, J.; et al. High-performance, tensile-strained Ge p-i-n photodetectors on a Si platform. *Appl. Phys. Lett.* **2005**, *87*, 103501. [CrossRef]
40. Klinger, S.; Berroth, M.; Kaschel, M.; Oehme, M.; Kasper, E. Ge-on-Si p-i-n Photodiodes With a 3-dB Bandwidth of 49 GHz. *IEEE Photonics Technol. Lett.* **2009**, *21*, 920–922. [CrossRef]
41. Zhao, X.; Moeen, M.; Toprak, M.S.; Wang, G.; Luo, J.; Ke, X.; Li, Z.; Liu, D.; Wang, W.; Zhao, C.; et al. Design impact on the performance of Ge PIN photodetectors. *J. Mater. Sci. Mater. Electron.* **2019**, *31*, 18–25. [CrossRef]

Review

Review of Highly Mismatched III-V Heteroepitaxy Growth on (001) Silicon

Yong Du [1,*], Buqing Xu [1,2], Guilei Wang [1], Yuanhao Miao [1,3,*], Ben Li [3], Zhenzhen Kong [1,2], Yan Dong [1], Wenwu Wang [1,2] and Henry H. Radamson [1,3,4,*]

1. Key Laboratory of Microelectronic Devices & Integrated Technology, Institute of Microelectronics, Chinese Academy of Sciences, Beijing 100029, China; xubuqing@ime.ac.cn (B.X.); wangguilei@ime.ac.cn (G.W.); kongzhenzhen@ime.ac.cn (Z.K.); dongyan2019@ime.ac.cn (Y.D.); wangwenwu@ime.ac.cn (W.W.)
2. Institute of Microelectronics, University of Chinese Academy of Sciences, Beijing 100049, China
3. Research and Development Center of Optoelectronic Hybrid IC, Guangdong Greater Bay Area Institute of Integrated Circuit and System, Guangzhou 510535, China; liben@giics.com.cn
4. Department of Electronics Design, Mid Sweden University, Holmgatan 10, 85170 Sundsvall, Sweden
* Correspondence: duyong@ime.ac.cn (Y.D.); miaoyuanhao@ime.ac.cn (Y.M.); rad@ime.ac.cn (H.H.R.); Tel.: +86-010-8299-5745 (Y.D.)

Abstract: Si-based group III-V material enables a multitude of applications and functionalities of the novel optoelectronic integration chips (OEICs) owing to their excellent optoelectronic properties and compatibility with the mature Si CMOS process technology. To achieve high performance OEICs, the crystal quality of the group III-V epitaxial layer plays an extremely vital role. However, there are several challenges for high quality group III-V material growth on Si, such as a large lattice mismatch, highly thermal expansion coefficient difference, and huge dissimilarity between group III-V material and Si, which inevitably leads to the formation of high threading dislocation densities (TDDs) and anti-phase boundaries (APBs). In view of the above-mentioned growth problems, this review details the defects formation and defects suppression methods to grow III-V materials on Si substrate (such as GaAs and InP), so as to give readers a full understanding on the group III-V hetero-epitaxial growth on Si substrates. Based on the previous literature investigation, two main concepts (global growth and selective epitaxial growth (SEG)) were proposed. Besides, we highlight the advanced technologies, such as the miscut substrate, multi-type buffer layer, strain superlattice (SLs), and epitaxial lateral overgrowth (ELO), to decrease the TDDs and APBs. To achieve high performance OEICs, the growth strategy and development trend for group III-V material on Si platform were also emphasized.

Keywords: III-V on Si; heteroepitaxy; threading dislocation densities (TDDs); anti-phase boundaries (APBs); selective epitaxial growth (SEG)

1. Introduction

As the big data is coming, continuing rapid development of Internet business, communication network moves toward the direction of high speed and large capacity. To meet the data information transmission requirements of efficient, speedy, and integrated data, very large-scale integrated circuits (VLSI) were developed via continuing miniaturization of the transistor characteristic size according to Moore's law [1]. Si is always considered as the backbone material in the micro- and nano electronic industry owing to its natural abundance, high mobility, larger wafer size, low cost, and mature manufacturing technologies, etc. [2]. However, as the device characteristic size reaches to the sub-7 nm technology node, Si based integrated circuits are suffering from the physical and technological limitations in speed, power consumption, integration, and reliability, which further affect the device performance [3]. At present, two main technical roadmaps were expected to prolong the

Citation: Du, Y.; Xu, B.; Wang, G.; Miao, Y.; Li, B.; Kong, Z.; Dong, Y.; Wang, W.; Radamson, H.H. Review of Highly Mismatched III-V Heteroepitaxy Growth on (001) Silicon. *Nanomaterials* **2022**, *12*, 741. https://doi.org/10.3390/nano12050741

Academic Editor: Cesare Malagù

Received: 22 December 2021
Accepted: 17 February 2022
Published: 22 February 2022

Publisher's Note: MDPI stays neutral with regard to jurisdictional claims in published maps and institutional affiliations.

Copyright: © 2022 by the authors. Licensee MDPI, Basel, Switzerland. This article is an open access article distributed under the terms and conditions of the Creative Commons Attribution (CC BY) license (https://creativecommons.org/licenses/by/4.0/).

Moore's law: (I) "non-silicon" high-mobility materials, such as SiGe, Ge, GeSn, GaAs, InAs, and InGaAs, were gradually extended into CMOS technology; (II) Si-based OEICs were proposed to integrate both photonic devices (such as the laser, optical modulator, optical waveguide, and photodetector) and electronic devices (transistors) on the sole Si wafer, which owns the advantages of faster transmission speed, larger transmission capacity, and low power consumption [4].

For high-mobility "non-silicon" materials, group III-V semiconductors can provide higher electron mobility (electron mobility of GaAs and InAs can reach up to 9000 cm^2/(Vs) and 40,000 cm^2/(Vs), respectively), and are ideal channel material for ultra-high speed and low-power devices, such as the high electron mobility transistor (HEMT) [5,6]. For example, to overcome the downscaling limit of conventional CMOS technology, monolithic integrations of various III-V devices, such as the sub−80 nm E–mode InGaAs/InAs HEMTs [7], InP-based HEMT [8], and AlGaN/GaN HEMT [9], have been proposed, enabling dense three-dimensional (3D) integration, low-power consumption, and high-speed applications [10]. On the other hand, for Si-based OEIC, the Si-based light source is the ultimate obstacle to achieve owing to the fact that Si is an indirect band-gap semiconductor material, and its emission efficiency is very low, which makes it unavailable as the active gain medium for Si-based high-efficient light sources. In contrast, most group III-V materials are definitely suitable for the optoelectronic devices in light-emitting/absorbing devices, including light-emitting diodes (LEDs), lasers, and detectors [11–14], owing to their direct bandgap properties, indicating their stronger photon emission and absorption efficiency in comparison than indirect semiconductors such as Si, Ge [15,16], and GeSn [17]. Thus, taking advantage of the excellent properties of III-V compounds, Si-based III-V CMOS devices and III-V photoelectric devices can further greatly improve the data transmission speed and amount, which effectively reduce integrated electricity and power consumption [18].

To realize the monolithic integration of III-V devices on the Si platform, it is critical to develop the heteroepitaxy technique for group III-V materials on Si [19]. Growth challenges for high-quality III-V heteroepitaxy on Si will cause APBs and TDDs/cracks [20,21]. APBs are caused by a polarity difference between III-V material and Si (surfaces for the III-V material and single Si are polar and non-polar), suggesting that it is prerequisite to prevent the formation of APBs. In case the APBs nucleated at the interface between III-V and Si, which can propagate through whole III-V epilayer, this leads to the devices' manufacturing on Si an impossibility [22]. Another important issue is the TDDs, an issue which is attributed to the mismatch of the lattice constant and thermal expansion coefficient between group III-V material and Si. As a result, both APBs and TDDs can lead to surface roughness, which act as the nonradiative recombination centers and leakage current to destroy the device performance [23]. Hence, the defects management strategy was proposed to decrease the TDDs and APBs for group III-V material, thus improving the device performance. Wafer bonding technologies, such as adhesive bonding [24,25], direct bonding [26], and fusion bonding [27,28], were adopted to form the advanced heterogeneous integration substrate platform. However, wafer bonding induces a high manufacturing cost and low integration density [29]. In addition, it is difficult to realize the graphics technology of alignment and passive devices in subsequent processing [30]. In this regard, growing high-quality III-V semiconductors on Si is the key pathway towards monolithic integration of III-V devices on Si-based OEICs.

The purpose of this review article is providing the types of defects and the mechanism of defects formation in silicon-based III-V heteroepitaxy and the detect solution. Particularly, we update recent advances in the epitaxial growth of large lattice-mismatched III-V materials on Si substrates, especially for GaAs and InP, which are both important materials for optic-device applications. This paper is arranged as follows: Section 2 introduces the fundamental challenges in III-V hetero-epitaxy on the (001) silicon wafer, and we also highlight their defect formation mechanism. Section 3 provides a brief review of growth strategies for the defect solution, including the miscut substrate, buffer layer, Strain super-lattice layers (SLSs), Aspect ratio trapping (ART), and epitaxial lateral overgrowth (ELO). Section 4 elaborates on recent approaches on growing high-quality III-V materials on Si. This includes global hetero-epitaxial thin film growth and selective-area hetero-epitaxy. Finally, we summarize the current status and discuss the potential future of III-V-on-Si heteroepitaxy.

2. Basic Challenges of III-V Hetero-Epitaxy on Si (001)

Heteroepitaxial growth represents a growth where materials with different lattice constants are grown in a stacked order, which is usually named "metamorphic growth" [31]. The relaxed lattice constant of the epitaxial layer is generally different from that of the substrate. To grow high-quality III-V layers on Si, fundamental challenges, such as the lattice mismatch, thermal mismatch, and substrate polarity difference, are the main limitations. Figure 1 shows the bandgap (wavelength) and lattice constants (lattice misfit) for the most commonly used group III-V and group-IV materials [32]. Below each semiconductor material, there are also annotation numbers for their own electron and hole mobilities, from which we can see that III-V semiconductor materials own higher electron mobility than Si, which are more suitable for the high mobility CMOS device. Meanwhile, direct bandgap property of III-V semiconductors made it more conducive to optoelectronic devices compared to the indirect gap of IV materials. However, there is a huge challenge to grow the III-V layer on the Si substrate owing to the highly mismatched nature of III-V and Si. In III-V semiconductors, GaAs (4.1%) and InP (8.0%) have close lattice constants to IV relatively, especially the Ge, which are more likely to realize the heteroepitaxy on the Si substrate. In addition, Ge has the close lattice constant and thermal expansion coefficients to GaAs, which are often used as a buffer layer to grow III-V on Si. This huge mismatch can bring out many defects such as: APBs, TDDs, stacking faults. In this section, the definition of mismatch and the mechanism of defect caused by mismatch will be introduced.

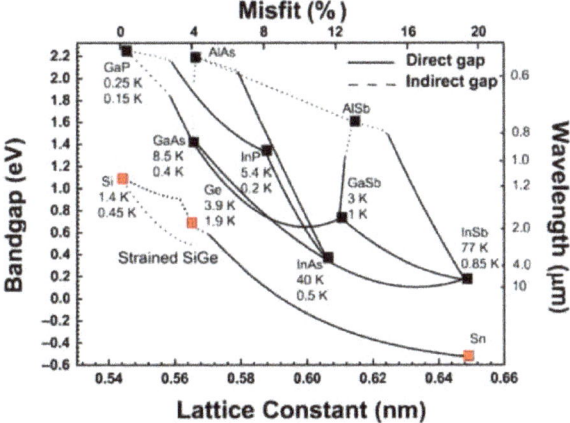

Figure 1. Bandgap (wavelength), lattice constants (lattice misfit), and mobilities for the most commonly used group III-V and group-IV materials. Reprinted with permission from ref. [32]. Copyright 2014 Springer Nature.

Electrical and optical properties of a semiconductor heavily depend on the crystal quality and, hence, defects in the crystal structure. There are several types of defects that can occur in semiconductor crystals, such as structural defects or compositional defects. Considering the spatial extension as a criterion, defects can be classified as 0 D point defects, such as vacancies, 1D line defects, such as misfit dislocations (MDs) or threading dislocations (TDs), 2D planar defects, such as APBs and stacking faults, 3D defects, such as voids. A detailed overview of defects is given by [33,34]. Figure 2 depicts three defect types relevant in this work. Figure 2a depicts a misfit dislocation forming at the interface to compensate for different lattice constants of the materials. Figure 2b shows the APBs' defect. Homopolar III-III or V-V bonds can form due to the atomic steps grown on the non-polar Si substrate, which lead to the formation of APB. Figure 2c is the stacking faults that can occur during the III-V growth. If the stacking sequence changes in every layer, a zinc-blende (ZB) ABC stacking can be switched to a Wurtzite (WZ) ABAB stacking [35,36], which can impact the optical band gap since some semiconductors exhibit different band gaps for different crystal structures [37] or even change the band gap from indirect to direct or vice versa [38,39]. The heteroepitaxial growth of mismatched III/V on Si introduces additional challenges; hence, the mechanisms of challenges and the defect will be discussed below.

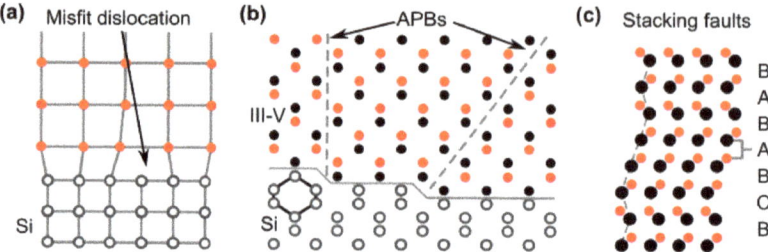

Figure 2. Schematic diagram of three defects (**a**) Misfit dislocation due to lattice mismatch, (**b**) APB at atomic steps of the substrate, (**c**) Stacking faults in the III-V material.

2.1. Lattice Mismatch

One important source of strain in heteroepitaxy is the difference in the lattice constant between different materials, referred to as lattice mismatch. This mismatch introduces strain in the epitaxial layer since it is forced to adapt to the lattice constant of the substrate when it is being deposited on. Eventually, after exceeding a critical thickness, the energy stored as strain will become so huge that the layer will relax. For example, at room temperature, the lattice constants of Si and GaAs were 0.543 nm and 0.565 nm, respectively, and the lattice mismatch was 4.1%. The strain in the heteroepitaxial layer resulting from mismatch is given by [4]:

$$\alpha_m = \frac{\alpha_s - \alpha_0}{\alpha_0} \quad (1)$$

where α_m is the mismatch strain in the epilayer; α_o and α_s are the substrate and overlayer lattice parameters, respectively.

If α_o is greater than α_s, it is a tensile strain; otherwise, it is compressive strain. In an epitaxial layer grown on a foreign substrate, the layer is subjected to biaxial strain in the plane of the substrate (normally the (001), if it is unrelieved, the biaxial strain will translate to a strain in the vertical direction according to:

$$\varepsilon^\perp = \varepsilon^\parallel \frac{1}{R_B} = \varepsilon^\parallel \frac{C_{11}}{2C_{12}} \quad (2)$$

2.2. Thermal Expansion Coefficient Mismatch

Most materials not only have specific lattice constants but also specific coefficients of thermal expansion (CTE). This is highly relevant in heteroepitaxy since epitaxy is normally carried out at a temperature of several hundreds of degrees, which means that the lattices of two different materials will contract to different extents upon cool-down. Going from growth temperature to room temperature, there will be an amount strain introduced in the epitaxial layer according to [40]:

$$\varepsilon_{th} = \int_{T_G}^{RT} \alpha_S - \alpha_0 dT \quad (3)$$

where α_s and α_0 are the thermal expansion coefficients of the substrate and the epitaxial overlayer, respectively, and T_G the temperature at which growth takes place. Since the grown layer is normally more or less relaxed during growth, the introduced thermal strain may lead to formation of dislocations.

When III-V thin film is deposited on a thick substrate, the layer will undergo a formation of misfit dislocations and threading dislocations. After the growth is completed, during the process of lowering the temperature of the wafer, the difference in CTE between the two causes the shrinkage ratio of the two materials to be different, resulting in thermal strain. We assume that during the growth process, it is completely relaxed. After the wafer is cooled down to room temperature, larger CTE (III-V materials) causes greater contraction than the Si substrate, so tensile strain is generated in III-V epitaxy. Nevertheless, the strain caused by the thermal expansion mismatch can be solved through buffer thickness. However, the thermal cracks emerge easily if a thick buffer accumulates too much strain energy when temperature changes. For example, CTE of GaAs (6.6×10^{-6} K^{-1}) is larger than Si (2.3×10^{-6} K^{-1}); the thermal mismatches between Ge, GaAs, and Si are 103%, 105%. Thickness for III-V films on Si is typically below 10 μm [41]. Therefore, huge thermal strain is generated in the thick III-V layer when the temperature drops to room temperature, resulting in thermal cracks through the III-V epitaxial layer. Similar to other defects, the presence of thermal cracks introduces destructive effects on the quality of the III-V epilayer and performance of optoelectronic devices, such as light scattering centers, the electrical leakage path, and a limitation on the total thickness of the epilayer [19]

2.3. Anti-Phase Boundary

Most materials have their own crystal structure and surface primarily. The V group (Si, Ge) has a diamond crystal structure, while III-arsenides and III-phosphides have a zincblende crystal structure which makes the different types of atomic stacking. For example, the diamond crystal structure has its ABAB... atomic stacking, but the zincblende crystal structure has its ABCABC... atomic stacking. When the III-V layer is grown on the Si substrate, the different types of atomic stacking make the APB defect formation, which arises from the polar on nonpolar nature of the III-V/Si heteroepitaxy and monatomic step of the (001) Si surface [42]. For instance, in the GaAs zincblende structure without defects, Ga atoms should be alternately connected with As atoms. Once the coordination of some atoms in the structure changes so that Ga atoms are no longer connected with As atoms, a two-dimensional structural defect will be formed at the interface where the changes occur, named APB. APBs arise as the existence of steps with odd atomic thickness on the surface of element semiconductor substrates (Si or Ge) and the uneven coverage of group III or V sources during silicon surface pretreatment [43].

In the process of substrate processing, it is impossible to obtain the (001) substrate with a perfect crystalline orientation. In this way, there are certain atomic steps on the actual substrate surface, which is a general monatomic layer height. The causes of APBs are shown in Ge substrate epitaxial GaAs. In the metal-organic chemical vapor phase epitaxy (MOCVD) system, arsenide (As) is pretreated with an arsenic atom (ideal) to grow GaAs on the Ge substrate (001) with the mono-atomic step surface. Due to the presence of the monatomic Ge step, the As atom and Ga atom are arranged alternately in the direction of (001), and GaAs interface with two orientations, and the As-As bond and Ga-Ga bond appear above the step, forming APBs. Figure 3a shows the single-layer steps (or odd layer height steps) to produce two domains in the III-V overlayer with opposite sub lattice allocation, whereas double-layer (or even-numbered) steps do not [44]. Although APBs do not involve partial dislocations, they can still interact with TDDs [45]. APBs are regarded as the non-radiative recombination centers for the optoelectronic devices, which will reduce the life of a few carriers in the device, and increase the scattering of most carriers, thus affecting the device performance. To characterize the influence of APBs on the optical properties, photoluminesce quenching and spectral broadening were usually adopted [46].

Besides, the APBs defect can be observed under SEM or AFM. As an example, irregular and curved boundaries were clearly observed for the SEM image of the as-grown GaAs/Si(100) sample (Figure 3b) [47]. APB is a plane defect, which can prevent the manufacture of Si-based III-V devices. Therefore, achieving APB-free III-V/Si heteroepitaxy is a fundament for following III-V devices' fabrication.

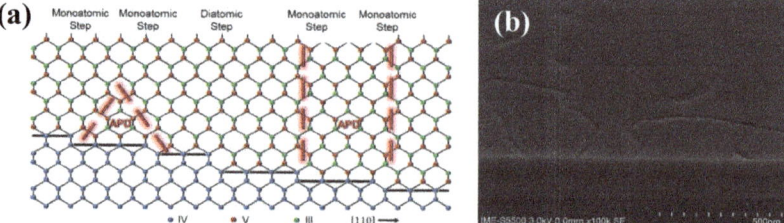

Figure 3. (**a**) (Color online) Schematic showing nonpolar/polar interface between Ge and GaAs. Monoatomic steps on the Ge surface result in APBs, planes of As-As, or Ga–Ga bonds. The APD can either self-annihilate (left) or rise to the surface (right). Diatomic steps on the Ge surface (center) do not result in APD formation. Reprinted with permission from ref. [44]. Copyright 2016 American Vacuum Society. (**b**) SEM plan view images of GaAs/Ge/Si (100) sample with APBs. Reprinted with permission from ref. [47]. Copyright 2021 Springer Nature.

2.4. Threading Dislocation Density

Heteroepitaxy of III/V materials on Si substrates results in the huge strain energy, which is released in the thickness of epitaxy via the formation of MDs along the heterointerface and TDs toward the surface. Thick epitaxy can release the mismatch strain but generates a large number of line defect dislocations. In addition, because of the mismatch TEC of III-V and Si, thick III-V epitaxy also accumulates much strain energy upon temperature cool down, inducing thermal cracks that emerge easily. These thermal cracks case the defects and surface roughness in the epitaxial layer; usually the dislocation density near the interface is as high as 10^9–10^{11}/cm^2 [48,49].

Dislocations are line defects representing a break of symmetry along a line, called the dislocation line, which are defined by a line vector, a Burgers vector describing the distortion of the lattice along the line, and a glide plane on which the dislocation moves. Dislocations can generally be subdivided into edge dislocations and screw dislocations. The fundamental difference between these two dislocation types is that whereas the edge dislocation is perpendicular to the dislocation line vector, the screw dislocation has a Burger vector parallel to the line vector. According to the angle between Burgers and the dislocation line, 90° (edge), 0° (screw), and 60° units are the important dislocations, and the 60° unit is the main dislocation which occurs mostly at the edge of island growth during initial epitaxy. Hence, the defect formation and glide mechanism are discussed. For heteroepitaxy to begin, a two-dimensional film Tc (a few nanometers) was grown on the substrate, allowing plastic relaxation to start. Because of the lager lattice mismatch, TDs will originate from the interface and glide along the slip planes to the surface with the increase in the epitaxy. When many dislocations appear in the same area, dislocation lines are formed by upward extension of multiple obvious dislocations. The entanglement of dislocation lines changes the direction of the dislocation movement. When multiple dislocations are entangled into one, the total number of dislocation lines will decrease, thus reducing the penetrating dislocation generated by upward growth and extension. However, the dislocation entanglement generates new dislocations in different directions, some of which annihilate with epitaxial growth and some penetrate to the surface, increasing the surface dislocation density. In addition, the surface dislocation mainly consists of proliferating dislocation and penetrating dislocation, forming a "dislocation half-loop". Figure 4 shows a sketch of MD formation by the glide of an existing TD from the substrate (I) and by dislocation half-loop formation (II). This "dislocation half-loop" has a great contribution to the strain release [50].

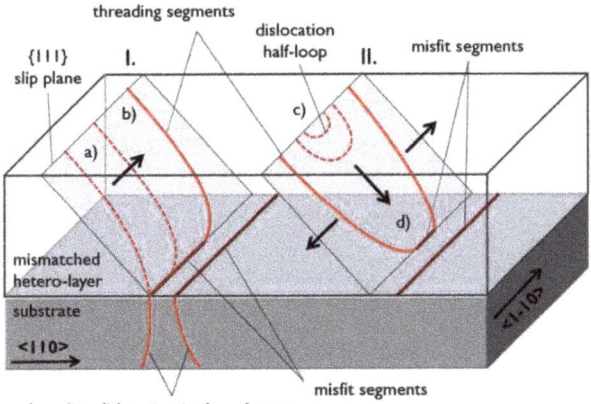

Figure 4. Schematic for the formation of misfit dislocation via threading dislocation glide: (**I**) TDs bend over and glide along the slip planes, (**a**,**b**) and half-loop formation; (**II**) half-loop nucleation at the surface and gliding down to the interface, (**c**,**d**). Reprinted with permission from ref. [50]. Copyright 2018 IOP Publishing.

TDs are one-dimensional crystal dislocations in semiconductor film, which has a serious impact on the properties of semiconductors. The TD is the scattering or absorption center of the carrier or light, which reduces the free path of the electron and greatly reduces the mobility of the carrier. For example, in optoelectronic devices, TDs are the center of non-radiative recombination because the intermediate bandgap energy level in the dislocation core is highly efficient at capturing minority carriers, resulting in a minority load in the material. These defects will form a non-radiative composite center, greatly reducing device lifetime and luminous efficiency. In the case of a semiconductor laser, only a large number of minority carrier reversals are realized in the active layer to obtain an effective gain, and a laser is generated, and it is seen that the reduction in minority carrier lifetime is disadvantageous [51]. In the laser structure, if the minority carrier lifetime is reduced due to dislocations, more injected minority carriers will form a non-radiative recombination before the number of population inversions are sufficient; then, the quality of the laser will fall. Early research work pointed out that for lasers, when the TDD is exceeded, the laser will not work properly due to the reduced lifetime of minority carriers [52]. Therefore, the necessary means to prevent the dislocation from extending upward and reducing TDD in the hetero-epitaxial layer is the main problem of laser fabrication on the basis of the current stage.

TDD is a quantitative parameter which describes the quality of the epitaxial layer. It can be measured by the three common approaches: (1) Etch-pit density (EPD) measurement [53]; (2) X-ray diffraction (XRD) measurement [54]; (3) Transmission electron microscopy (TEM) [55]. In the EPD method, TDD is obtained by calculating the pits at the crystalline region by optical observation or atomic force microscopy (AFM), which is a very easy, quick, and cheap process, but it tends to underestimate the TDD. XRD provides a non-destructive measurement of TDD in the range from 10^5 to 10^9 cm^{-2}. It is possible to calculate the TDD by measuring FWHM of rocking curve widths, because dislocations broaden the rocking curve. TEM measurement enables direct observation of TDs and quantitative analysis in the layer.

2.5. Stacking Faults

Stacking faults (SFs) are planar defects (PDs) representing a disruption in the crystallographic stacking order. In crystals with the Face-Centered Cubic (FCC) type lattice, they normally occur on {111} planes since these have the lowest SF energy. SFs can occur either as an insertion or removal of a crystallographic plane. This may happen either during deposition or by the gliding of a plane from its natural position to another. Joseph et al. [44] investigated the SFs originating from defects or contamination on the surface prior to growth, especially at low T_{sub}, which caused pits on the surface along [110] direction, as shown in Figure 5.

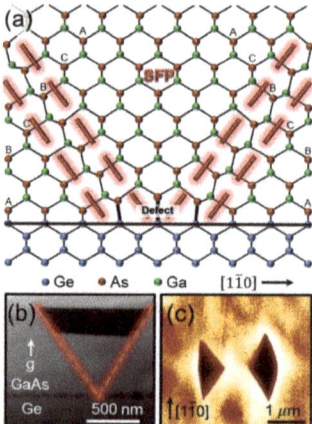

Figure 5. (Color online) (**a**) Schematic down [110] direction showing a SFP that originates from defect or contamination on the Ge surface; (**b**) XTEM with g = 002; (**c**) AFM image for the surface pits. Reprinted with permission from ref. [44]. Copyright 2016 American Vacuum Society.

3. Defect Solution for III-V Hetero-Epitaxy on (001) Silicon Wafer

3.1. Surface Treatment for Si Substrate

The atomic-level Si substrate platform is a basis for the III-V semiconductor devices' manufacture. It is because rough or particle substrates can cause the stacking faults during the heteroepitaxy. To avoid stacking faults, a very clean surface for the Si substrate is very important. The ex-situ process [56] (including cycled HF dip and O_2 plasma treatments) was developed, and film thickness variation (around 0.3 nm) is well reproduced (Figure 6).

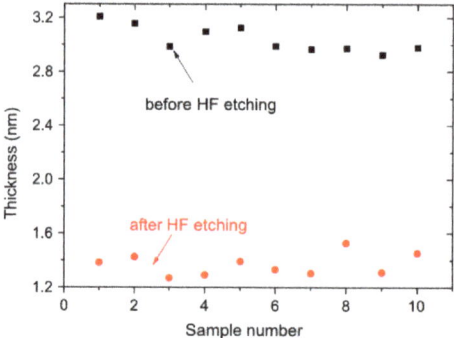

Figure 6. Thickness variation before and after HF 1% bath taken from a single Si substrate. Reprinted with permission from ref. [56]. Copyright 2015 Elsevier BV.

3.2. Process Optimization for III-V Heteroepitaxy Growth

3.2.1. Miscut Si Substrate

There are two main difficulties in heterogeneous growth of silicon-based III-V materials: APBs and TDs. As mentioned in Section 2.3, APBs' defect arises from polar mismatch between the III-V materials and the Si substrate, two alternating (2 × 1) and (1 × 2) dimerization on the monatomic steps of the Si (001) surface. In order to avoid anti-phase disorder in the III/V layer, it is important to nucleate on a (001) Si surface with bi-atomic steps. Double steps on the Si (001) surface are desired in order to suppress APBs in subsequently grown III-V epilayers. At present, it is universally acknowledged that the use of miscut Si substrates with various angles from 2° to 6° is effective in suppression of the formation of APBs [57,58]. A miscut Si substrate can make the Si-Si dimmers parallel to the upper terrace, and the double-atomic steps can form predominantly. The step structures of Si (001) and their energetic were studied theoretically by Chadi [59]. To obtain miscut substrates, thermal treatment is usually adopted to initiate the double-step formation, which was verified on Si substrates with a miscut in <110> directions [60]. The high-temperature treatment in As atmosphere using the miscut substrate can make the surface of the silicon substrate form the diatomic step, existing as (1 × 2) surface reconstruction, and the direction of the As-As dimer or Si-Si is parallel to the direction of surface step. This form is called single-domain surface, which is a stable surface structure. The III-V family layer obtained on this structure is a single-phase structure, thus inhibiting the APBs. However, the formation of a double-atomic step does not always guarantee the APB-free III-V epitaxial layer on Si. To ensure that the Si substrate surface is almost all diatomic steps or only a few single atomic steps, the crucial keys are the diatomic step validation of the Si substrate surface and process optimization. Sakamtoto et al. [61] verified the formation of diatomic steps of the Si surface by high temperature annealing and etched by anisotropy, respectively. Carved and reflection high energy electron diffraction (RHEED) are two ways to prove the Si surface formation of diatomic steps. The mechanism of mono-atomic step transformation to diatomic step transformation on the Si surface under cyclic annealing at high temperature was analyzed by Kawabe [62].

3.2.2. Bulk Hetero-Epitaxial Growth of III-V Thin Films on Si Substrate

TDD is another problem originating from the large lattice mismatch between the III-V and Si substrate. The effective suppression and reduction in TDs can be considered from two directions.

(1). The buffer layer and dislocation barrier layer with a strain field structure are the common method to reduce TD because the strain field generated by them can bend the direction of dislocation extension, thus effectively reducing the penetration depth of dislocation.

Low temperature buffer layer technology is a widely used scheme for heterogeneous epitaxy of silicon-based III-V materials, which can effectively inhibit the generation of dislocation at the interface [63,64]. The low temperature (LT) buffer layer is critical to the quality of III-V materials. III-V materials are generally island nucleated on the Si surface at a low temperature, which is the key factor affecting the nucleation density. When there is high growth temperature of the buffer layer (e.g., 650 °C), nucleation density is small, and large compressive strain causes many defects in the core. In contrast, lowering the temperature decreases the migration performance of surface nucleating atoms and the initial nucleus, thus increasing the nucleation density and reducing the size of the nucleus. The relaxation of the III-V layer at the interface releases strain and reduces the defect density at the top III-V layer [65]. Inserting buffer materials which the lattice constants between Si and III-V groups is an improved method to low TDD. The buffer layer can also be a material with a component gradient or gradual component gradient. For GaAs/Si heteroepitaxy, a wide variety of methods using Ge [66], SiGe [19], GeSnSi [67], InGaP [68] were developed. Among these materials, Ge has been most widely used because of its complete miscibility with Si, well-developed Ge-on-Si growth technology, and nearly the same lattice constant with GaAs [69].

The graded SiGe component buffer layer can effectively disperse dislocation into different component epitaxial layers to obtain a high-quality top epitaxial layer. These buffer layers can provide a high surface base for III-V epitaxial growth owing to the little mismatch between them, which can improve the quality of the III-V epilayer. In addition, the control of initial nucleation conditions of the buffer layer is also the most critical part to obtain high-quality top layer materials. For example, the lattice constant of GaP is very close to that of Si. After obtaining the high-quality GaP/Si materials, the gradient layer $GaAs_xP_{1-x}$ can be used to obtain the transition to GaAs [70]. However, a very thick Ge buffer layer or graded SiGe buffer for III-V growth on Si causes difficulties for interconnection between the III-V Optical device and existing CMOS devices because of huge step height. Therefore, in order to obtain both lower TDD and a thin structural layer, multiplied superlattice layers are introduced.

Strain superlattice layers (SLs) commonly consist of multiple pairs of two lattice-mismatched layers alternately under compression and tension. If the thickness of each SLs' layer is less than a certain critical thickness (30 nm), which otherwise creates misfit dislocations, each SL accommodates elastic strains caused by lattice mismatch. The strain field of SLs can bend over and force the dislocations propagating upward to move laterally toward the edge of the sample, leading to dislocation coalescence and annihilation. Note that the SLs should have enough lattice mismatch and thickness to generate strain required for bending dislocations. SLs are used to filter dislocation, and dislocation density can be reduced an order of magnitude [71] when $In_xGa_{1-x}As/GaAs$ and $GaAs_{1-x}P_x/GaAs$ SLs structures were inserted between the silicon substrate and III-V material. However, SLs will introduce new strain in the epitaxial layer, which will cause mismatching dislocations from the III-V/Si interface to slip and interact, merge, or vanish. These SLs can make the propagating TDs bend over to interfaces and serpentine back and forth between the different superlattice interfaces, which increases the chance of coalescence and annihilation with other dislocations.

In addition, instead of two-dimensional SLs, the self-assembled QDs can be better used as DFLs to decrease TDD. Because the strain-driven self-organized QDs produce a large three-dimensional (3D) strain field around themselves, dislocations around QDs can be bent over and annihilated in a similar way to SLSs' DFLs. Consequently, 3D QDs islands can promote the propagating dislocations to bend over more easily due to the stronger Peach–Koehler forces [72]. Yang et al. [73] proposed and demonstrated the employment of InAs QDs as 3D DFLs in GaAs-based material. Then, Shi et al. [74] reported a four-fold reduction in density of TDs in the InP/Si system by using self-organized InAs/InAlGaAs QDs as DFLs. A number of TDs generated from InP/GaAs and GaAs/Si interfaces propagate toward the top surface, leading to the TDD of 1.3×10^9 cm^{-2}. However, the growth process of quantum dots has relatively high requirements. How to control the growth conditions of quantum dots, or the best quantum dots, is needed to be solved.

(2). During the process of epitaxial growth, controlling and optimizing the growth condition of the epitaxial layer are another method to decrease TDD. Heteroepitaxy growth is a complex process science as it involves issues, e.g., nucleating, temperature, thickness, annealing, so a systematic investigation for III-V heteroepitaxy on Si is necessary. For example, too high initial temperature can induce the forming of 3D islands in in initial nucleation. A high temperature annealing process [75,76] can make defects slip and disappear and too-thick epitaxy and can increase the bow on the wafer [77]. Meanwhile, another measure includes a buffer layer with thermal cycle treatment [78], and other methods have also been developed to decrease the TDD.

Currently, the heteroepitaxy of III-V materials on Si substrates consists of primarily two methods, one is the global area epitaxy and the other is selective epitaxial growth (SEG). Global area epitaxy generally includes silicon-based III-V direct epitaxy of a group of materials and epitaxy using a buffer layer. SEG is a more effective method to reduce

TDD, which can limit the defects in the patterned channel, obtaining a high quality III-V heteroepitaxial layer.

3.2.3. Selective Epitaxial Growth (SEG)

Selective epitaxial growth (SEG) is introduced for the integration of different materials on the same plane or for the realization of high-quality III-V semiconductor layers. This technique is based on a certain purpose graphed Si (or Ge) substrate, locally epitaxial on the III-V layer, graphed as an insulating medium (generally SiO_2). The graphic substrate has the advantages of releasing strain caused by heat mismatch, strong repeatability, and ease of combination with other epitaxial methods, which makes it another promising method. There are two mechanisms for dislocations reduction using graphic substrates: one is aspect ratio trapping (ART); the other is epitaxial lateral overgrowth (ELO).

ART technology is the solution of the epitaxial high-quality III-V family layer in silicon graphics grooves with a height/width ratio greater than 2. It is a method to limit the dislocation at the bottom of the groove by using the limiting effect of the SiO_2 side wall on the dislocation in the groove on the Si graphics substrate through the selection epitaxy. In the groove of this size, the growth plane changes from the original (001) plane to a crystal plane composed of {111} and {113} when the group III-V material was grown [79]. Defects, such as dislocations, also extend along the crystal plane and are limited when dislocations meet the groove insulation wall, thus obtaining a top layer with almost no defects (Figure 7).

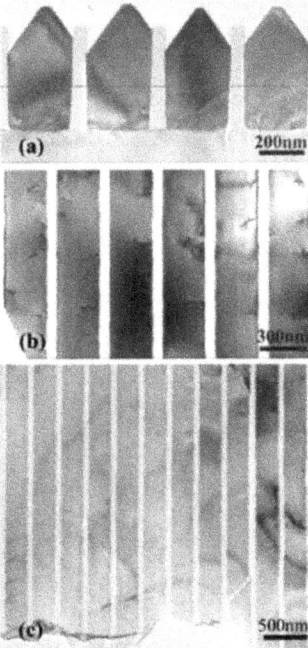

Figure 7. (a) Cross-sectional TEM of silicon-based GaAs in the groove in the direction of [110] (the groove width is 270 nm, and the depth to width ratio is 1.8); (b) the plane TEM image, the defect is near the insulation wall; (c) TEM image of the thinned sample. Reprinted with permission from ref. [79]. Copyright 2007 American Institute of Physics.

ART epitaxy technology has the following advantages: (1) easy to integrate with a variety of high mobility group III-V materials and device structures; (2) it can use the selective epitaxy to achieve the epitaxy growth of Ge materials and group III-V materials between different grooves to achieve the monolithic integration of the Si base; (3) the graphic substrate of the scheme can be prepared by STI (shallow trench isolation) templates in traditional Si-based microelectronic technology, which is convenient for future large-scale integration; (4) the scheme can be directly in the groove to achieve a high-quality group III-V nano-scale on Si, compared to other nano-material preparation methods; the scheme is more convenient for the next generation of Si based high mobility device preparation and application. The scheme can combine the excellent optoelectric properties of III-V group materials with Si, and has great application potential in the future Si-based monolithic optoelectronic integration.

Epitaxial lateral overgrowth (ELO) is a technique developed to overcome the difficulties with obtaining a high-quality epitaxial layer on a foreign substrate. The idea is to use a substrate of a first material with a thin layer of a second material as a starting point. The layer of second material will be full of defects due to the previously outlined mechanisms. On top of this layer, a mask, normally a dielectric such as SiO_2 or Si_3N_4, is deposited, and openings in the mask are defined by lithography and etched. Growth is then conducted selectively in the openings with no nucleation on the mask (shown in Figure 8a). Once the grown material reaches the height of the mask, it starts growing laterally across the mask without nucleation on it, as shown in Figure 8b. In the laterally grown parts, propagating defects such as threading dislocations and stacking faults will be blocked by the mask and consequently cannot propagate into the layer above the mask.

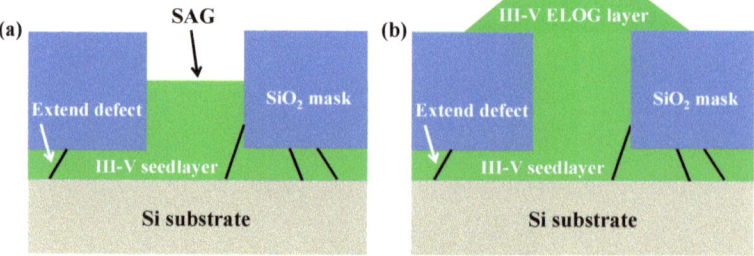

Figure 8. Principle of SAG (**a**) and ELO (**b**) applied to heteroepitaxy of III-V on Si.

It was shown that the angle between the mask openings and the crystallographic direction greatly influences the lateral and vertical growth rates as well as the bounding facet plane [80]. Recently, it was also shown that image forces acting on dislocations close to the mask sidewalls in the openings cause dislocations to bend towards the mask sidewalls, thereby enhancing the filtering effect so that virtually no dislocation propagation though the mask openings occurs [81].

Above all, traditional Si based III-V materials' heteroepitaxy technology described above is still facing a series of problems. For example, the demand of TDD values should be lowered to 10^{-6} cm^{-2}, the RMS surface roughness as low as below 0.5 nm, and compatibility with the traditional CMOS process make it difficult to realize the large-scale integration application of Si-based III-V group devices in the future. Therefore, how to solve the defect of the highly heteroepitaxy mismatch of Si-based III-V group materials is the problem that most scholars are solving at present.

4. Latest Approach of Heteroepitaxy of Si-Based III-V Group Materials

4.1. III-V Thin Films Hetero-Epitaxial Grow on Si Wafer-Scale

Fabricating the optoelectronic devices, a high-quality structure of III-V layers grown on the Si substrate is prerequisite. However, defects such as the APBs and the TDDs, propagating from the heterointerface to the surface, seriously affect the performance of the device. Bulk III-V structure layers are the basis of larger scale optoelectronic devices' fabricating. This section will introduce several methods to achieve TDD-lowering and an APB-free III-V layer on Si substrates, mainly focusing on global epitaxial growth on the bulk Si substrate.

4.1.1. APB-Free of III-V on Miscut Ge/Si Substrates

The APB defect is very obvious, which is derived from the different atomic step between the III-V and Si substrate. However, substrates with a sufficient miscut exhibit a double-stepped terrace structure that can significantly reduce the APB. In 1986, Kawabe [82] grew APB-free of GaAs/Si films on a mis-oriented Si (001) substrate toward (110), which has a better structural quality and luminescence efficiency. Then, the affection of different crystal directions on APB elimination was studied [83,84]. It was suggested that misorientation toward (100) is optimum, since it produces steps in the vertical (110) directions, and this assists the formation of edge-type which is fit dislocations that accommodate the misfit more efficiently [85,86]. Then, the influence of different angles of the Si substrate on the inhibition of APB is also discussed. Wanarattikan et al. [87] grew GaAs layers with two-step growth on miscut Ge (001) substrates mis-oriented by big angles between 4° and 6° towards [110] direction. They found that APBs were limited at the 20–30 nm GaAs/Ge interface, while APBs-free 480 nm GaAs regions can be significantly obtained on the 6° miscut Ge (001) substrates with the RMS of 0.9 nm. A higher quality of GaAs with four times the FWHM of the GaAs epilayer than that grow on the normal Si substrate. However, the large angle substrate is not only difficult to manufacture, but also incompatible with the existing silicon manufacturing technology. Figure 9 shows the model of APBs' generation and self-annihilation mechanisms In Figure 9a, an incomplete pre-layer at the initial Ge/Si surface induces the APBs' generation when III-V epitaxy grows on the axis (001) Ge/Si substrate. The Ga and As atoms can be adsorbed on Ge atoms, forming Ga–Ga and As–As bonds along the [1] direction. Instead, the miscut substrate can offer a short terrace length between steps, which is conducive to APDs' annihilation at an initial growth stage. Figure 9b,c are the model of APDs' formation and annihilation.

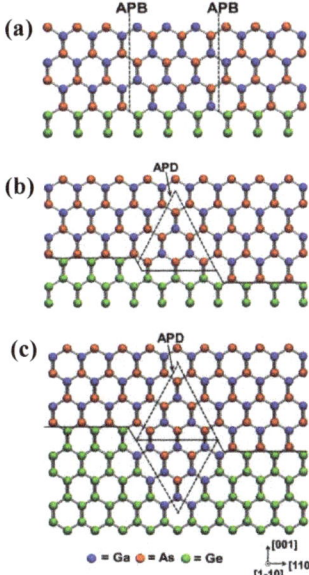

Figure 9. The model of generation and self-annihilation mechanisms of APBs: (**a**) an on-axis Ge (001) substrate and (**b**,**c**) miscut Ge (001) substrates. Reprinted with permission from ref. [87]. Copyright 2015 Elsevier.

Recently, one notably called "exact" Si (001) substrate with a slight mis-orientation (<0.5°) was made to grow APB-free III-V epilayers [88]. Figure 10 shows the GaAs layer was grown on different types of Si substrate. In Figure 10a, a high density of randomly oriented APBs on the GaAs surface with the RMS roughness of 1.6 nm was obtained when grown on a normal Si substrate. Figure 10b shows the "quasi-normal" Si (001) substrate with a 0.15° after the surface preparation procedure by annealing, presenting a 2 × 1 surface structure and predominant double steps. Based on the double steps of the "quasi-normal" Si (001) substrate, a 150 nm GaAs overlayer was deposited. Figure 10c shows the AFM image of the GaAs surface. We can see the APB-free surface of GaAs with a 0.8 nm RMS value. Above all, it is not necessary to use a large miscut substrate; instead, using this "quasi-nominal" substrate can make the GaAs layers more compatible with the existing silicon manufacturing technology.

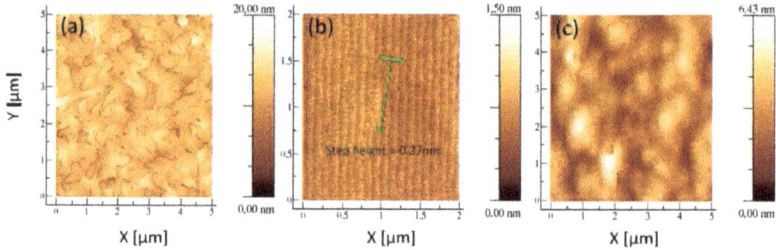

Figure 10. AFM image of (**a**) GaAs grown on un-optimized Si (001): High density of randomly oriented APBs; RMS roughness 1.6 nm; (**b**) 0.15° Si (001) surface after optimized preparation (800–950 °C annealing under H_2). The surface is therefore mainly double-stepped; (**c**) APBs-free GaAs layer grown on optimized 0.15° Si (001): RMS roughness is 0.8 nm. Reprinted with permission from ref. [88].

As for the growth of InP, there is little research on bulk heteroepitaxy due to the large lattice mismatch (8%). M. Grundmann et al. [89] studied the existence or the lack of APB in the InP, providing the information about the presence of single or double atomic steps on the Si, respectively. They found the APBs-free InP could be obtained if it used the 3.8° miscut Si substrate. APBs could be decreased by the miscut substrate, but there were still hillocks on the InP layer.

4.1.2. TDD-Reduction of III-V by Inserting Buffer Layers

TDD is a common problem in heteroepitaxy, which is caused by mismatch of the lattice constant and CTE between different materials. The TDDs extend directly through the epitaxy layer from the interface surface, which greatly affects the performance of devices. For a high-quality III-V layer monolithically grown on Si, achieving a low density of TD is a key issue. In particular, the TDs penetrating an active region of optoelectronic devices significantly degrade their performance.

The forming of TDD begins the initial stage of III-V growth on Si, since the growth begins with the formation of the island on the Si surface. A simple two-step growth has been most widely adopted in III-V heteroepitaxy. The two-step growth starts with low-temperature (LT) growth in the initial stage, then growing the overlayer at typical high temperature (HT). During this growth method, the defects are introduced in the LT step because it can introduce a higher density of islands, which is better for islands coalescing into a continuous layer at HT growth [90]. Although the conventional two-step growth can be used in the process of heteroepitaxy, for Si based III-V heteroepitaxy, the large mismatch of the lattice constant and CET results in a large penetrating dislocation and thermal strain in the epitaxy material. Hence, a wide variety of methods have been extensively studied, including the buffer layer, annealing, three-step growth, and superlattices (SLs). Based on the III part of the growth principle of III-V heteroepitaxy on Si, the big challenge is a large lattice mismatch between them, which induces the quantitative TDD. Therefore, inserting a buffer layer in which the lattice constant and CET are matched with Si and III-V is a popular scheme of Si based III-V heteroepitaxy. This method can effectively suppress the dislocation extension from the bottom to the surface. As we know, Ge is most widely used because of its nearly the same lattice constant and thermal expansion matching between

GaAs and Si. There is much research on optimization of growth process parameters. Yu et al. [64] investigated the growth of GaAs epitaxy on Si substrates with a Ge buffer. Before growing GaAs on the Ge buffer, an arsenic pre-layer was deposited with graded temperature ramped from 300 to 420 °C. Their results display that the TDD of GaAs was significantly reduced by inserting the Ge buffer. They demonstrated a graded-temperature arsenic pre-layer to improve the surface roughness to 1.1 nm, and a low V/III ratio of 20 to suppress the interdiffusion between Ge and GaAs, earning an APB-free GaAs epitaxy with the TDD of 2×10^7 cm^{-2}. Zhou et al. [91] also grew 450 nm GaAs films on miscut Ge-on-Si substrates by MOCVD using a two-step epitaxial method. They found that a 3 nm initial thin buffer layer is critical for the suppression of anti-phase boundaries and threading dislocations. The polishing process is essential to remove the ultrathin LT- GaAs, obtaining a smooth surface for HT-GaAs layer growth. Finally, high-quality GaAs top layers with a low TDD of 2.25×10^5 cm^{-2} and the RMS less than 1 nm were obtained. Figure 11a shows the cross-sectional TEM images of GaAs/Ge/Si. Threading dislocations are restricted at the Ge/Si interface, as shown in Figure 11b. At the same time, heteroepitaxy of GaAs on the Ge surface is not the source for threading dislocation because of the little mismatch between Ge and GaAs. In Figure 11c, APBs were inhibited in the initial thin LT-GaAs buffer layer owing to the double-atomic Ge steps and high temperature annealing (>700 °C) under arsine.

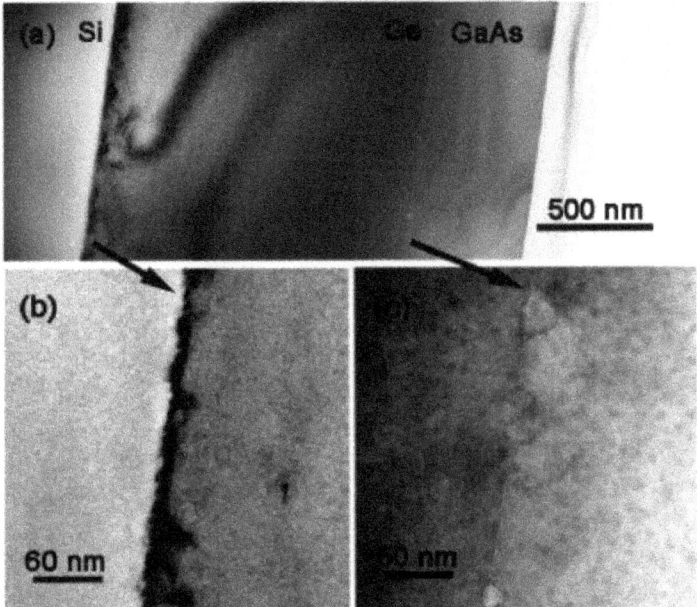

Figure 11. (a) Cross-sectional TEM images of GaAs/Ge/Si; (b,c) are the interface of Ge/Si and GaAs/Ge, respectively. Reprinted with permission from ref. [91]. Copyright 2014 IOP Publishing.

The growth of GaAs is very sensitive to roughness and strain of the buffer layer. Therefore, it is necessary to optimize the Ge buffer layer before III-V epitaxy. Bogumilowicz et al. [77] investigated the effect of the Ge buffer layer with different thickness on the threading dislocations in GaAs epitaxial layers. First, a range of 0.36 and 1.38 μm thickness of the Ge buffer was grown on the miscut Si substrate. The results displayed that increasing the thickness of the Ge buffer results in a decline RMS value of 0.5 nm. Based on this optimized Ge buffer, a smoother 0.27 μm GaAs was obtained with a RMS less than 1 nm and low defect density of 3×10^7 cm^{-2}. However, a thicker Ge + GaAs epitaxial stack produced a linear increase in the wafer curvature, which causes a bow of the substrate. This bow may introduce huge strain inside the wafer, which further deteriorates the surface roughness of GaAs and the following device performance. Figure 12 shows the surface morphology of GaAs layers grown on different thicknesses of Ge-buffered Si (001) substrates. The thicknesses of the Ge buffer layer are: (a) 357 nm, (b) 764 nm, (c) 1377 nm, respectively. From the scale and the crystallographic directions, Figure 12b presents low APB density and surface roughness; the APBs' linear density decreased rapidly as the thickness of Ge changed: 0.4 µm^{-1} for the 357 nm Ge buffer down to 0.1 µm^{-1} for the 357 nm Ge buffer and less than 0.1 µm^{-1} for 1377 nm Ge. Subsequently, Du et al. [92] also confirmed this conclusion on the influence of Ge thickness variation on the TDD of the GaAs epitaxial layer.

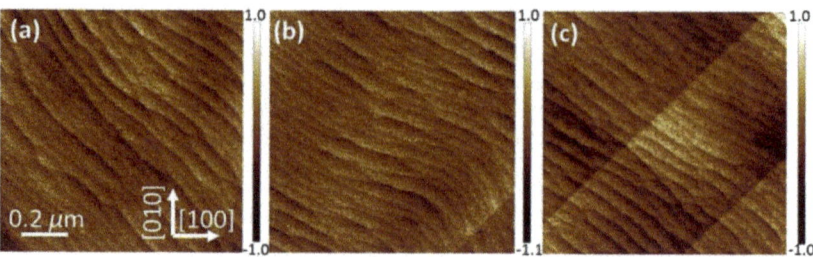

Figure 12. AFM images of GaAs layers grown on Ge-buffered Si (001) substrates. The Ge buffer thickness increased from (**a**) 357 nm up to (**b**) 764 nm and finally (**c**) 1377 nm. Reprinted with permission from ref. [77]. Copyright 2016 Elsevier BV.

4.1.3. TDD-Reduction of III-V by Thermal Annealing

However, the engineered Ge buffer on the Si substrate always exists with large strain, which is difficult for the following GaAs growth. Therefore, the graded $Si_{1-x}Ge_x$ buffer layer was used for GaAs to grow on the Si substrate, owing to offering efficient strain relaxation, and therefore a final Ge cap layer serves as a virtual substrate for GaAs growth. For a graded $Si_{1-x}Ge_x$ buffer grown on Si, a slow increase in Ge composition layers can induce a low number of "glissile" TDs. These effective "glissile" dislocations can increase the segment length of misfit dislocation, which accelerates the strain release. Thereby, the nucleation of new TDs is minimized. Meanwhile, the composition gradient dislocation can bend over and slip during the multilayer and then obtain the upper epitaxial layer with low TDD. Boeckl et al. [93] applied UHVCVD technology to prepare the Ge_xSi_{1-x} buffer layer on the Si substrate, and obtained a GaAs/Si epitaxial layer with a penetrating dislocation density of 10^6 cm^{-2} magnitude. After that, substantial efforts were devoted to achieving artificial $Ge/Ge_xSi_{1-x}/Si$ substrates [94,95]. However, a final Ge layer of composition 100% typically takes 10 µm of epitaxial growth when it is almost fully relaxed theoretically. A thicker buffer layer will not only result in a material consumption, but also be an incompatibility with the small CMOS devices. More important, thermal strain will be introduced during the high temperature ramping down, which increases the roughness of the final product surface [96]. In addition, in order to obtain a smooth surface in rough Ge/GeSi buffers for III-V growth, a chemical-mechanical polishing (CMP) process was used, which can decrease the TDD, but increase the fabrication cost [97].

The thermal annealing (TA) method is indispensable to reduce defect density during growth, enabling thermally activated dislocation migration and thus the annihilation of dislocations. Indeed, the TA-induced reduction of TDD in III-V/Si has been substantially investigated [98–101]. Barrett et al. [101] investigated the post growth annealing (PGA) effect on growing GaAs films on Si (001). He studied the effect of an ex situ post-growth annealing temperature range of 550–700 °C and time on the dislocation density of the GaAs layer. They found that the APB density was reduced ten times when the annealing temperature is above 650 °C. Figure 13 shows the plots of the APB density for different annealing conditions. Obviously, APB density decreases rapidly to a nonzero value after the higher temperature annealing at 650 °C and 700 °C, but for low annealing temperature, the APB density is still large even with a longer annealing time. The mechanism may be explained by the energetics of APB habit planes. High annealing temperature has sufficient energetics to propel the APB slip on {110} type planes [100].

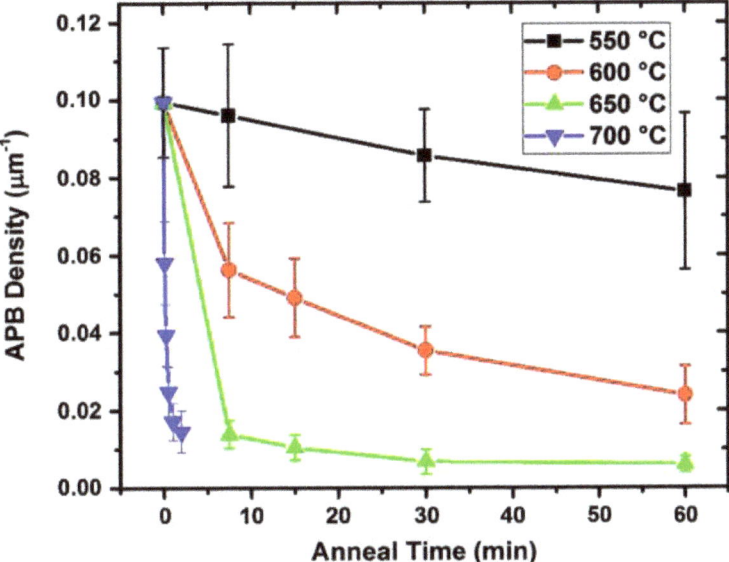

Figure 13. Plot of surface APB densities in the GaAs layer for different annealing conditions. APB density consistently decreases with increasing anneal time and temperature. Reprinted with permission from ref. [101]. Copyright 2019 Springer Nature.

Yet, previously reported annealing temperatures are either thermocouple target temperatures or ambient temperatures in the furnaces. Compared with post annealing, thermal cyclic annealing (TCA) is more conducive to defect elimination and strain relaxation. Callahan et al. [99] investigated the thermal cycle annealing (TCA) effects on the defect reduction in GaAs/Si, and reported that the dislocation density was considerably reduced to 2×10^6 cm^{-c} as the annealing temperature and cycling number increased. His results revealed that the thermally induced stress as a driving force of dislocation motion contributed to the dislocation annihilation, such as coalescence. Meanwhile, based on their numerical analysis, an excellent quality of GaAs layers with a low TDD of 10^5 cm^{-2} would be realized if the thermal cycle annealing is carried out more than 1000 times. Recently, Shang et al. [102] grew a GaAs layer through an in situ thermal cycle annealing (TCA) in the chamber to investigate the effect of TCA times on the reduction in TDD of the GaAs-on-Si template. Figure 14 shows the plot of the TDD with a different TCA process. We can see in Figure 14a that increasing the times of the TCA can reduce the TDD of the GaAs obviously, but a higher TCA of 735 °C enables a minimum TDD of 3.7×10^7 cm^{-2} after 12 cycles of TCA. The mechanism is that times of TCA can prompt the TDs slip, offset or propagate to the edge of the wafer, resulting in a low thermal mismatch strain. However, a higher annealing temperature above 745 °C causes catastrophic degradation of the GaAs surface owing to the formation of a Ga droplet, as shown in Figure 14b. Figure 14c shows the comparison of ECCI images of the surface of GaAs before and after cycles of TCA. It is clearly seen that the TDD was reduced from 4.18×10^8 cm^{-2} to 3×10^7 cm^{-2} after 16 cycles of TCA.

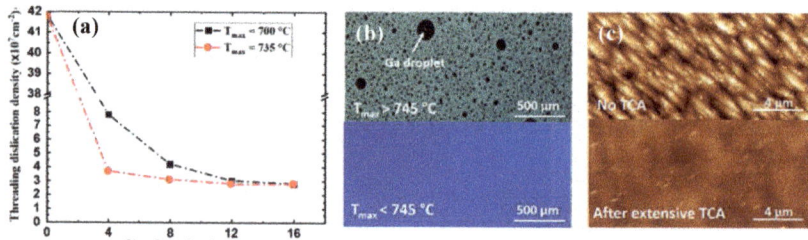

Figure 14. (a) TDD of the TCA process of GaAs-on-Si template. The minimum TDD achievable via TCA at the given GaAs thickness is about 3.7×10^7 cm^{-2}; (b) Nomarski microscope image of the GaAs surface after annealing above and below 745 °C temperature. Gallium droplets are observed when T max is higher than 745 °C; (c) ECCI images of the as-grown GaAs with no TCA (top) and after 16 cycles of TCA (bottom). The TDD is about 4.18×10^8 cm^{-2} with no TCA and decreases to 3.7×10^7 cm^{-2} after extensive TCA. Reprinted with permission from ref. [102].

For the InP/Si, TA has also been applied to improve the crystal quality [103]. However, the effect of thermal annealing on the defect reduction is not as dramatic as in GaAs/Si because the difference in CTE between InP and Si is relatively small; thus, the dislocation motion by thermally driven strain is limited.

4.1.4. TDD-Reduction of III-V by Multi-Step Epitaxial Growth

Multi-step epitaxial growth is a modified conventional two-step growth method, which inserts the Intermediate temperature (IT) layer between the LT and HT layer, and was commonly employed in recent years [104–106]. The two-step growth is a low temperature nucleation layer and high temperature growth layer. The purpose of the low temperature nucleation layer is to ensure sufficient time for the initial three-dimensional fusion to reduce the surface roughness and promote the fusion between the dislocations, thereby limiting the dislocation movement and reducing the penetration depth. However, the instability of the initial nucleation layer in low temperature growth makes harsh growth conditions for the high temperature growth; therefore, it is difficult to grow III-V materials stably with low surface roughness and defect density. Multi-step epitaxial growth such as three-step or four-step, which insert intermediate temperature growth, helps to prevent nuclear island forming in a metastable state from being reconstructed or damaged at high temperature. Wanarattikan et al. [87] studied the effect of the process of two-step growth and one-step growth on GaAs buffer layers using miscut Ge substrates. They designed the two-step growth with: low temperature growth at 470 °C and high temperature growth at 580 °C. Their results presented that compared with the one-step growth process at a temperature of 550 °C, two-step growth of the GaAs process exhibited a lower TDD value by about three times; the lowest APB density is 2.7×10^7 cm^{-2}. In following, a multi-step growth process was also studied. Wang et al. [105] demonstrate the three-step growth of GaAs on the Si (001) substrate in a low-pressure metal organic chemical vapor deposition reactor compared with two-step growth. To decrease the TDD further, TCA was also introduced for comparison. They designed their three-step growth process as: a 70 nm-thick initial LT-GaAs nucleation layer was grown at 420 °C, a 300 nm MT-GaAs epilayer grown at 630 °C, and then a 1.5-µm-thick HT-GaAs epilayer grown at 685 °C. Table 1 is different characteristic data of GaAs/Si samples grown with different procedures. Compared with the two-step growth, the TDD and RMS values of GaAs were reduced obviously by three-step growth. Meanwhile, the combination of three-step growth with two TCA steps can further improve the surface morphology and crystal quality of metamorphic GaAs. A TDD of only 1.1×10^7 cm^{-2}, EPD of 3×10^6 cm^{-2}, and the smallest RMS of 1.8 nm can be obtained via this Combinatorial method.

Table 1. Different characteristic data of GaAs/Si samples grown with different procedures. Reprinted with permission from ref. [105] Copyright 2013 American Vacuum Society.

Samples	Growth Procedure	DCXRD ω−Scan FWHM (arcsec)	RMS Roughness in 10×10 μm^2 (nm)	TDD (cm^{-2})
A1	Two−step growth	327.5	2.9	4.4×10^7
A2	Two−step growth + one TCA step	210.2	2.5	1.9×10^7
B1	Three−step growth	298.2	2.4	3.7×10^7
B2	Three−step growth + one TCA step	184.3	2.0	1.4×10^7
B3	Three−step growth + two TCA steps	164.2	1.8	1.1×10^7

According to the above, although three-step growth can reduce the RMS of the GaAs surface to 1.8 nm, it still cannot meet the requirements of device preparation. In 2021, Du et al. [47] also investigated the three-step growth of GaAs on both 0°— and 6°—miscut Si substrates with an engineered Ge buffer. First, a flatter Ge buffer layer was obtained through CMP, which is more favorable for GaAs growth. The conventional two-step growth process was low temperature at 460 °C, high temperature at 670 °C. The results of the two-step growth displayed a foggy surface of GaAs with the RMS of 3.4 nm. Then, an intermediate temperature at 600 °C was inserted between low and high temperature growth of GaAs to impede the defects to propagate to the HT layer. Figure 15a–d show the GaAs surface morphology of a comparison of the two-step with three-step growth on 0° and 6° miscut Si substrates. The three-step growth process can obviously eliminate the pits (TDs) on both substrates, but APB strips still exist on 0° miscut Si substrates. In other words, APB-free GaAs film with a low TDD of 7.4×10^7 cm^{-2} and RMS of 1.27 nm could be obtained on 6°− miscut Si substrates by three-step growth.

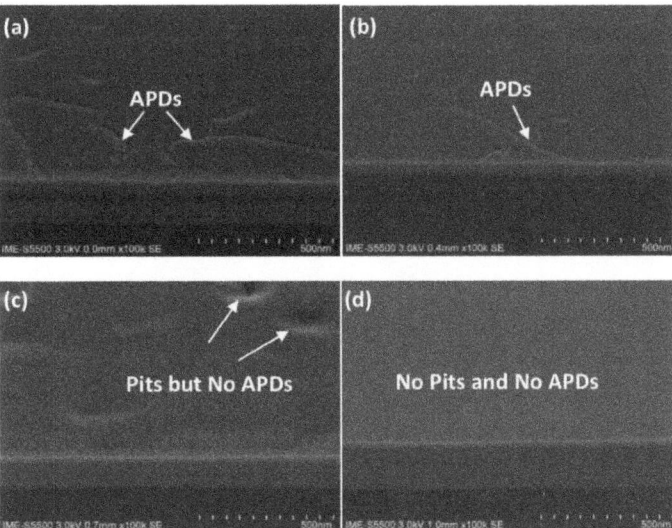

Figure 15. The SEM plan view images of GaAs (a) two-step growth on 0° miscut Si substrates and (b) three-step growth on 0° miscut Si substrates; (c) two-step growth on 6° miscut Si substrate, (d) three-step growth on 6° miscut Si substrate. Reprinted with permission from ref. [47]. Copyright 2021 Springer Nature.

For the InP/Si, direct growth of InP on Si produces a much higher TDD than that of GaAs on Si; a two-step growth is difficult to obtain a flat InP surface [107]. However, our group is developing the high-quality InP epitaxial layer on a 200 mm miscut Si platform using the multi-step growth technique, which has an APB-free InP-on-Si substrate. These breakthrough results will be submitted later.

4.1.5. TDD-Reduction of III-V by Inserting Strained-Layer Superlattices Layer

Recently, strain-layer superlattices (SLs) were employed as dislocation filter layers (DFLs) to filter the dislocations' density by bending the dislocation direction with the strong strain field around the quantum well (QW) or 3D quantum diots (QDs) [108–110]. The detail mechanism of dislocation being filtered and eliminated by SLs is explained in part III. Ternary-binary SLSs DFLs are widely used in III-V/Si heteroepitaxy, including InGaAs/GaAs, InAlAs/GaAs, InGaAs/InP, and so on [111–114]. For instance, Tang et al. [115] compared InAlAs/GaAs and InGaAs/GaAs (SLSs) as dislocation filter layers to grow 1.3 µm InAs/GaAs quantum dot laser structures on Si substrates. Two types of SLSs are designed as: five-period of 10 nm $In_{0.15}Ga_{0.85}As$/10 nm GaAs and five-period of 10 nm $In_{0.15}Al_{0.85}As$/10 nm GaAs, respectively. Figure 16a–c shows the cross-sectional TEM of low magnification of two different SLs layers. We can see that free dislocations of GaAs layers are visible after the insertion of InAlAs/GaAs SLSs in Figure 16b, but a few dislocations are also exist in GaAs layers after the insertion of InGaAs/GaAs SLSs in Figure 16a. TDD reduction of the GaAs after insertion of different types of SLSs was also characterized by TEM and EPD in Figure 16c, respectively. After three sets of InAlAs/GaAs SLSs, the GaAs sample shows an average defect density of about 2.0×10^6 cm^{-2} while the one with InGaAs/GaAs SLSs has about 5.0×10^6 cm^{-2}. In addition, photoluminescence (PL) also verified that the sample with InAlAs/GaAs SLSs is about two times stronger than that with InGaAs/GaAs SLSs, which means that InAlAs/GaAs SLSs are more effective in blocking the propagation of threading dislocations than InGaAs/GaAs SLSs under the similar growth conditions.

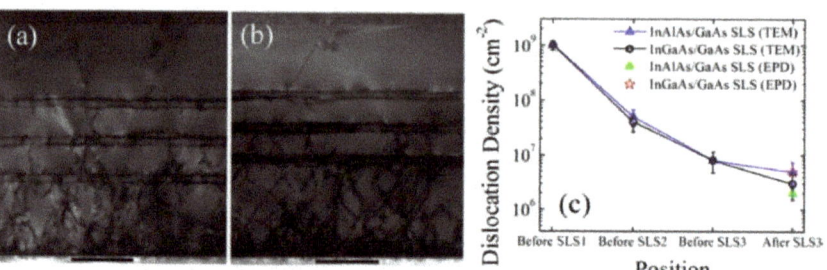

Figure 16. Cross-sectional TEM dark field images of (**a**) InGaAs/GaAs SLS and (**b**) InAlAs/GaAs SLS. (**c**) Reduction in dislocation induced by the SLS layers measured at different positions. Reprinted with permission from ref. [115].

The changing of composition of SLSs materials can affect the band potential barrier, which has an important impact on defects' limitation. Later, Tang et al. [116] investigated the indium composition and thickness of $In_xGa_{1-x}As$/GaAs SLSs for 1.3 µm QD lasers on Si. They designed the efficacy of indium composition x which were 0.16, 0.18, and 0.20, and the thickness of GaAs were 8, 9, and 10 nm. To improve the effectiveness of InGaAs/GaAs DFLs, two different growth methods were introduced in Figure 17a,b. In growth method I in Figure 17a, a GaAs spacer layer was grown during the period of heating up to 610 °C right after the deposition of InGaAs/GaAs SLSs at 420 °C. In contrast, in growth method II, the GaAs spacer layer was grown after in-situ annealing of the SLSs at 610 °C in Figure 17b. In Figure 17c, the PL peak intensity of the QD laser structure with growth method II was at least three times higher than that with growth method I. This improvement can be attributed to the high-temperature growth of the GaAs spacer layer and in-situ annealing of SLSs. It is also revealed that the optimized indium composition and GaAs thickness in SLSs were 0.18 and 10 nm, respectively. In Figure 17d, it was shown that the employment of three sets of $In_{0.18}Ga_{0.82}As/GaAs$ SLSs DFLs effectively blocked and annihilated the TDs. In addition, the UCSB team [113] grew the GaAs layer on the GaP-engineered Si substrate using $In_{0.1}Ga_{0.9}GaAs/GaAs$ strain super-lattices (SLSs). The $In_{0.1}Ga_{0.9}GaAs/GaAs$ SLSs can further reduce the penetration dislocation density in the GaAs buffer layer to 7.3×10^6 cm^2.

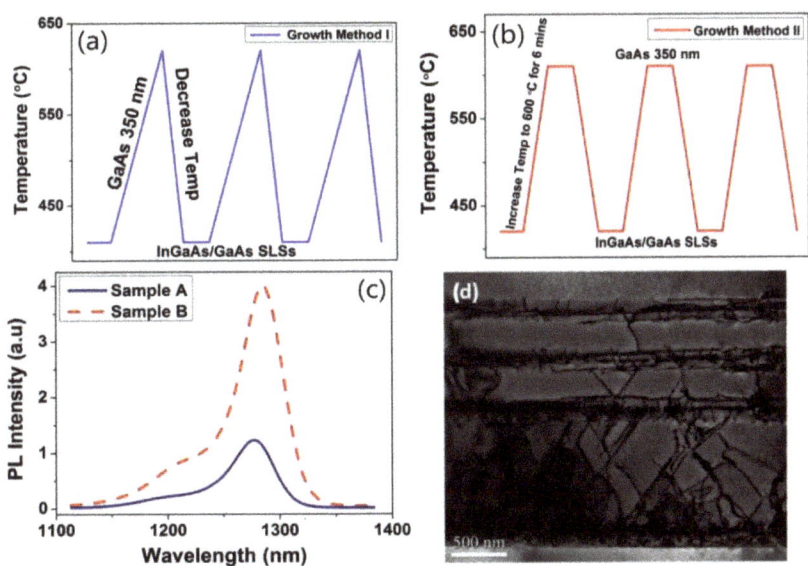

Figure 17. (**a**) Growth method I; (**b**) Growth method II; (**c**) PL spectra measured at room temperature for the two sample; (**d**) Dark-field cross-sectional TEM image of optimized In0.18Ga0.82As/GaAs SLSs DFLs. Reprinted with permission from ref. [116]. Copyright 2016 IEEE.

In the InP/Si platform, the DFLs based on InGaAs/InP, In(Ga)AsP/InP, (In)GaP/InP [117,118] were also commonly adopted. In 2020, Klamkin et al. [107] reported their advanced InP-on-Si virtual substrate which is optimized by inserting $In_{0.73}Ga_{0.27}As$ (13 nm)/InP (19 nm) 10-pair SLSs on the GaAs-on -V-grooved Si (GoVS) template. In this report, InP buffer layers were first grown on the GoVS template using multi-step growth, followed by four sets of InGaAs (13 nm)/InP (19 nm) 10-pair SLSs with 300-nm-thick InP spacer layers. Figure 18a shows the cross-sectional STEM image of the InP-on-Si template and the extracted dislocation density at various growth stages. Six lines are the different growth stages. First, at low temperature growth of InP on the GoVS substrate, many dislocations are visible at the interface of GaAs and InP; the TDD is in the order of 10^{10} cm^{-2}. After the three-step growth of InP (line 2), a large number of TDs are annihilated and coalesced, leading to a reduced defect density of approximately 1.5×10^9 cm^{-2}. In following, a higher set of SLSs is inserted to filter dislocations, which can be seen in the image that the TDs decrease after the multi-SLSs insertion. Figure 18b shows the plot of the TDD value with the various growth stages. The dislocation filtering efficiency is enhanced for the higher set of SLSs; the final InP surface TDD is reduced to 1.17×10^8 cm^{-2} after four sets of InGaAs/InP DFLs (line 6). The final InP surface morphology was also characterized by ECCI in Figure 18c. APB-free and low TD were present, but few SFs and pinholes appear. The counted densities for TDs, SFs, and pinholes were 6.9×10^7 cm^{-2}, 1.1×10^7 cm^{-2}, and 3.5×10^7 cm^{-2}, respectively. Such pinholes are mainly due to the fact that higher SLSs also introduce new dislocations. It was revealed that InGaAs/InP SLSs can obviously reduce the TDD to 10^7 cm^{-2}, but they formed a rough surface with many hillocks.

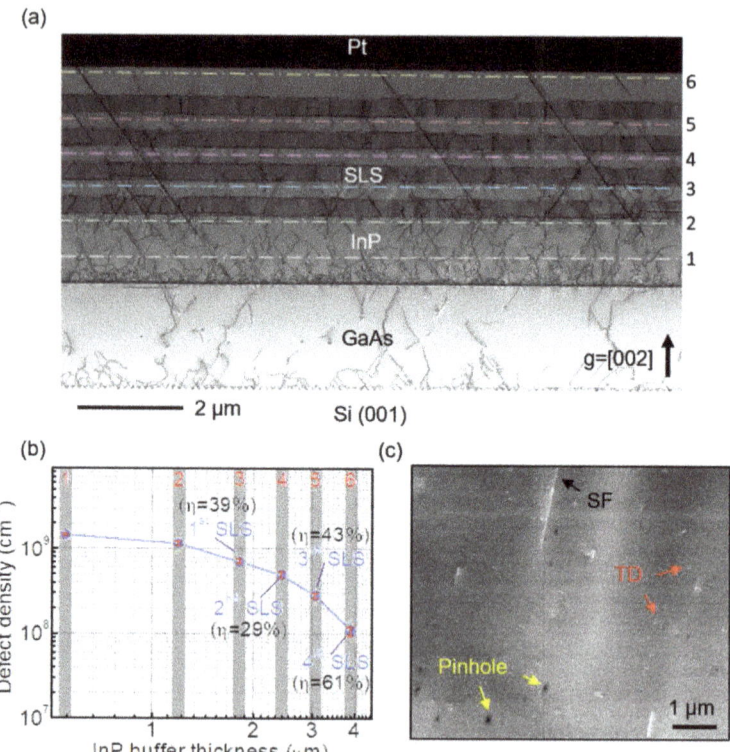

Figure 18. (**a**) Cross-sectional STEM image of the InP-on-Si template to demonstrate the generation and propagation of threading dislocations and stacking faults; (**b**) Extracted dislocation density as a function of the InP buffer thickness at various growth stages. (**c**) Typical ECCI image of the InP surface, where different kinds of defects can be identified and counted. Reprinted with permission from ref. [107]. Copyright 2020 American Institute of Physics.

In addition, the self-assembled QDs can be used as DFLs to filter the TD of the InP/Si layer. Because the strain-driven self-organized QDs produce a large three-dimensional strain field around themselves, dislocations around QDs can be bent over and annihilated in a similar way to SLSs DFLs. Shi et al. [119] grew the InP layer on the GaAs-on-Si substrate by inserting optimized multiple InAs/InP QDs as DFLs. They inserted two periods of five-layer InAs/InP QDs dislocation filters to obtain a smoother surface before the subsequent QD stack growth during the HT-InP layer growth. A RMS roughness of 2.88 nm of a binary InP layer can be obtained, minimizing the generation of large InAs islands. Figure 19a shows the cross sectional of InP on planar Si inserted with two periods of five-layer InAs/InP QD DFLs. The structure of InP and the InAs/InP DFLs layer are observed clearly. Figure 19b shows the effect of InAs/InP DFLs on defect elimination by TEM images. It can be seen that the TDD is bent and eliminated by the first five-layer InAs/InP QD DFLs, but sufficient defects can still propagate upward to the top surface. After the second stage of QD DFLs, very few TDs can be detected, and most of the defects are propelled or pinned by the stacked QDs, leading to either annihilation or coalescence of the TDs. Finally, a low defect density of 3×10^8 cm^{-2} was achieved for the InP-on-Si substrate.

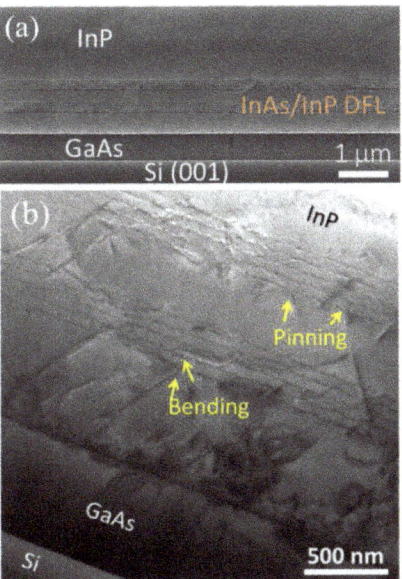

Figure 19. (a) Cross-sectional SEM of InP layer grown on planar GaAs/Si by inserting two periods of five-layer InAs/InP QD DFLs; (b) cross-sectional TEM images of InP grown on planar GaAs/Si by inserting two periods of five-layer InAs/InP QD DFLs. Reprinted with permission from ref. [119]. Copyright 2018 AIP Publishing.

Adopting SLSs' dislocation filter to reduce the TDD also was used with the selective epitaxial technology recently. For instance, Norman et al. [120] obtained the GaAs/Si epitaxial layer by SEG; the $In_{0.15}Ga_{0.85}As$ SLSs dislocation filter was grown on the V-groove graphic substrate. ECCI shows a low dislocation density as 2×10^7 cm^{-2}.

4.2. III-V Thin Films Selective Epitaxial Growth on Si Wafer-Scale

However, the miscut Si substrates are not popular in current industrial process flows because of the high consumption and are incompatible with advanced Si manufacturing technologies. Thus, some researchers explore other methods to reduce the problem of APBs in epitaxial GaAs layers on nominally on-axis Si (001) wafers.

Selective epitaxial growth (SEG), allowing the epitaxial layer to grow on the predefined region by substrate patterning, offers additional control over the strain relaxation process to control the dislocation. Aspect ratio trapping (ART) is the most common method of SEG owing to the simplicity of design. This epitaxial technology, through a high depth-width ratio, limits the dislocation and other defects originating from the Si surface to the bottom of the groove by using the SiO_2 sidewall, so as to obtain high-quality, dislocation-free III-V materials at the top, which greatly reduces the dislocation density in the materials. The ART template can be made by STI technology from traditional CMOS processes, which can realize the monolithic integration of III-V group materials and Ge materials on the Si substrate.

4.2.1. Aspect Ratio Trapping Technology (ART)

The original concept of ART for epitaxial III-V on the silicon substrate was proposed by Fitzgerald in 1991 [121]. At that time, this idea was called the "epitaxial necking effect": They point out that the {111} crystal plane family is the slip plane in the zinc-blende lattice structure, and the TD is mainly along the {111} plane, which develops a 45° with plane (001). So, when the width of the selected area is less than the thickness of the epitaxial material, the TD will reach the edge of the material and terminate. The method was firstly revealed in Ge/Si hetero-epitaxy [122] and then applied to III-V/Si epitaxy. Bai et al. [123] introduced the ART method directly to GaAs epitaxy on silicon. They first deposited SiO$_2$ on silicon with a certain thickness and then etched along the [110] direction to reach a certain surface of the silicon substrate width of grooves. The SiO$_2$ side wall of the groove limits the development of TD from the GaAs-silicon interface; part of GaAs has almost no defects, as shown in Figure 20. In Figure 20a, a lot of dislocations are visible at the interface of GaAs/Si, and gradually limited within the SiO$_2$ trenches, then completely terminated within the first 200 nm of GaAs growth. The schematic of initial coalesced GaAs growth and coalesced GaAs growth is shown in Figure 20b.

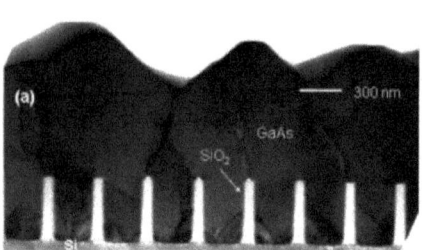
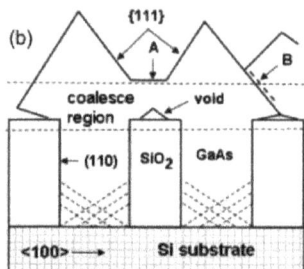

Figure 20. (a) Cross-sectional TEM image of GaAs grown on ART patterned Si; (b) coalesced GaAs grown and the schematic illustration of coalesced GaAs growth. Reprinted with permission from ref. [123]. Copyright 2008 American Institute of Physics.

Although the above SiO$_2$ mask can limit the development of TDD, the APBs are still generated, which bring defects to the materials and limits the photoelectric properties. Therefore, TDD and APB are further reduced by introducing a buffer layer. Li et al. [124] investigated the growth of GaAs layers on polished Ge/Si by selective ART. They first grew the Ge layer on the patterned SiO$_2$ substrate, then deposited GaAs on the SEG Ge buffer layer. Figure 21 shows the layer structure. Their results indicated that APB-free GaAs can be obtained only on a polished SEG Ge buffer layer on the exact (001) Si substrate. Figure 21b shows the APB surface of GaAs when grown on a non-polished SGE Ge buffer layer. In Figure 21c, an APB-free of 1 μm GaAs layer was obtained with the full-width at half-maximum (FWHM) is only 140 arcsec. The significant APB reduction in the GaAs layer was attributed to the nature of SEG-based Ge growth, which results in a virtual miscut Ge surface after CMP. However, hard-to-control asymmetry of GaAs facets and the thicker structural layer are not conducive to device integration which remains a problem.

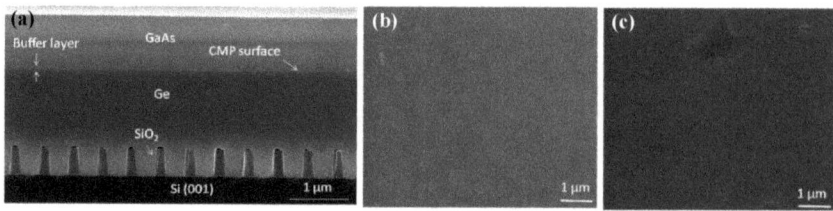

Figure 21. (a) Cross-sectional SEM image of GaAs overgrown on polished SEG Ge buffer layer on patterned Si (001) substrate. (b) Plane view of GaAs surface grown on exact oriented (001) Ge substrate and (c) on polished SEG Ge/Si substrate. Reprinted with permission from ref. [124]. Copyright 2009 Elsevier BV.

In order to resolve this problem, a thin buffer layer is grown in the groove to selectively continued III-V materials. Wang et al. [125] demonstrated the SEG method of high-quality InP layers in submicron trenches on normal Si substrates using a thin Ge buffer layer.

Figure 22 shows the cross-sectional TEM images of the SEG InP layer in 100 nm STI trenches. {111} and {311} facets are visible after the Si process. Then, a thin Ge buffer layer was deposited to form a relatively round surface. This rounded Ge surface removes facets, and the SEG InP grows following the Ge surface in a step flow growth mode; thus, a different crystal orientation can be avoided, which can solve the problem of voids' formation. Meanwhile, an annealing process can prompt the single surface steps of Ge to migrate and merge into double steps, which is essential to avoid any APB formation. In addition, many threading dislocations' TDs are confined at the side of the trench; an APB-free and low TDD InP layer is obtained at the top of the trenches.

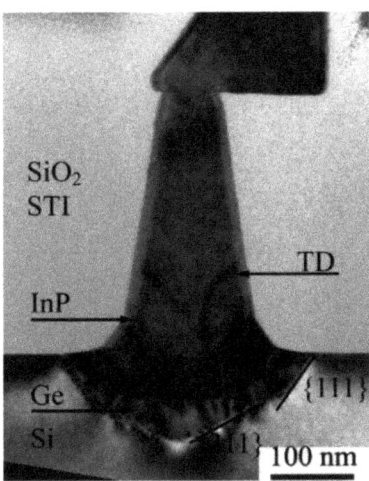

Figure 22. Bright field cross-section TEM images of InP grown in the 100 nm wide STI trenches. {111} and {311} Si facets were obtained after Si etch with HCl vapor. TDs are confined in the bottom of the trenches. Reprinted with permission from ref. [125]. Copyright 2010 American Institute of Physics.

Even though a pre-epitaxial Ge buffer layer is helpful to solve the APB, the quality degradation of III-V materials still cannot be completely eliminated because the diatomic steps cannot form naturally, spontaneously on the Si declination substrate and Ge buffer layer surface. Although the formation of diatomic steps can be promoted by certain pretreatment, it is still not guaranteed that all epitaxial interfaces are diatomic steps. When the III-V materials are deposited in the position without diatomic steps, there will always be possible APBs. The density of defects such as twin planes traveling along the trench direction is fairly high. In 2012, the IMEC group innovatively developed a method to construct natural diatomic step surfaces by pre-etching the silicon substrate at the bottom of the ART method SiO_2 groove into a "V" groove consisting of two {111} faces using an alkaline solution, which can effectively suppress the generation of APB in the III-V epitaxial layer [126]. Growing III-V materials on V-grooved (111) Si surfaces can greatly enhance the quality of epitaxial III-V materials in the ART process [127–134]. There are many advantages by the use of {111} Si V-grooves in the ART growth process. First, APBs can vanish in the V-grooves by the crystallographic alignment between the Si and III-V materials; secondly, compared with the Si (001) plane, little defects will generate when III-V materials nucleate on the Si (111) plane; thirdly, it can selectively grow the active region in any location on the silicon substrate, and the size and position of the active area can be controlled manually. Figure 23 shows the schematic diagrams of the III-V lattice in the "V-shape" of Si. Figure 23a shows III-V lattice in the V-shape of Si with {111} facets along the [110] direction [125], which have the same polarity, but in Figure 23b, a single step on the Si (111) surface is equal to the interplanar spacing of Si {111} planes; such steps might not lead to the formation of APBs in the III−V material.

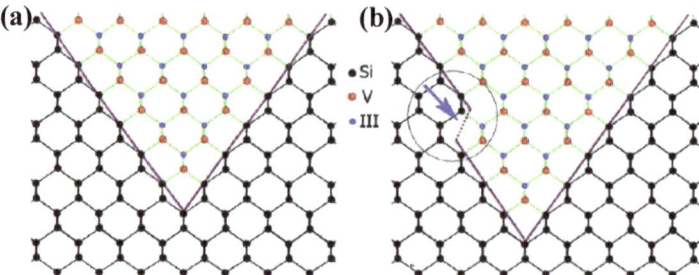

Figure 23. Schematic diagrams of III−V lattice in: (**a**) the "V-shape" of Si with {111} facets; (**b**) a monatomic step on a (111) plane. Reprinted with permission from ref. [126]. Copyright 2012 American Chemical Society.

Tommaso et al. [128] grew GaAs fins in sub-100 nm trenches patterned on Si (001) substrates using the ART approach. They demonstrated the trench bottom geometries in "V" shaped with a consequence of the NH_4OH etch. A 75 nm deep of the "V" shaped groove is formed with the presence of small {113} and (001) facets, which can minimize the interfacial energy and prevent the formation of APBs. Figure 24a–c display bright field STEM images of GaAs-on-V-grooved-Si in directions both perpendicular and parallel to the trenches. All TDs (meandering lines) are found annihilated on the oxide walls and confined at the trench bottom. Few {111} planar defects can be identified, and none of them reach the surface, suggesting the upper part of the inspected GaAs portion is free of defects.

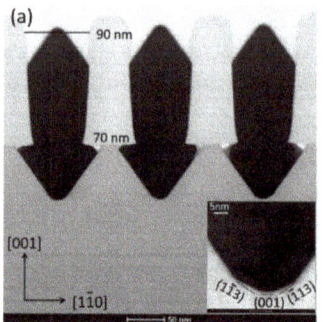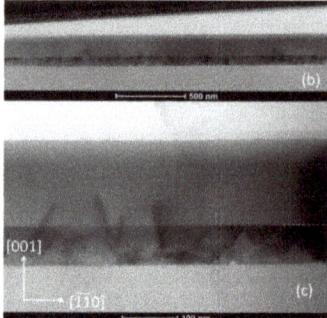

Figure 24. (**a**) Bright field STEM image of GaAs on V-grooved Si. Inset: High magnification bright field STEM image of the trench bottom [113] direction; (**b**,**c**) show low and high magnification bright field-STEM images of a cross section parallel to the trenches Reprinted with permission from ref. [128]. Copyright 2015 American Institute of Physics.

However, the defects of III-V material are also related to the structure of the groove and the growth process. The different aspect ratio also affects the limitation of material defects. Kunert et al. [50] reported that the GaAs fins selectively grow in a V-shaped trench with the aspect ratio of 7.5, 3, 1. Figure 25 shows a cross sectional SEM of the GaAs selectively grown in different ARs. In the case of the ARs being 7.5 and 3, all dislocation defects are trapped and confined inside the STI region in Figure 25b,c. However, for the ARs of 1 in Figure 25d, TD defects are also found above the trench, which indicates that an AR of 1 is not sufficient to block all dislocation. In fact, in these narrow trenches, the InP layer is very defective, with an extremely rough and discontinuous surface. As the surface treatment for wide and narrow trenches is identical, we must conclude that the geometrical confinement within the narrow trenches induces a transition from 2D to 3D growth.

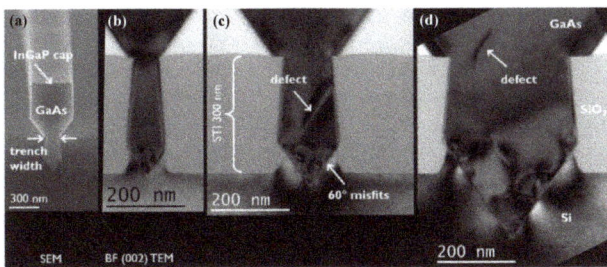

Figure 25. (**a**) a slightly tilted cross SEM image of a GaAs nano-ridge was deposited in 100 nm wide trenches capped with an InGaP cap; cross-sectional bright field TEM images of GaAs were grown in (**b**) 40 nm; (**c**) 100 nm; (**d**) 300 nm trench. Reprinted with permission from ref. [50]. Copyright 2018 IOP Publishing.

In the III-V compounds' semiconductor, InP global epitaxy on the Si substrate is rarely reported due to the difficulty of a huge 8% lattice mismatch. However, based on the advantages of ART technology, growing InP on silicon by the ART approach is common [132,133]. In the first place, the creation of {111}-oriented V-grooved pockets in trenches by this ART approach not only can prohibit the formation of APBs, but also promote strain relaxation via the formation of planar defects such as stacking faults. By designing the shape of the Si substrate, a highly twinned region is forming at the InP/Si interface, enabling a growth of active regions closer to the Si substrate. This twinned region is conducive to the strain relaxation when InP is deposited inside the V-grooved Si pockets. Moreover, InP can serve as a buffer layer for the growth of other III-V semiconductor compounds, such as InGaAs, InGaP. In the past few years, we summarize four growth schemes that were investigated by some researchers in Figure 26. Selective epitaxial growth of InP is deposited in different types of trench (V-grooved or rounded shape trench); different buffer layers (Ge, GaAs) are deposited for the growth of InP.

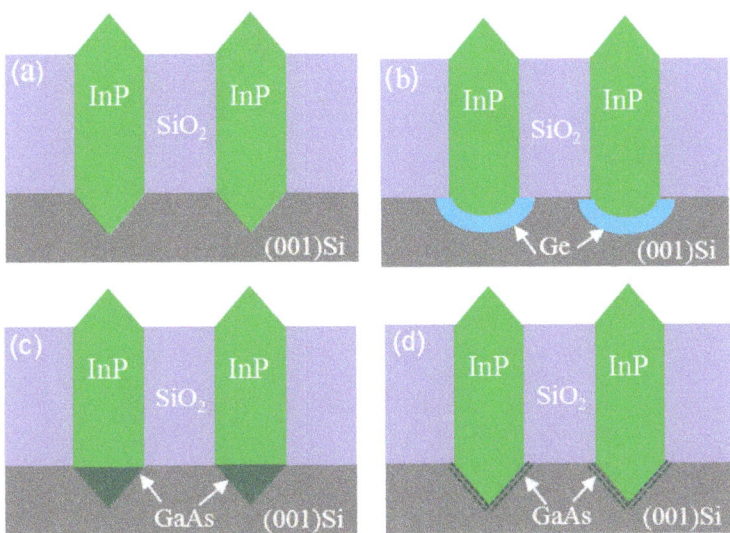

Figure 26. Four epitaxial schemes of InP on patterned Si using ART: (**a**) direct InP epitaxy on V-grooved Si; (**b**) InP on rounded Ge surface; (**c**) InP on a GaAs intermediate buffer filling up the V-grooves; (**d**) InP on a few–nanometer–thick GaAs stress relaxing layer.

Although the creation of {111}-oriented V-grooved pockets in trenches by this ART approach can prohibit the formation of APBs, the weakness of this method is that it is impossible to capture the (111) steering defect along the parallel direction of the groove. This defect can result in stacking fault, twins on the sidewalls, in the upper InP layer, which makes it impossible to obtain large-size plane InP layer. Merckling et al. [134] studied the impact of starting geometry at the bottom STI on the crystalline alignment of the InP layer. They explored the starting geometry at the bottom as rounded etch with Ge buffer versus a crystalline <111> V-groove structure in the Si; the model is in Figure 27a. Rounded Ge structure exhibits different crystallographic facets such as {001}, {111}, and {113}, instead only {111} crystallographic planes on V-groove structure. Different crystallographic facets in rounded Ge surface may provide a nucleation surface with a unique polarity, which is inconducive to grow better nucleation uniformity. Instead, Si V-groove {111} enclosure will provide a surface with a unique polarity. The quality of InP layer was quantitatively characterized by HRXRD. The extracted FWHM value shows broad diffraction peak from InP grown on rounded Ge/Si at 1690 arcsec, while a much narrower diffraction peak from InP grown on "V" shape Si is 540 arc sec. Meanwhile, scanning spreading resistance microscopy (SSRM) was used to measure the local electrical resistance of InP for different process in Figure 27b,c. It is clear that low resistance (high conductivity) of the Ge buffer region is observed in the bottom of the trench, but a much thinner and a higher resistance of the V-groove InP/Si interface. Their SSRM results displayed that the InP grew on a rounded-Ge surface exhibits a low resistance (high conductivity) in the ~10^6 ohms range in Figure 27b, while the resistance is clearly improved to ~5×10^7 ohms range by using a V-groove starting surface in Figure 27c. In other words, higher quality of InP layer was achieved by the use of a V-grooved Si starting surface. In addition, more TDs were generating from the round Si/Ge interface, periodically decorated with misfit dislocations, propagating in the III-V layer.

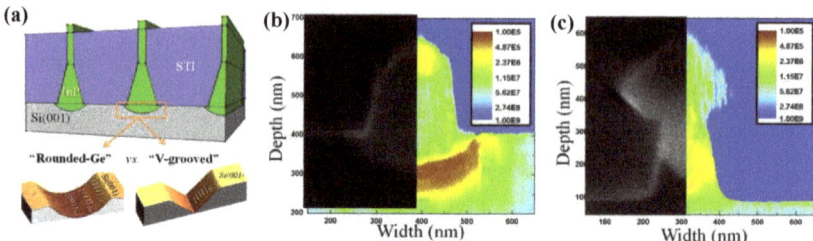

Figure 27. (a) Schematics of different starting surface (rounded-Ge and V-grooved Si) for the III-V growth on Si; Room temperature SSRM patterns in W = 80 nm trenches of InP epitaxially grown on (b) rounded-Ge and (c) V-groove Si starting surface. Reprinted with permission from ref. [134]. Copyright 2014 American Institute of Physics.

By mimicking the metamorphic InP/GaAs buffer on planar Si [135], GaAs and InP have the same crystal structure (sphalerite structure), which can reduce dislocation defects for the InP layer by using GaAs as the buffer layer. Figure 26c illustrates the model of a GaAs intermediate buffer layer. Li et al. [133] grew the high-quality uncoalesced thin films InP by MOCVD in SiO_2 trenches on Si (001) via the ART method. Addition of a V-grooved Si surface to the ART process can more effectively trap misfit defects and APBs at a GaAs/Si intermediate interface as shown in Figure 28. They first directly grew InP in a blanket Si substrate with a different thickness of the GaAs layer. The 30-nm-GaAs buffer was not sufficient to allow misfit dislocations to be trapped in the ART structure. APBs and stacking faults were also observed from the Si surface to the upper InP layer as seen in Figure 28a. In Figure 28b, a thicker 200nm single GaAs buffer can make the stacking faults and twins originate from the InP/GaAs interface along {111}, which effectively inhibited the defects and APBs. Then, V-grooved surfaces at the trench bottom were formed for InP ART. A

V-grooved feature would effectively increase the desired aspect ratio in the ART process and enable GaAs growth inside a pre-defined Si {111} enclosure, which can be expected to promote initial GaAs nucleation uniformity. From Figure 28c, it can be seen that high crystallographic quality of the InP is above the dotted line, where InP is essentially defect free. Figure 28d is the reference of InP global epitaxy on the exact (001) Si substrate, for which various mixed defects exist with high TDDs in the InP layer.

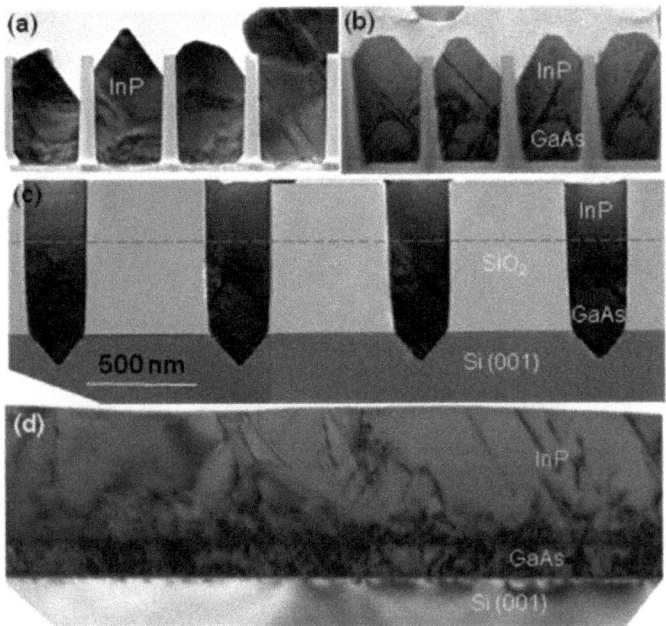

Figure 28. Cross-sectional TEM images of (**a**) InP/GaAs/Si with a 30 nm GaAs buffer layer; (**b**) InP/GaAs/Si with 200 nm intermediate GaAs layer; (**c**) InP/GaAs/Si structure with V-grooved Si surface and modified growth conditions and (**d**) same growth as (**c**) but grown on a Si (001) substrate. All images are to the same scale. Reprinted with permission from ref. [133]. Copyright 2009 IOP Publishing.

Bulk growth and quantum well (QW) are also induced as a dislocation filter to reduce TD in the ART technology further. Zhou [136] grew InP in nanoscale V-grooved trenches on the Si (001) substrate using InGaAs/InP multi-quantum-well by metal organic chemical vapor via the ART method. To obtain the best InGaAs quantum well potential barrier, a 60% In composition of InGaAs/InP MQW was deposited on InP/GaAs buffer layers in nanoscale V-grooved trenches. Figure 29 shows the cross-sectional TEM image of the sample. Highly uniform InGaAs/InP MQWs were visible over different trenches in Figure 29a. Few defects propagated through the GaAs/InP buffer layer to the MQW region in Figure 29b. From the magnification of InGaAs/InP MQW and the InP contact layer part, a clearly identified line and no crystalline defects are observed in Figure 29c. The four periods of 3 nm/6 nm InGaAs/InP QWs are observed with fine periodicity, flat and sharp interfaces in an atomic size, which indicated the highest quality of materials in Figure 29d.

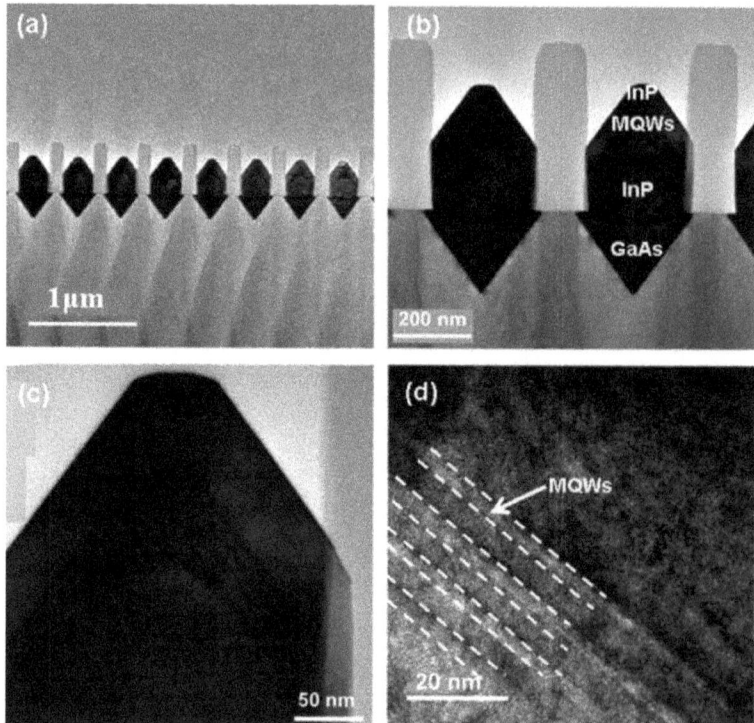

Figure 29. (**a**,**b**) Low magnification cross-sectional TEM image of an InGaAs/InP MQW grown on InP/GaAs buffer layers in nanoscale V-shaped trenches of Si (001) substrate; (**c**) A magnified cross-sectional TEM image of the top part of the sample; (**d**) A high-resolution TEM image taken from the region of side MQW. Reprinted with permission from ref. [136]. Copyright 2016 American Institute of Physics.

The use of {111}-oriented V-grooved pockets in SiO$_2$ trenches by the ART approach is effective to restrict the generation of APBs and TDs, but the V-groove etched by the KOH solution is not only easy to cause serious damage to the crystal arrangement of the Si surface, but also easy to cause surface contamination by wet etching. In 2019, Wei et al. [130] grew the III-V materials on the V shape bottom of the {111} plane Si substrate by MBE. They first obtained the U-shaped pattern with ridges along [110] direction on the Si (001) substrate in Figure 30a; then, the homoepitaxy of the 550 nm Si layer was conducted by MBE at 600 °C. After that, the (111)-faceted Si hollow structures were achieved on the U-shape patterned Si (001) substrate in Figure 30b. The GaAs layers were deposited on the V shape Si substrate using a two-step method, as shown in Figure 30c. This outgrown (111)-faceted Si hollow structure is considered as a diatomic step to grow the ABP-Free GaAs layer on the GoVS template. More importantly, the thermal strain is released and attributed to this hollow structure. Figure 30d shows a perfect surface with the RMS at 1.3 nm and a low TDD of 7.0×10^6 cm^{-2}. This novel process can solve the problem of incompatibility between the miscut Si substrate and conventional Si substrate used in the CMOS process, which has a great commercial value.

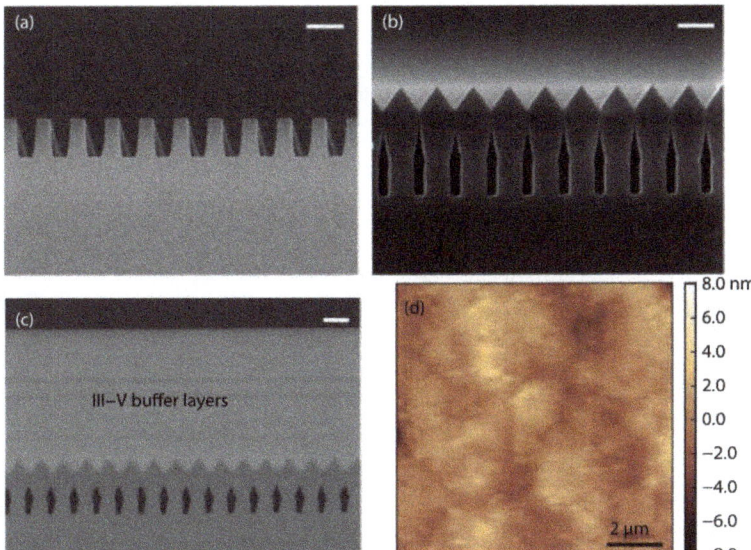

Figure 30. (**a**) Cross-sectional SEM images of U-shape patterned Si substrate; (**b**) Cross-sectional SEM images of homoepitaxy of 550 nm Si on (111)-faceted Si hollow; (**c**) III-V buffer layers on (111)-faceted Si hollow substrate; (**d**) AFM image of III-V buffer layers on Si substrate. Reprinted with permission from ref. [130]. Copyright 2019 IOP Publishing.

From the above Figure 30, it is amazing that the presence of an extra free surface can absorb ~10% thermal strain between Si and GaAs layers, which can obviously reduce the TDD. As we know, the forming of the thermal crack is the thermal strain caused by the large CET of III-V and Si during the cool-down process. Controlling thermally induced strain during the growth is very important to the device fabrication. In order to prevent the crack formation, Saravanan et al. [137] grew GaAs films on a Si/porous Si/Si (SPS) substrate at a low temperature of 450 °C. Their result shows a biaxial tensile stress as low as 1.69 kbar for GaAs/SPS at the 77 K photoluminescence spectra, but with a higher TDD of 10^{11} cm^{-2}. Inserting the buffer layer can decrease the TDD, while thicker buffer layers (~4 μm) can also induce thermal crack. Therefore, the strain compensated layer was introduced to reduce the thermal strain. Takano et al. [138] grew GaAs epilayers on Si substrates by inserting $In_xGa_{1-x}As$ buffer layers via low-pressure metalorganic vapor-phase epitaxy. The TDD of 4.8×10^6 cm^2 for the GaAs layer was achieved by insertion of an InGaAs strained layer. This incomplete strain relaxation in InGaAs layers can compensate the tensile strain due to the large TEC between GaAs and Si materials. However, the compositional buffer layer has some drawbacks, such as several microns in thickness, high cost, and large usage of material. In following, a thinner superlattice (SL) buffer was applied as the strain compensated layer to minimize the thermal crack, and the SL layer can not only facilitate strain relaxation by interaction with misfit dislocations, but also compensate the thermal strain energy from the mismatch of III-V to Si [107,116,136].

Recently, the use of a patterned substrate to minimize the thermal crack was widely developed. As for pre-patterned growth, cracks can be avoided by growth on a small area due to strain relaxation near the pattern edge. As seen from Figure 31, the thermal strain of GaAs is released and attributed to this Si hollow structure; approaches using the Ge on the micropillar patterned Si (001) substrates were also proposed. Zhang et al. [139] grew APB-free GaAs film on employed the {113}-faceted Ge/Si (001) hollow substrate by MBE. First, the fabrication of the {113}-faceted Ge/Si (001) hollow substrate is as follows: the U-shape grating pattern was defined on an on-axis Si (001) substrate by using deep ultraviolet lithography and reactive ion etching; then, depositing the 60 nm Si buffer on the patterned substrate, a 600 nm Ge layer was grown to achieve the {113}-faceted Ge hollow structures, as shown in Figure 31a. Second, the typical two-step GaAs was grown on the {113}-faceted Ge/Si (001) hollow substrate in Figure 31b. Figure 31c shows the magnification of {113}-faceted Ge and the GaAs interface; few APBs were found at the bottom of the Ge {113} sawtooth structure, but APBs are not observed in the top GaAs layer. Similarly, the {113}-faceted Ge structure can annihilate the APBs at the interface with the {111}-faceted Si structure, which can be considered as a miscut substrate. However, the Ge {113} crystal plane has less miscut than the Si {111} crystal plane, which brings out a trapezoidal shape of APB in ten nanometers of the bottom of the Ge {113}. In following, the 7-μm-thick GaAs layer, which is far beyond the typical value of the cracking thickness, can be grown on the {113}-faceted Ge/Si hollow substrate. They characterized the thermal strain issue by a high-resolution XRD reciprocal space mapping (RSM) performed around (004) and (224) reflections, as shown in Figure 31d,e. From the peak position of RSM, the in-plane strain $\varepsilon_{\parallel}$ of the GaAs and Ge layer were calculated from the extracted calculation of the in-plane lattice constant and out-of-plane lattice constant of the GaAs and Ge. Compared with the lattice constant of bulk GaAs and Ge, the residual thermal strain of Ge is about 89.8% lower than that of the Ge layer on normal Si substrates [140]. The residual thermal strain of GaAs is 29.4% lower than that of the GaAs layer grown on the Ge/Si template [141]. Their results indicated that this hollow structure plays an essential role in thermal strain reduction. With a 400 nm GaAs deposition, a smooth GaAs surface with a RMS of 0.67 nm was acquired, and a low TDD of $5.7 \times 10^6/cm^2$ was obtained by following InGaAs/GaAs quantum-well DFLs inserting.

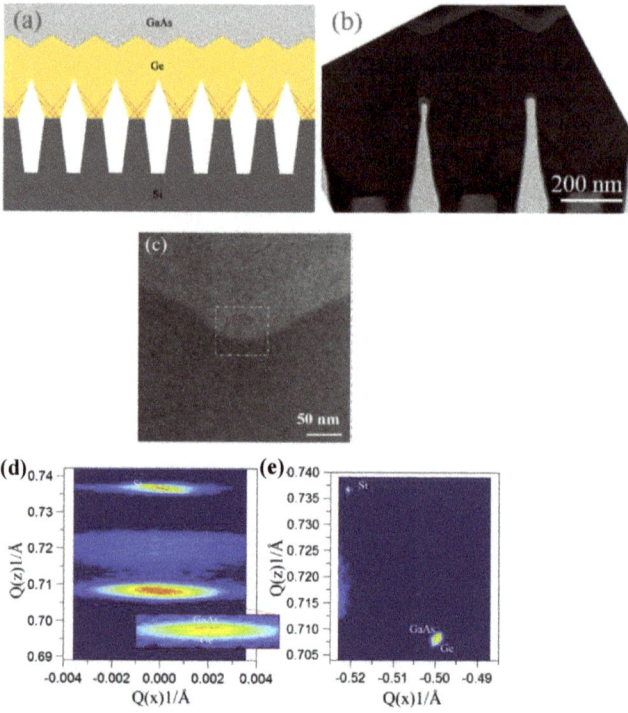

Figure 31. (a) The schematic illustration of the GaAs grown on the Ge/Si hollow substrate; (b) Cross-sectional TEM image of 600 nm thick Ge layer grown on the hollow patterned (001) Si; (c) Zoom-in HADDF image of the interface at the bottom region; high-resolution XRD reciprocal-space mapping around (d) (004) and (e) (224). Inset of (c) Higher resolution (004) map from triple-axis mode. Reprinted with permission from ref. [139].

4.2.2. Epitaxial Lateral Overgrowth (ELO)

Although the ART technique discussed above can effectively solve the APB problem, and block the propagation of crystalline defects using the groove mask, this technique also limits the maximum achievable dimension of III-V epitaxial layers, which are more amenable in some device applications [142]. In order to provide a large film plane for the preparation of III−V devices, the ART technique needs to be continuously optimized so that the SEG III−V materials are epitaxial and grown out of closely spaced trenches, and extend laterally above the oxidation strips until the materials from adjacent grooves merge together to form a continuous high quality epitaxial layer. Recently, an alternative ART technique called Lateral aspect ratio trapping (LART) was developed. As schematically illustrated in Figure 32a, micrometer-scale III-V crystals can be directly grown above the buried oxide layer by this LART method [143]. Compared to the conventional "aspect ratio trapping" approach, the LART approach is changing the direction of the groove, enabling the growth front to the lateral direction. The lateral oxide trenches can be created by dry etch terminated on the buried oxide; then, anisotropic wet etching was processed to form "V" shaped Si. This {111}-oriented Si bevel structure not only prohibits the formation of APBs in the initiating lateral growth of InP layers, but also the propagation of TDD was effectively blocked by the wide lateral SiO_2 trenches, which reveal a high "aspect ratio" in the lateral direction. Figure 32b,c shows a tilted-view SEM image of the SEG InP layer grown on (001) SOI substrates using the LART method. Figure 33b shows an InP-epi "wing" grown using the lateral ART approach, and Figure 33c displays the two symmetrical InP-epi wings. From the SEM images, we can see the Si pedestal sandwiched between the top oxide spacer, and the buried oxide layer features two {111}-oriented surfaces. Starting from the nucleation sites provided by the {111} Si facets, the InP crystal evolves laterally along the [110] direction into wing-structures with two {111} facets. The angle between the two {111} facets is around 110° which indicates a zincblende crystal structure Then, the InP stripes continue laterally to grow inside the long nano-scale SOI trenches resulting in dislocation-free InP crystals right atop the buried oxide layer. In addition, the dimension of III-V nano-ridges can be controlled by changing the thickness of the Si layer on SOI, which is limited by the size of photolithography in the conventional ART approach. This in-plane and close placement of the III-V layer with the Si device layer also facilitates the integration of III-V light emitters with Si-based photonic components.

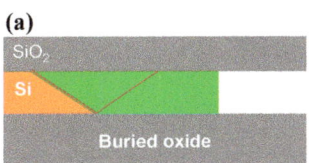 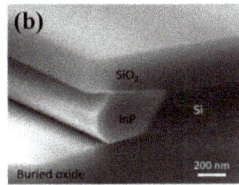 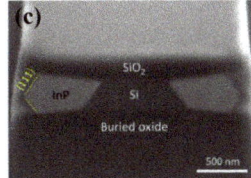

Figure 32. (a) The scheme of lateral ART. The red dotted lines denote the confinement of the majority of crystalline defects at the InP/Si interface and the trapping of residual TDs by the top and bottom SiO_2 layers. (b) Tilted view SEM image of one InP sandwiched between the top oxide spacer and the buried oxide layer. (c) Cross-sectional SEM images of two symmetrical InPs grown using the lateral ART approach. Reproduced with permission from ref. [143]. Copyright 2019 American Institute of Physics.

The above ART methods or lateral ART approach often produce sub-micrometer dimensions of APB-III-V layers, but to expand the dimension of the epitaxial III-V further to tens and even hundreds of micrometers, ART technology needs to be improved. Epitaxial lateral overgrowth (ELO) is an innovative technique to provide large regions of device materials. In the ELO process, a III-V buffer layer is first grown on the Si substrate, then polished by CMP. Microsized SiO_2 stripe patterns are selectively etched to expose the III-V buffer layer for regrowth. Then, the III-V epitaxial layer is vertically regrown through the opened region of the mask and thereafter can be laterally grown over the mask. Most of TDs in the buffer layer are blocked by the bottom of the mask, but a small number of TDs around the opened region will propagate upwards, as shown in Figure 33a. Therefore, a high crystalline quality and thicker layer can be made by this ELO. For the GaAs on Si, there are many pioneering works reported [144,145]. Tsaur et al. [146] first obtained the single-crystal GaAs layers by means of ELO seeded within stripe openings in a SiO_2 mask over GaAs layers grown on Ge-coated Si substrates. TEM and scanning cathodoluminescence studies indicated that the laterally overgrown GaAs layers have a dislocation density. However, the early reports on ELO for GaAs-on-Si suffered from the limited defect-free region and the mechanical weakness of the laterally grown parts, both of which became severe as the ELO layer was further grown. In 2015, Yunrui et al. [147] demonstrated that the GaAs coalescence layer grew on the patterned 1.8 μm GaAs buffer layer by epitaxial lateral overgrowth using metal-organic chemical vapor deposition. A 410-nm-thick coalesced ELO GaAs layer was obtained with the low root-mean-square surface roughness of 6.29 nm. Figure 33b,c show the SEM images of the growth and evolution of the ELO GaAs layer on the opening trenches. As shown in Figure 33b, the growth front of GaAs is faceted, with (111) B on both sides and (001) on the top. With the deposition time, the overall 410 nm ELO GaAs layer was shown in Figure 33c.

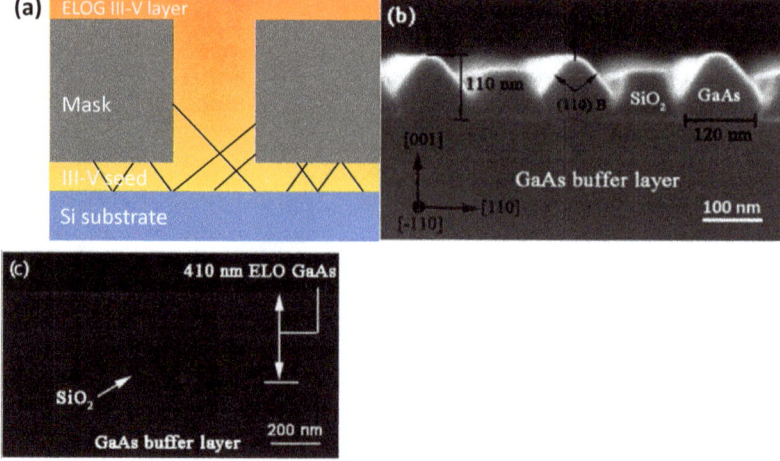

Figure 33. (**a**). Schematics of lateral epitaxial overgrowth (ELO); (**b**) cross-sectional SEM images of the films after the first selective growth stage; (**c**) cross-sectional SEM images of the ELO GaAs layer. Reprinted with permission from ref. [147]. Copyright 2015 American Institute of Physics.

Compared with GaAs-on-Si, ELO techniques are more widely employed [148–150] on InP-on-Si growth, because it is very difficult to obtain low TDDs and large-scale via the direct growth of InP on Si. For the optimization of InP ELO on the (001) Si substrate, Metaferia et al. [151] investigated the ELO of InP from mesh and line openings on the masked InP seed layer on the Si (001) wafer. Their results showed that the coalesced region produced the TDD in a range from 6×10^6 cm^{-2} to 4×10^7 cm^{-2} depending on the thickness of the ELO layer. In their work, the ELO InP layer was grown on a 1.5-μm-thick InP/Si substrate with a 40-nm-thick SiO_2 mask. The mesh and line masks (opening and masking width of 200 nm and 3 μm, respectively), tilted 15° and 30° off the [110] direction, are compared. The quicker coalescence in the mesh opening resulted in better surface roughness at the early growth stage (~10-μm-thick InP layer), but, after an extended growth (~100 μm-thick InP layer), both mesh and line opening cases exhibited a similar surface roughness (RMS roughness of 16~25 nm). It was shown that regardless of both angles (15° and 30°) and masks (mesh and line opening), the TDDs were measured to be a similar value. The TDD of the 10 μm-thick and 100 μm-thick InP layer was measured to be $2\sim4 \times 10^7$ cm^{-2} and $6\sim7 \times 10^6$ cm^{-2}, respectively. Han et al. [152] displayed two improved ELO strategies to achieve large-dimension III-V materials with a close proximity to the Si substrates in Figure 34. The first scheme is called conformal growth; III-V films lateral selective epitaxial growth starts from III-V seeds on silicon. In the conformal growth scheme [153], after the III-V films were deposited on the Si substrate, a thin layer of the growth mask is then patterned to form the III-V seeds. Then, the regrown III-V stripes were laterally selectively grown along the initial III-V seeds to extend the large dimension size. Another scheme is named corrugated epitaxial lateral overgrowth (CELO) [154]; short III-V thin film segments were processed as a III-V seeds by bulk III-V film deposition and patterned; then, a thin patterned oxide mask was deposited on III-V seeds for the following homogenous selective regrowth of the InP process. In both strategies, crystalline defects are confined within the seed mesa, while the laterally overgrown III-V is free of any TDs. Figure 34b shows the cross-sectional SEM in [110] view of CELOG InP on Si; defects are confined in the lateral trenches, and no TDS can be found in the CELOG InP layer. These two improved processes can be used for large-dimension III-V layer manufacturing, which could be furthered processed to the subsequent growth of lasers' structures.

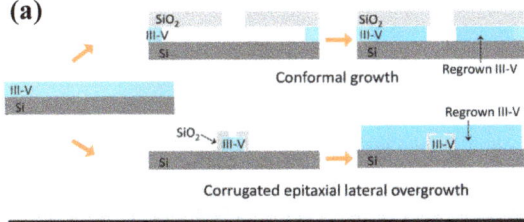

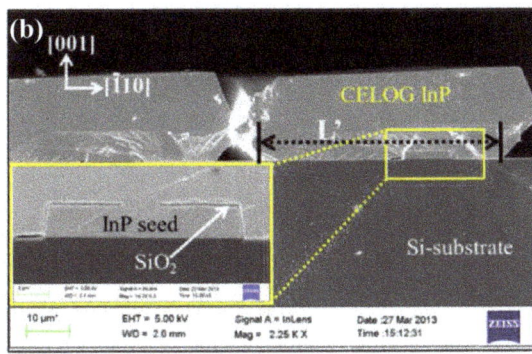

Figure 34. (a) Schematic of two improved ELO process; (b) cross-sectional SEM in [110] direction view of CELOG InP on Si. Reprinted with permission from ref. [154]. Copyright 2015 IOP Publishing.

Following the above results, in 2020, Omanakuttan et al. [155] fabricated and studied the high crystalline quality of InP by the self-aligned corrugated epitaxial lateral overgrowth (CELOG) method. Figure 35a–d display the InP-seed mesa fabrication process flow. A schematic of the InP-seed mesa was patterned on Si processed for CELOG and is shown in Figure 35e. Through a series of processes: mask deposition, photolithography, and etching, then the CELOG of InP/Si can be obtained by a schematic of Figure 35f. The defective InP seed acts as the CELOG InP growth origination; the threading dislocations propagating from the InP seed can be eliminated by creating dislocation loops for increasing the InP layer thickness. Their results displayed that RMS surface roughness of 2.95 nm obtained for the uniform InP CELOG layer. A higher intensity band edge emission in the cathodoluminescence spectra near-BE band at 892 nm (1.39 eV) and enhanced carrier lifetime (710 ps) of InP are observed above the CELO InP/Si interface compared to the defective seed InP layer on Si, which are attributed to the reduced TDD realized ($\sim3 \times 10^8$ cm^{-2}) by the CELO method. For the application of the photonic integrated circuit (PIC), a large dimension and uniform InP layer with ultralow dislocation on Si are desired. The CELOG can be extended to the formation of other large dimensions of III-V/Si heterostructures.

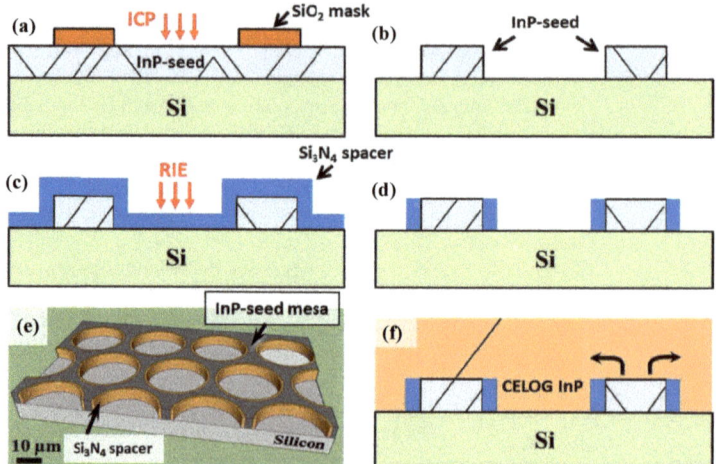

Figure 35. (**a**–**d**) Process of InP-seed mesa fabrication for CELO; (**e**) Schematic of the InP-seed mesa pattern on Si before CELO. The thickness of InP-seed layer is 2 µm, and the distance between the two adjacent circular rings of diameter 30 µm is 5 µm; (**f**) Schematic of the CELO InP/Si cross-section. Reprinted with permission from ref. [155].

In the laterally grown parts, propagating defects such as threading dislocations and stacking faults will be blocked by the mask and consequently cannot propagate into the layer above the mask. ELOG was used to grow GaN [156], as well as InGaAs [157] on Si.

The InAs semiconductor material is a very good candidate for ultra-high-speed electronic and optoelectronic devices, due to its narrow band gap, small electron effective mass, and very high electron mobility (~33,000 cm^2/V s at 300 K). However, the large epitaxial strain generated by the high lattice mismatch (11.4%) and the difference in thermal expansion between InAs and Si, leads to a roughness surface and high TDD, which degrade the electrical and optical properties. At present, growth of thin-film InAs is usually difficult, but direct growth of vertical nanowires can usually accommodate a larger lattice mismatch as compared to thin-film growth [158]. Regarding InAs nanowires, various strategies including vertical VLS growth [159], vertical SEG growth [160,161] have been proposed. Among these techniques, vertical VLS growth has an obvious disadvantage: using Au as the catalyst leads to the formation of deep-level traps with silicon and degrades the device performance, which makes it forbidden in the CMOS fabrication process, so vertical SAE growth was widely developed. In the process of InAs NWs' growth, the morphology of NWs is essential to homogeneous optical and electrical properties. Hertenberge et al. [160] obtained very high yields of ~90 percent of vertically (111)-oriented InAs nanowires selectively grown on the patterned Si(111) substrate. Then, Bjork et al. [161] studied (111)Si (axial) and (1-10)Si (radial) growth of InAs NWs by varying growth duration, temperature, group-III molar flows, V/III ratio, mask material, mask opening size, and inter-wire distance. To achieve uniform nucleation and a high vertical yield of wires, an As-terminated surface and an optimized TMIn flow and V/III ratio are required. Below 520 °C and 540 °C, respectively, the <111> and the <1-10> growth is surface kinetically limited. Their results also indicated that by placing wires in large arrays, it is possible to stop the <1-10> growth rate completely in favor of the <111> growth rate. Recently, Grégoire et al. [162] report that the vertical and high aspect ratio InAs NWs with a hexagonal shape were grown on both GaAs (111)B and Si(111) patterned substrates by selective epitaxial growth (SEG). The morphology and the quality of InAs NWs' arrays grown on GaAs(111)B and Si(111) were characterized by SEM and photoluminescence measurements in Figure 36a–d, respectively. For NWs grown on GaAs(111)B in Figure 36a, a strong peak at 0.445 eV with a full width at half-maximum (FWMH) of 32meV was observed, while a low peak at 0.413 eV with a FWHM of 34 meV grown on Si was observed, as shown in Figure 36b. A lower PL intensity recorded from InAs NWs grown on Si compared to GaAs is probably due to the lower NW density and diameter. By decreasing the NW diameter, the principal PL peak shifted to a higher energy, confirming the presence of both WZ and ZB phases in InAs NWs grown by SAG-HVPE. These NW arrays exhibited strong PL intensity and optical absorption, which is encouraging for future optical devices.

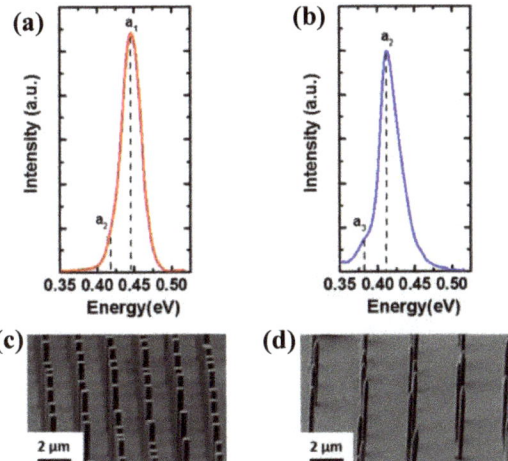

Figure 36. PL spectra at 10 K of InAs NWs grown on (**a**) GaAs (111) B and (**b**) Si (111); (**c,d**) Corresponding SEM images. The pitches are, respectively, 2 and 2.5 µm. Reprinted with permission from ref. [162]. Copyright 2021 American Chemical Society.

In the III−V semiconductor, III-antimonide, such as GaSb, has attracted a tremendous amount of research interest. GaSb (0.72 eV) is a direct bandgap material, which can be used as a photodetector of the mid-IR spectrum. Moreover, GaSb exhibited attractive characteristics for p-MOSFETs due to their high bulk hole mobility (over 1000 cm^2/Vs). Similarly, it has great challenges to grow GaSb on Si substrate, due to the large lattice mismatch and thermal mismatch. Recently, there were reports of a GaSb-on-insulator (-OI) by direct wafer bonding [163] and epitaxial layer transfer [164], but these methods remain costly due to the limited sizes of available III−V substrates. However, a one-dimensional (1D) GaSb nanowire (NW) manufactured by selective epitaxial growth (SEG) provides extra benefits for many advanced utilizations, such as the better stress relaxation, capability of the advanced gate stacking integration, more efficient light adsorption and trapping. For the GaSb NWs' application in electronic devices, a well-controlled nanowire-like morphology with a small diameter (several tens of nanometers) is essential. Yang et al. [165] grew very thin and uniform GaSb NWs with diameters down to 20 nm by the use of a sulfur surfactant. This GaSb NWs were configured into transistors and exhibited impressive electrical properties with the peak hole mobility of ~200 cm^2/Vs, better than any mobility value reported for a GaSb nanowire device. To control the GaSb crystal shape and dimension deeply, an alternative to the SEG technique known as template-assisted selective epitaxy (TASE) was developed [166,167]. Borg et al. [167] investigated a monolithic integration of high-mobility horizontal GaSb NWs on the SOI substrate by TASE. They found that a high degree of morphological control allows for GaSb nanostructures with critical dimensions down to 20 nm. Figure 37a displays a SEM image of an exemplary array of

GaSb nanostructures outside the templates. The exposed front GaSb surface, at which the crystal grows, typically forms a large {111} facet often with two smaller and opposing {110} facets (see Figure 37c,d). Because of the polar nature of the GaSb zinc-blende crystal structure, nonequivalent 180° rotations of the crystal lattice can occur upon nucleation on the nonpolar Si. This results in two distinct orientations of the growth facets as illustrated in Figure 37d. Meanwhile, the GaSb growth is governed by excess Sb present on the GaSb surface, leading to a strong growth rate dependence on the V/III ratio and temperature. Hall/van der Pauw measurements are conducted on 20-nm-thick GaSb nanostructures, revealing high hole mobility of 760 cm^2/(V s).

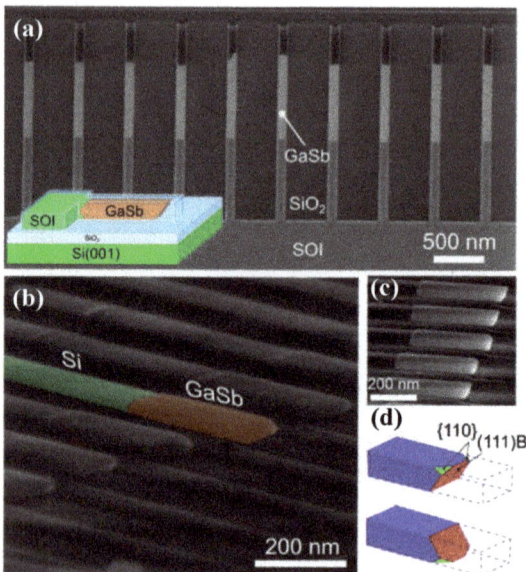

Figure 37. (**a**) SEM image of an array of horizontal GaSb nanostructures integrated coplanar with a SOI layer using TASE. The inset shows an illustration of the layer structure. (**b**,**c**) Tilted SEM images of GaSb nanostructures. (**d**) Schematic illustration highlighting the typical faceting observed at the growth front of the GaSb crystals. Reprinted from ref. [167]. Copyright 2017 American Chemical Society.

In summary, the APB problem can be solved by a miscut Ge/Si substrate, {113}-faceted Ge/Si (001) hollow substrate in the global epitaxial method, and {111}-oriented V-grooved in the ART method. However, high quality III–V heteroepitaxy with low TDD on Si is crucial to the monolithic integration of III–V devices on Si-based PICs. Table 2 is the summary of the reported TDD and RMS values of GaAs, InP in the growth of the substrate, epitaxy method, buffer materials, process in these years. In terms of material growth quality, the TDD of the GaAs layer as low as 10^7 cm^{-2} can be achieved by inserting the superlattice via global epitaxy. Then, the 8-inch GaAS-O-I substrate can be realized in our group soon. However, few research projects report on the preparation of other III–V epitaxy methods on the Si substrate. For selective epitaxial growth (SEG), many III–V semiconductors such as InP, InCaP, InAs, GaSb, etc. can be grown by the ART method. A TDD of 10^6 cm^{-2} of the InP layer was obtained in the V-grooves patterned Si substrate. In terms of application, the high-quality and stress-controlled large-size III–V epitaxial layer grown by global epitaxy is more conducive to the preparation of Si-based OEIC. On the contrary, the III-V nanowire arrays fabricated by the SEG method can produce the monolithic integration of III-V nanodevices on silicon substrates. Especially for a smaller size of CMOS electronic devices, several tens of nanometers of III-V NWs are considered optimal for high-performance tunnel-FETs.

Table 2. Summary of reported TDD value of III-V materials in terms of substrate, epitaxy method, buffer, procedure.

Year	Substrate	Epitaxy Method	Buffer	Procedure	III-V Materials	TDD (cm^{-2})	RMS (nm)	Refs.
2011	0° Si	Global	Ge	two-step growth	GaAs	2×10^7	1.1	[70]
2011	4° Si	Global	Ge	two-step growth	GaAs	1.8×10^7	—	[64]
2013	0° Si	Global	—	three-step growth+ TCA	GaAs	1.1×10^7	0.73	[113]
2014	Ge	Global	—	two-step growth	GaA	2.7×10^7	0.7	[93]
2015	6° Ge	Global	—	two-step growth	GaAs	—	0.6	[97]
2016	0° Si	Global	Ge	two-step growth	GaAs	3×10^7	0.5	[98]
2018	0° Si	Global	Ge	In$_{0.18}$Ga$_{0.82}$As/GaAs SLSs	GaAs	2.3×10^6	—	[121]
2019	0° Si	ART	Si	(111)-faceted Si hollow	GaAs	7.0×10^6	1.3	[138]
2020	0° Si	ART	Ge	{113}-faceted Ge/Si hollow substrate.	GaAs	5.7×10^6	0.67	[145]
2021	0° Si	Global	CMP-Ge	three-step growth	GaAs	7.4×10^7	1.27	[53]
2018	0° Si	Global	Ge	InAs/InP QD DFLs	InP	3×10^8	2.88	[127]
2011	0° Si	ELO	—	two-step growth	InP	4×10^8	—	[158]
2019	0° Si	CELOG	—	two-step growth	InP	3×10^8	2.95	[162]
2020	0° Si	Global	—	In$_{0.73}$Ga$_{0.27}$As/InP SLSs	InP	4.5×10^7	2.38	[115]

5. Conclusions and Outlooks

High-quality III-V heteroepitaxy on the Si substrate is crucial to the Si-based optoelectronic integration circuits (OEICs). In this work, we reviewed the three major challenges of the Si based IIII-V heteroepitaxy: (1) anti-phase boundaries (APB); threading dislocation (TD); (3) stacking faults (SF). Meanwhile, the mechanism and theoretical model of three kinds of defects are also deeply analyzed, which is convenient for readers to understand its development process in detail. In order to solve each issue, a wide variety of strategies is discussed. The bulk and high quality of III-V structure layers are the basis of larger scale optoelectronic devices' fabricating. For global epitaxy, the offcut Si substrates can easily suppress the formation of APBs, but it is incompatible with the integration of CMOS manufacturing. To reduce the TDDs, a lot of methods, including the buffer layer, annealing, multi-step growth, defect filter layers, and so on, were developed. In this, the thick layer and high annealing process can lead the new defects as stacking faults and thermal cracks, resulting in the low TDD of 10^6 cm^{-2} for GaAs, and a higher TDD (~10^7 cm^{-2}) of InP-on-Si due to a larger lattice mismatch (~8%) between InP and Si. For selective epitaxial growth (SEG), {111}-oriented V-grooved pockets in SiO$_2$ trenches by the ART approach are effective to restrict the generation of APBs and TDs, but the solution-etched V-groove is not always uniform and easily damages the crystal arrangement of the Si surface, which obviously affects the defects' distribution. In addition, the ART method also limits the maximum achievable dimension of the epitaxial III-V materials. Epitaxial lateral overgrowth (ELO) is an innovative technique to provide large regions of device materials, especially for plane-InP growth on Si. Moreover, other III-V semiconductors, such as InAs, GaSb, which have high mobility but larger mismatches with the Si substrate, can be fabricated in small-size Nanowires (NWs) by the ART method, representing potential applications as the channel material integrated on Si substrates to MOSFETs devices.

Even though major obstacles for III-V-on-Si heteroepitaxy, such as APBs, TD, stacking faults, are resolved now, the quality of III-V layers on on-axis Si is still unsatisfactory for the deployment on the PICs or ICs. Currently, we can obtain the low TDD of 10^6 cm^{-2}, but it still challenging to reduce the TDD to native substrates (~10^4 cm^{-2}). Therefore, further effects need to be devoted to improving the quality of the III-V layer on Si to realize the integrated III-V on the Si-based platform fully. The Epitaxial Lateral Overgrowth (CELO) is an innovative process for large-scale high-quality Si-based III-V epitaxial layers; systematic research and process optimization are still needed to improve the materials' quality and device performance further. In addition, the bounding of III-V-O-I can thin the LT low-quality layer, which can fabricate the high-performance optoelectronic devices. Therefore, based on the current research, efforts should continue to be devoted to resolving the TDDs and the compatibility of integration techniques with large-volume CMOS manufacturing.

Author Contributions: Conceptualization, Y.D. (Yong Du) and H.H.R.; methodology, Y.D. (Yong Du), B.X., B.L. and Z.K.; data curation, Y.D. (Yong Du) and G.W.; writing—original draft preparation, Y.D. (Yong Du); writing—review and editing, Y.D. (Yong Du), G.W., Y.M., W.W., Y.D. (Yan Dong) and H.H.R.; supervision, H.H.R. All authors have read and agreed to the published version of the manuscript.

Funding: This work was supported by the construction of the high-level innovation research institute from the Guangdong Greater Bay Area Institute of Integrated Circuit and System (Grant No. 2019B090909006) and the projects of the construction of new research and development institutions (Grant No. 2019B090904015), in part by the National Key Research and Development Program of China (Grant No. 2016YFA0301701), the Youth Innovation Promotion Association of CAS (Grant No. 2020037), and the National Natural Science Foundation of China (Grant No. 92064002).

Conflicts of Interest: The authors declare no conflict of interest.

References

1. Radamson, H.H.; Zhu, H.L.; Wu, Z.H.; He, X.B.; Lin, H.X.; Liu, J.B.; Xiang, J.J.; Kong, Z.Z.; Xiong, W.; Li, J.; et al. State of the Art and Future Perspectives in Advanced CMOS Technology. *Nanomaterials* **2020**, *10*, 1555. [CrossRef] [PubMed]
2. Fisher, G.; Seacrist, M.R.; Standley, R.W. Silicon Crystal Growth and Wafer Technologies. *Proc. IEEE* **2012**, *100*, 1454–1474. [CrossRef]
3. Radamson, H.H.; Luo, J.; Simoen, E.; Zhao, C. *CMOS Past, Present and Future*; Woodhead Publishing: Cambridge, UK, 2018; ISBN 9780081021392.
4. Wang, T.; Zhang, J.J.; Liu, H.T. Quantum dot lasers on silicon substrate for silicon photonic integration and their prospect. *Acta. Phys. Sin.* **2015**, *20*, 64. [CrossRef]
5. Riel, H.; Wernersson, L.E.; Hong, M.; Alamo, J.A. III-V compound semiconductor transistors-from planar to nanowire structures. *MRS Bull.* **2014**, *39*, 668–677. [CrossRef]
6. Convertino, C.; Zota, C.; Schmid, H.; Ionescu, A.M.; Moselund, K. III-V heterostructure tunnel field-effect transistor. *J. Phys. Condens. Matter* **2018**, *30*, 96–102. [CrossRef] [PubMed]
7. Kim, D.H.; Alamo, J.A. Scaling behavior of $In_{0.7}Ga_{0.3}As$ HEMTs for logic. In Proceedings of the 2006 IEEE International Electron Devices Meeting, San Francisco, CA, USA, 11–13 December 2006; pp. 1–4. [CrossRef]
8. Suemitsu, T. GaAs- and InP-Based High-Electron-Mobility Transistors. In *Comprehensive Semiconductor Science and Technology*; Elsevier, Ltd.: Amsterdam, The Netherlands, 2011; Volume 5, pp. 85–109. ISBN 978-0-444-53153-7.
9. Aggerstam, T.; Lourdudoss, S.; Radamson, H.H.; Sjödin, M.; Lorenzini, P.; Look, D.C. Investigation of the interface properties of MOVPE grown AlGaN/GaN high electron mobility transistor (HEMT) structures on sapphire. *Thin Solid Film.* **2006**, *515*, 705–707. [CrossRef]
10. Deshpande, V.; Djara, V.; O'Connor, E.; Hashemi, P.; Morf, T.; Balakrishnan, K.; Caimi, D.; Sousa, M.; Fompeyrine, J.; Czornomaz, L. Three-dimensional monolithic integration of III-V and Si (Ge) FETs for hybrid CMOS and beyond. *Jpn. J. Appl. Phys.* **2017**, *56*, 04CA05. [CrossRef]
11. Kosten, E.D.; Atwater, J.H.; Parsons, J.; Polman, A.; Atwater, H.A. Highly efficient GaAs solar cells by limiting light emission angle. *Light Sci. Appl.* **2013**, *2*, e45. [CrossRef]
12. Sobolev, M.M.; Soldatenkov, F.Y.; Shul'pina, I.L. Misfit dislocation–related deep levels in InGaAs/GaAs and GaAsSb/GaAs p–i–n heterostructures and the effect of these on the relaxation time of nonequilibrium carriers. *J. Appl. Phys.* **2018**, *123*, 161588. [CrossRef]
13. Yang, Q.; Wu, Q.M.; Luo, W.; Yao, W.; Yan, S.Y.; Shen, J. InGaAs/graphene infrared photodetectors with enhanced responsivity. *Mater. Res. Express.* **2019**, *6*, 116208. [CrossRef]
14. Abouzaid, O.; Mehdi, H.; Martin, M.; Moeyasrt, J.; Salem, B.; David, S.; Souifi, A.; Chauvin, N.; Hartmann, J.M.; Bouraoui, L.; et al. O-Band Emitting InAs Quantum Dots Grown by MOCVD on a 300 mm Ge-Buffered Si (001) Substrate. *Nanomaterials* **2020**, *10*, 2450. [CrossRef] [PubMed]
15. Zhao, X.; Moeen, M.; Toprak, M.S.; Wang, G.; Luo, J.; Ke, X.; Li, Z.; Liu, D.; Wang, W.; Zhao, C.; et al. Design impact on the performance of Ge PIN photodetectors. *JMSE* **2020**, *31*, 18–25. [CrossRef]
16. Zhao, X.; Wang, G.; Lin, H.; Du, Y.; Luo, X.; Kong, Z.; Su, J.; Li, J.; Xiong, W.; Miao, Y.; et al. High Performance pin Photodetectors on Ge-on-Insulator Platform. *Nanomaterials* **2021**, *11*, 1125. [CrossRef]

17. Miao, Y.; Wang, G.; Kong, Z.; Xu, B.; Zhao, X.; Luo, X.; Lin, H.; Dong, Y.; Lu, B.; Dong, L.; et al. Review of Si-Based GeSn CVD Growth and Optoelectronic Applications. *Nanomaterials* **2021**, *11*, 2556. [CrossRef] [PubMed]
18. Yang, J.; Bhattacharya, P.; Mi, Z.; Qin, G.X.; Ma, Z.Q. Quantum dot lasers and integrated optoelectronics on silicon platform. *Chin. Opt. Lett.* **2008**, *6*, 727–731. [CrossRef]
19. Lourdudoss, S. Heteroepitaxy and selective area heteroepitaxy for silicon photonics. *Curr. Opin. Solid State Mater. Sci.* **2012**, *16*, 91–99. [CrossRef]
20. Yang, V.K.; Groenert, M.; Leitz, C.W.; Pitera, A.J.; Currie, M.T.; Fitzgerald, E.A. Crack formation in GaAs heteroepitaxial films on Si and SiGe virtual substrates. *J. Appl. Phys.* **2003**, *93*, 3859–3865. [CrossRef]
21. Tang, M.C.; Park, J.S.; Wang, Z.C.; Chen, S.M.; Jurczak, P.; Seeds, A.; Liu, H.Y. Integration of III-V lasers on Si for Si photonics. *Prog. Quantum Electrons* **2019**, *66*, 1–18. [CrossRef]
22. Georgakilas, A.; Stoemenos, J.; Tsagaraki, K.; Komninou, P.; Flevaris, N.; Panayotatos, P.; Christou, A. Generation and annihilation of antiphase domain boundaries in GaAs on Si grown by molecular beamepitaxy. *J. Mater. Res.* **1993**, *8*, 1908–1921. [CrossRef]
23. Fang, S.F.; Adomi, K.; Iyer, S.; Morkoç, H.; Zabel, H.; Choi, C.; Otsuka, N. Gallium arsenide and other compound semiconductors on silicon. *J. Appl. Phys.* **1990**, *8*, 1908–1921. [CrossRef]
24. Keyvaninia, S.; Muneeb, M.; Stankovi, S.; Van Veldhoven, P.J.; Van Thourhout, D.; Roelkens, G. Ultra-thin DVS-BCB adhesive bonding of III-V wafers, dies and multiple dies to a patterned silicon-on-insulator substrate. *Opt. Mater. Express* **2013**, *3*, 35–46. [CrossRef]
25. Stankovic, S.; Jones, R.; Sysak, M.N.; Heck, J.M.; Roelkens, G.; Thourhout, D.V. Hybrid III-V/Si distributed-Feedback Laser Based on Adhesive Bonding. *IEEE Photonics Technol. Lett.* **2012**, *24*, 2155–2158. [CrossRef]
26. Yokoyama, M.; Iida, R.; Ikku, Y.; Kim, S.; Takagi, H.; Yasuda, T.; Yamada, H.; Osamu, L.; Fukuhara, N.; Hata, M. Formation of III-V-on-insulators structures on Si by direct wafer bonding. *Semicond. Sci. Technol.* **2013**, *28*, 094009. [CrossRef]
27. Tanabe, K.; Watanabe, K.; Arakawa, Y. III-V/Si hybrid photonic devices by direct fusion bonding. *Sci. Rep.* **2012**, *2*, 349. [CrossRef]
28. Cheng, Y.T.; Lin, L.; Najafi, K. Localized silicon fusion and eutectic bonding for MEMS fabrication and packaging. *J. Microelectromechanical Syst.* **2000**, *9*, 3–8. [CrossRef]
29. Jones, R.; Doussiere, P.; Driscoll, J.B.; Lin, W.; Yu, H.; Akulova, Y.; Komljinovic, T. Heterogeneously Integrated InP/Silicon Photonics: Fabricating fully functional transceivers. *IEEE Nanotechnol. Mag.* **2019**, *13*, 17–26. [CrossRef]
30. Isenberg, J.; Warta, W. Free carrier absorption in heavily doped silicon layers. *Appl. Phys. Lett.* **2004**, *84*, 2265–2267. [CrossRef]
31. Richardson, C.J.K.; Lee, M.L. Metamorphic epitaxial materials. *MRS Bull.* **2016**, *41*, 193–198. [CrossRef]
32. Liu, C.W.; Hannon, J.B. New materials for post-Si computing. *MRS Bull.* **2014**, *39*, 658–662. [CrossRef]
33. Yu, P.Y.; Cardona, M. *Fundamentals of Semiconductors*; Springer: Berlin/Heidelberg, Germany, 2010; ISBN 978-3-662-03313-5.
34. Holt, D.B.; Yacobi, B.G. *Extended Defects in Semiconductors*; Cambridge University Press: Cambridge, UK, 2007.
35. Staudinger, P.; Mauthe, S.; Moselund, K.; Schmid, H. Concurrent Zinc-Blende and Wurtzite Film Formation by Selection of Confined Growth planes. *Nano Lett.* **2018**, *18*, 7856–7862. [CrossRef]
36. Staudinger, P.; Moselund, K.E.; Schmid, H. Exploring the Size imitations of Wurtzite III-V Film Growth. *Nano Lett.* **2020**, *20*, 686–693. [CrossRef] [PubMed]
37. Li, K.; Sun, H.; Ren, F.; Ng, K.W.; Tran, T.T.; Chen, R.; Chang-Hasnain, C.J. Tailoring the Optical Characteristics of Microsized InP Nano needles Directly Grown on Silicon. *Nano Lett.* **2014**, *14*, 183–190. [CrossRef] [PubMed]
38. De, A.; Pryor, C.E. Predicted band structures of III-V semiconductors in the wurtzite phase. *Phys. Rev. B* **2010**, *81*, 155210. [CrossRef]
39. Assali, S.; Zardo, I.; Plissard, S.; Kriener, D.; Verheijen, M.A.; Bauer, G.; Meijerink, A.; Belabbes, A.; Bechstedt, F.; Haverkort, J.E.M.; et al. Direct Band Gap Wurtzite Gallium phosphide Nanowires. *Nano Lett.* **2013**, *13*, 1559–1563. [CrossRef]
40. Wu, P.H.; Huang, Y.S.; Hsu, H.P.; Li, C.; Huang, S.H.; Tiong, K.K. The study of temperature dependent strain in Ge epilayer with SiGe/Ge buffer layer on Si substrate with different thickness. *Appl. Phys. Lett.* **2014**, *104*, 241605. [CrossRef]
41. Yang, J.J.; Jurczak, P.; Cui, F.; Li, K.S.; Tang, M.C.; Billiald, L.; Beanland, R.; Sanchez, R.; Liu, H.Y. Thin Ge buffer layer on silicon for integration of III-V on silicon. *J. Cryst. Growth* **2019**, *514*, 109–113. [CrossRef]
42. Kroemer, H. Polar-on-nonpolar epitaxy. *J. Cryst. Growth* **1987**, *81*, 193–204. [CrossRef]
43. Li, Y.; Lazzarini, L.; Giling, L.J.; Salviati, G. On the sublattice location of GaAs grown on Ge. *J. Appl. Phys.* **1994**, *76*, 5748. [CrossRef]
44. Faucher, J.; Masuda, T.; Lee, M.L. Initiation strategies for simultaneous control of antiphase domains and stacking faults in GaAs solar cells on Ge. *J. Vac. Sci. Technol. B* **2016**, *34*, 041203. [CrossRef]
45. Niehle, M.; Trampert, A.; Rodriguez, J.B.; Cerutti, L.; Tournié, E. Electron tomography on III-Sb heterostructures on vicinal Si(001) substrates: Anti-phase boundaries as a sink for threading dislocations. *Scripta. Mater.* **2017**, *132*, 5–8. [CrossRef]
46. Brammertz, G.; Mols, Y.; Degroote, S.; Motsnyi, V.; Leys, M.; Borghs, G.; Caymax, M. Low-temperature photoluminescence study of thin epitaxial GaAs films on Ge substrates. *J. Appl. Phys.* **2006**, *99*, 093514. [CrossRef]
47. Du, Y.; Xu, B.Q.; Wang, G.L.; Gu, S.H.; Kong, Z.Z.; Li, B.; Yu, J.H.; Bai, G.B.; Wang, W.W.; Radamson, H.H.; et al. Growth of high-quality epitaxy of GaAs on Si with engineered Ge buffer using MOCVD. *J. Mater. Sci.* **2021**, *32*, 6425–6437. [CrossRef]
48. Linder, K.K.; Philips, J.; Qasaimeh, O.; Bhattacharya, P.; Jiang, J.C. In(Ga)As/GaAs self-organized quantum dot light emitters grown on silicon substrates. *J. Cryst. Growth* **1999**, *201/202*, 1186–1189. [CrossRef]

49. Wang, T.; Liu, H.Y.; Lee, A.; Pozzi, F.; Seeds, A. 1.3 μm InAs/GaAs quantum-dot lasers monolithically grown on Si substrates. *Opt. Express* **2011**, *19*, 11381–11386. [CrossRef]
50. Kunert, B.; Mols, Y.; Baryshniskova, M.; Waldron, N.; Schulze, A.; Langer, R. How to control defect formation in monolithic III/V hetero-epitaxy on (100) Si: A critical review on current approaches. *Semicond. Sci. Technol.* **2018**, *33*, 093002. [CrossRef]
51. Sze, S.M.; Ng, K.K. *Physics of Semiconductor Devices*; Wiley-Interscience: New York, NY, USA, 2006; ISBN 9780471143239.
52. Yamaguchi, M.; Amano, C. Efficiency calculations of thin-film GaAs solar cells on Si substrates. *J. Appl. Phys.* **1985**, *58*, 3601. [CrossRef]
53. Clawson, A. Guide to references on III-V semiconductor chemical etching. *Mater. Sci. Eng. Rep.* **2001**, *31*, 1–438. [CrossRef]
54. Radamson, H.H.; Hallstedt, J. Application of high-resolution X-ray diffraction for detecting defects in SiGe(C) materials. *J. Phys. Condes. Matter* **2005**, *17*, S2315–S2322. [CrossRef]
55. Radamson, H.H.; Joelsson, K.B.; Ni, W.-X.; Hultman, L.; Hansson, G.V. Characterization of highly boron-doped Si, Si1-x Gex and Ge layers by high-resolution transmission electron microscopy. *J. Cryst. Growth* **1995**, *157*, 80. [CrossRef]
56. Madiomanana, K.; Bahri, M.; Rodriguez, J.B.; Largeau, L.; Cerutti, L.; Mauguin, O.; Castellano, A.; Patriarche, G.; Tournie, E. Silicon surface preparation for III-V molecular beam epitaxy. *J. Cryst. Growth* **2015**, *413*, 17–24. [CrossRef]
57. Houdre, R.; Morko, H. Properties of GaAs on Si grown by molecular beam epitaxy. *Crit. Rev. Solid State Mater. Sci.* **1990**, *16*, 91–114. [CrossRef]
58. Choi, D.; Harris, J.S.; Kim, E.; McIntyre, P.C.; Cagnon, J.; Stemmer, S. High-quality III-V semiconductor MBE growth on Ge/Si virtual substrates for metal-oxide-semiconductor device fabrication. *J. Cryst. Growth* **2009**, *311*, 1962–1971. [CrossRef]
59. Swartzentruber, B.S.; Kitamura, N.; Lagally, M.G.; Webb, M.B. Behavior of steps on Si(001) as a function of vicinality. *Phys. Rev. B* **1993**, *47*, 13432. [CrossRef]
60. Laracuente, A.R.; Whitman, L.J. Step structure and surface morphology of hydrogen-terminated silicon: (001) to (114). *Surf. Sci.* **2003**, *545*, 70–84. [CrossRef]
61. Sakamoto, T.; Hashiguchi, G. Si (001)-2×1 single-domain structure obtained by high temperature annealing. *Jpn. J. Appl. Phys.* **1986**, *25*, 1A:L78. [CrossRef]
62. Kawabe, M.; Ueda, T. Self-annihilation of antiphase boundary in GaAs on Si (100) grown by molecular beam epitaxy. *Jpn. J. Appl. Phys.* **1987**, *26*, 944–946. [CrossRef]
63. Masami, T.; Hidefumi, M.; Mitsuru, S.; Yoshio, I. Continuous GaAs Film Growth on Epitaxial Si Surface in Initial Stage of GaAs/Si Heteroepitaxy. *Jpn. J. Appl. Phys.* **1993**, *32*, L1252–L1255. [CrossRef]
64. Yu, H.W.; Chang, E.Y.; Yananoto, Y.; Tillack, B.; Wang, W.C.; Kuo, C.I.; Wong, Y.Y.; Nguyen, H.Q. Effect of graded-temperature arsenic prelayer on quality of GaAs on Ge/Si substrates by metalorganic vapor phase epitaxy. *Appl. Phys. Lett.* **2011**, *99*, 171908. [CrossRef]
65. Bolkhovityanov, Y.B.; Pchelyakov, O.P. GaAs epitaxy on Si substrates: Modern status of research and engineering. *Physics-Uspekhi* **2008**, *51*, 437–456. [CrossRef]
66. Buzynin, Y.; Shengurov, V.; Zvonkov, B.; Buzynin, A.; Denisov, S.; Baidus, N.; Drozdov, M.; Pavlov, D.; Yunin, P. GaAs/Ge/Si epitaxial substrates: Development and characteristics. *AIP Adv.* **2017**, *7*, 015304. [CrossRef]
67. Radamson, H.H.; Noroozi, M.; Jamshidi, A.; Thompson, P.E.; Östling, M. Strain engineering in GeSnSi materials. *ECS Trans.* **2013**, *50*, 527. [CrossRef]
68. Komatsu, Y.; Hosotani, K.; Fuyuki, T.; Matsunami, H. Heteroepitaxial growth of InGaP on Si with InGaP/GaPstep-graded buffer layers. *Jpn. J. Appl. Phys.* **1997**, *36*, 5425–5430. [CrossRef]
69. Groenert, M.E.; Leitz, C.W.; Pitera, A.J.; Yang, V.; Lee, H.; Ram, R.J.; Fitzgerald, E.A. Monolithic integration of room-temperature cw GaAs/AlGaAs lasers on Si substrates via relaxed graded GeSi buffer layers. *J. Appl. Phys.* **2003**, *93*, 362–367. [CrossRef]
70. Benediktovitch, A.; Ulyanenkov, A.; Rinaldi, F.; Saito, K.; Kaganer, V.K. Concentration and relaxation depth profiles of $In_xGa_{1-x}As/GaAs$ and $GaAs_{1-x}P_x/GaAs$ graded epitaxial films studied by x-ray diffraction. *Phys. Rev. B.* **2011**, *84*, 035302. [CrossRef]
71. Soga, T.; Hattori, S.; Takeyasu, M.; Mmeno, M. Characterization of epitaxially grown GaAs on Si substrates with III-V compounds intermediate layers by metalorganic chemical vapor deposition. *J. Appl. Phys.* **1985**, *57*, 4578. [CrossRef]
72. Peach, M.O.; Koehler, J.S. Forces exerted on dislocations and the stress fields produced by them. *Phys. Rev.* **1950**, *80*, 436–439. [CrossRef]
73. Yang, J.; Bhattacharya, P.; Mi, Z. High-performance $In_{0.5}Ga_{0.5}As$/GaAs quantum-dot lasers on silicon with multiple-layer quantum-dot dislocation filters. *IEEE Trans. Electron Devices* **2007**, *54*, 2849–2855. [CrossRef]
74. Shi, B.; Li, Q.; Lau, K.M. Self-organized InAs/InAlGaAs quantum dots as dislocation filters for InP films on (001) Si. *J. Cryst. Growth* **2017**, *464*, 28–32. [CrossRef]
75. Lee, J.W.; Shichijo, H.; Tsai, H.; Matyi, R.J. Defect reduction by thermal annealing of GaAs layers grown by molecular beam epitaxy on Si substrates. *Appl. Phys. Lett.* **1986**, *50*, 31–33. [CrossRef]
76. Choi, C.; Otsuka, N.; Munns, G.; Houdre, R.; Morkoc, H. Effect of in situ and ex situ annealing on dislocations in GaAs on Si substrates. *Appl. Phys. Lett.* **1987**, *50*, 992–994. [CrossRef]
77. Bogumilowicz, Y.; Hartmann, J.M.; Rochat, N.; Salaun, A.; Martin, M.; Bassani, F.; Baron, T.; David, S.; Bao, X.Y.; Sanchez, E. Threading dislocations in GaAs epitaxial layers on various thickness Ge buffers on 300 mm Si substrates. *J. Cryst. Growth* **2016**, *453*, 180–187. [CrossRef]

78. Okamoto, H.; Watanabe, Y.; Kadota, Y.; Ohmachi, Y. Dislocation Reduction in GaAs on Si by Thermal Cycles and InGaAs/GaAs Strained-Layer Superlattices. *Jpn. J. Appl. Phys.* **1987**, *26*, 1950–1952. [CrossRef]
79. Li, J.Z.; Bai, J.; Park, J.S.; Adekore, B.; Fox, K.; Carroll, M.; Lochtefeid, A.; Shellenbarger, Z. Defect reduction of GaAs epitaxy on Si (001) using selective aspect ratio trapping. *Appl. Phys. Lett.* **2007**, *91*, 021114. [CrossRef]
80. Sun, Y.T.; Lourdudoss, S. Effect of growth conditions on epitaxial lateral overgrowth InP on InP/Si (001) substrate by hydride vapor phase epitaxy. *Photonics Packag. Integr. III* **2003**, *4997*, 221–231.
81. Olsson, F.; Xie, M.; Lourdudoss, S.; Prieto, I.; Postigo, P.A. Epitaxial lateral overgrowth of InP on Si from nano-openings: Theoretical and experimental indication for defect filtering throughout the grown layer. *J. Appl. Phys.* **2008**, *104*, 093112. [CrossRef]
82. Kawabe, M.; Ueda, T. Molecular Beam Epitaxy of Controlled Single Domain GaAs on Si (100). *Jpn. J. Appl. Phys.* **1986**, *25*, L285. [CrossRef]
83. Lo, Y.H.; Wu, M.C.; Lee, H.; Wang, S. Dislocation microstructures on flat and stepped Si surfaces: Guidance for growing high-quality GaAs on (100) Si substrates. *Appl. Phys. Lett.* **1988**, *52*, 1386. [CrossRef]
84. Posthill, J.B.; Tarn, J.; Das, K.; Humphreys, T.; Parikh, N.R. Observation of antiphase domain boundaries in GaAs on silicon by transmission electron microscopy. *Appl. Phys. Lett.* **1988**, *53*, 1207. [CrossRef]
85. Strite, S.; UnIO, M.S.; Adomi, K.; Gao, G.-B.; Agarwal, A.; Rockett, A.; Morko, H. GaAs/Ge/GaAs heterostructures by molecular beam epitaxy. *J. Vac. Sci. Technol. B* **1990**, *8*, 1131. [CrossRef]
86. Yonezu, H. Control of structural defects in group III–V–N alloys grown on Si. *Semicond. Sci. Technol.* **2002**, *17*, 762–768. [CrossRef]
87. Wanarattikan, P.; Sakuntam, S.; Denchitcharoen, S.; Uesugi, K.; Kuboya, S.; Onabe, K. Influences of two-step growth and off-angle Ge substrate on crystalline quality of GaAs buffer layers grown by MOVPE. *J. Cryst. Growth* **2015**, *414*, 15–20. [CrossRef]
88. Alcotte, R.; Martin, M.; Moeyaert, J.; Cipro, R.; David, S.; Bassani, F.; Ducroquet, F.; Bogumilowicz, Y.; Sanchez, E.; Ye, Z.; et al. Epitaxial growth of antiphase boundary free GaAs layer on 300 mm Si (001) substrate by metal organic chemical vapour deposition with high mobility. *APL Mater.* **2016**, *4*, 046101. [CrossRef]
89. Grundmann, M.; Krost, A.; Bimberg, D. Observation of the first-order phase transition from single to double stepped Si (001) in metal organic chemical vapor deposition of InP on Si. *J. Vac. Sci. Technol. B* **1991**, *9*, 2158. [CrossRef]
90. Akiyama, M.; Kawarada, Y.; Kaminishi, K. Growth of single domain GaAs layer on (100)-Oriented Si substrate by MOCVD. *Jpn. J. Appl. Phys.* **1984**, *23*, 11. [CrossRef]
91. Zhou, X.L.; Pan, J.Q.; Liang, R.R.; Wang, J.; Wang, W. Epitaxy of GaAs thin film with low defect density and smooth surface on Si substrate. *J. Semicond.* **2014**, *35*, 073002. [CrossRef]
92. Du, Y.; Kong, Z.Z.; Toprak, M.S.; Wang, G.L.; Miao, Y.H.; Xu, B.Q.; Yu, J.H.; Li, B.; Lin, H.X.; Radamson, H.H.; et al. Investigation of the Heteroepitaxial Process Optimization of Ge Layers on Si (001) by RPCVD. *Nanomaterials* **2021**, *11*, 928. [CrossRef]
93. Andre, C.L.; Carlin, J.A.; Boeckl, J.J. Investigations of high-performance GaAs solar cells grown on Ge-Si$_{1-x}$/Ge$_x$-Si substrates. *IEEE Trans. Electron Devices* **2005**, *52*, 1055–1060. [CrossRef]
94. Ting, S.M.; Fitzgerald, E.A. Metal-organic chemical vapor deposition of single domain GaAs on Ge/Ge$_x$Si$_{1-x}$/Si and Ge substrates. *J. Appl. Phys.* **2000**, *87*, 2618–2628. [CrossRef]
95. Chen, K.P.; Yoon, S.F.; Ng, T.K.; Hanoto, H.; Lew, K.L.; Dohrman, C.L.; Fitzgeraid, E.A. Study of surface microstructure origin and evolution for GaAs grown on Ge/Si$_{1-x}$Ge$_x$/Si substrate. *J. Phys. D Appl. Phys.* **2009**, *42*, 035303. [CrossRef]
96. Samavedam, S.B.; Fitzgerald, E.A. Novel dislocation structure and surface morphology effects in relaxed Ge/Si-Ge (graded)/Si structures. *J. Appl. Phys.* **1997**, *81*, 3108–3116. [CrossRef]
97. Currie, M.T.; Samavedam, S.B.; Langdo, T.A.; Leitz, C.W.; Fitzgerald, E.A. Controlling threading dislocation densities in Ge on Si using graded SiGe layers and chemical-mechanical polishing. *Appl. Phys. Lett.* **1998**, *72*, 1718–1720. [CrossRef]
98. Ayers, J.E.; Schowalter, L.J.; Ghandhi, S.K. Post-growth thermal annealing of GaAs on Si(001) grown by organometallic vapor phase epitaxy. *J. Cryst. Growth* **1992**, *125*, 329–335. [CrossRef]
99. Jung, D.; Callahan, P.G.; Shin, B.; Mukherjee, K.; Gossard, A.C.; Bowers, J.E. Low threading dislocation density GaAs growth on on-axis GaP/Si (001). *J. Appl. Phys.* **2017**, *122*, 225703. [CrossRef]
100. Cho, N.-H.; Carter, C.B. Formation, faceting, and interaction behaviors of antiphase boundaries in GaAs thin films. *J. Mater. Sci.* **2001**, *36*, 4209–4222. [CrossRef]
101. Barrett, C.S.; Atassi, A.; Kennonet, E.L.; Weinrich, Z.; Haynes, K.; Bao, X.-Y.; Martin, P.; Jones, K.S. Dissolution of antiphase domain boundaries in GaAs on Si(001) via post-growth annealing. *J. Mater. Sci.* **2019**, *54*, 7028–7703. [CrossRef]
102. Shang, C.; Selvidge, J.; Hughes, E.; Norman, J.C.; Taylor, A.A.; Gossard, A.C.; Mukherjee, K.; Bowers, J.E. A Pathway to Thin GaAs Virtual Substrate on On-Axis Si (001) with Ultralow Threading Dislocation Density. *Phys. Status Solidi A* **2020**, *218*, 2000402. [CrossRef]
103. Ababou, Y.; Desjardins, P.; Chennouf, A. Structural and optical characterization of InP grown on Si (111) by metalorganic vapor phase epitaxy using thermal cycle growth. *J. Appl. Phys.* **1996**, *80*, 4997–5005. [CrossRef]
104. Tran, C.A.; Masut, R.A.; Cova, P.; Brebner, J.L.; Leonelli, R. Growth and characterization of InP on silicon by MOCVD. *J. Cryst. Growth* **1992**, *121*, 365–372. [CrossRef]
105. Wang, Y.F.; Wang, Q.; Jia, Z.G.; Li, X.Y.; Deng, C.; Ren, X.M.; Cai, S.W.; Huang, Y.Q. Three-step growth of metamorphic GaAs on Si(001) by low-pressure metal organic chemical vapor deposition. *J. Vac. Sci. Technol. B* **2013**, *31*, 051211. [CrossRef]

106. Hayafuji, N.; Miyashita, M.; Nishimura, T.; Kadoiwa, K.; Kumabe, H.; Murotani, T. Effect of Employing Positions of Thermal Cyclic Annealing and Strained-Layer Superlattice on Defect Reduction in GaAs-on-Si. *Jpn. J. Appl. Phys.* **1990**, *29*, 2371–2375. [CrossRef]
107. Shi, B.; Klamkin, J. Defect engineering for high quality InP epitaxially grown on on-axis (001) Si. *J. Appl. Phys.* **2020**, *127*, 033102. [CrossRef]
108. Yamaguchi, M.; Nishioka, T.; Sugo, M. Analysis of strained-layer superlattice effects on dislocation density reduction in GaAs on Si substrates. *Appl. Phys. Lett.* **1989**, *54*, 24–26. [CrossRef]
109. Qian, W.; Skowronski, M.; Kaspi, R. Dislocation density reduction in GaSb films grown on GaAs substrates by molecular beam epitaxy. *J. Electrochem. Soc.* **1997**, *144*, 1430–1434. [CrossRef]
110. Nozawa, K.; Horikoshi, Y. Low threading dislocation density GaAs on Si(100) with InGaAs/GaAs strained-layer superlattice grown by migration-enhanced epitaxy. *J. Electron. Mater.* **1992**, *21*, 641–645. [CrossRef]
111. Aleshkin, V.Y.; Baidus, N.V.; Dubinov, A.A.; Fefelov, A.G.; Krasilnik, A.F.; Kudryavteev, E.v.; Nekorkin, S.M.; Novikov, A.V.; Pavlov, D.A.; Samartsev, I.V.; et al. Monolithically integrated InGaAs/GaAs/AlGaAs quantum well laser grown by MOCVD on exact Ge/Si(001) substrate. *Appl. Phys. Lett.* **2016**, *109*, 061111. [CrossRef]
112. George, I.; Becagli, F.; Liu, H.Y.; Wu, J.; Tang, M.; Beanland, R. Dislocation filters in GaAs on Si. *Semiconductor Sci. Technol.* **2015**, *30*, 114004. [CrossRef]
113. Jung, D.; Norman, J.; Kennedy, M.; Herrick, R.; Shang, C.; Jan, C.; Gossard, A.C.; Bowers, J.E. Low Threshold Current 1.3 μm Fabry-Perot III-V quantum dot Lasers on (001) Si with Superior Reliability. In Proceedings of the Optical Fiber Communication Conference, San Diego, CA, USA, 1–15 March 2018; Optical Society of America: Washington, DC, USA, 2018. [CrossRef]
114. Dalfors, J.; Lundstrom, T.; Holtz, P.O.; Radamson, H.H.; Monemar, B. The effective masses in strained InGaAs/InP quantum wells deduced from magnetoexcitation spectroscopy. *Appl. Phys. Lett.* **1997**, *71*, 503–505. [CrossRef]
115. Tang, M.; Chen, S.; Wu, J.; Jiang, Q.; Dorogan, V.G.; Benamara, M.; Mazur, Y.I.; Salamo, G.J.; Seeds, A.; Liu, H. 1.3-μm InAs/GaAs quantum-dot lasers monolithically grown on Si substrates using InAlAs/GaAs dislocation filter layers. *Opt. Express* **2014**, *22*, 11528–11535. [CrossRef]
116. Tang, M.; Chen, S.; Wu, J.; Jiang, Q.; Kennedy, K.; Jurczak, P.; Liao, M.Y.; Beanland, R.; Seeds, A.; Liu, H.Y. Optimizations of Defect Filter Layers for 1.3-μm InAs/GaAs Quantum-Dot Lasers Monolithically Grown on Si Substrates. *IEEE J. Sel. Top. Quantum Electron.* **2016**, *22*, 50–56. [CrossRef]
117. Prost, W.; Khorenko, V.; Mofor, A.; Bakin, A.; Khorenko, E.; Ehrich, S.; Wehmann, H.-H.; Schlachetzki, A.; Tegude, F.-J. High-speed InP-based resonant tunneling diode on silicon substrate. In Proceedings of the 35th European Solid-State Device Research Conference, Grenoble, France, 12–16 September 2005; pp. 257–260. [CrossRef]
118. Shi, B.; Li, Q.; Wan, Y.; Ng, K.W.; Zou, X.B.; Tang, C.W.; Lau, K.M. InAlGaAs/InAlAs MQWs on Si Substrate. *IEEE Photonics Technol. Lett.* **2015**, *27*, 748–751. [CrossRef]
119. Shi, B.; Li, Q.; Lau, K. Epitaxial growth of high quality InP on Si substrates: The role of InAs/InP quantum dots as effective dislocation filters. *J. Appl. Phys.* **2018**, *123*, 193104. [CrossRef]
120. Norman, J.; Kennedy, M.J.; Selvidge, J.; Li, Q.; Wan, Y.; Liu, A.Y.; Callahan, P.G.; Echlin, M.P.; Pollock, T.M.; Lau, K.M.; et al. Electrically pumped continuous wave quantum dot lasers epitaxially grown on patterned, on-axis (001) Si. *Opt. Express* **2017**, *25*, 3927–3934. [CrossRef] [PubMed]
121. FITZGERALD, E.A.; CHAND, N. Epitaxial necking in GaAs grown on pre-patterned Si substrates. *J. Electron. Mater.* **1991**, *20*, 839–853. [CrossRef]
122. Langdo, T.A.; Leitz, C.W.; Currie, M.T.; Fitzgerald, E.A. High quality Ge on Si by epitaxial necking. *Appl. Phys. Lett.* **2000**, *76*, 3700–3702. [CrossRef]
123. Li, J.Z.; Bai, J.; Major, C.; Carroll, M.; Lochtefeld, A.; Shellenbarger, Z. Defect reduction of GaAs/Si epitaxy by aspect ratio trapping. *J. Appl. Phys.* **2008**, *103*, 106102. [CrossRef]
124. Li, J.Z.; Bai, J.; Hydrick, J.M.; Park, J.S.; Major, C.; Carroll, M.; Fiorenza, J.G.; Lochtefeld, A. Growth and characterization of GaAs layers on polished Ge/Si by selective aspect ratio trapping. *J. Cryst. Growth* **2009**, *311*, 133–3137. [CrossRef]
125. Wang, G.; Leys, M.R.; Loo, R.; Richard, O.; Bender, H.; Waldron, N.; Brammertz, G.; Dekoster, J.; Wang, W.; Seefeidt, M.; et al. Selective area growth of high quality InP on Si (001) substrates. *Appl. Phys. Lett.* **2010**, *97*, 121913. [CrossRef]
126. Paladugu, M.; Merckling, C.; Loo, R.; Richard, O.; Bender, H.; Dekoster, J.; Vandervorst, W.; Caymax, M.; Heyns, M.M. Site Selective Integration of III−V Materials on Si for Nanoscale Logic and Photonic Devices. *Cryst. Growth Des.* **2012**, *12*, 4696–4702. [CrossRef]
127. Guo, W.; Date, L.; Pena, V.; Bao, X.; Merckling, C.; Waldron, N.; Collasert, N.; Caymax, M.; Sanchez, E.; Vancoille, E.; et al. Selective metal-organic chemical vapor deposition growth of high quality GaAs on Si (001). *Appl. Phys. Lett.* **2014**, *105*, 062101. [CrossRef]
128. Orzali, T.; Vert, A.; O'Brien, B.; Herman, J.; Vivekanand, S.; Hill, R.J.; Karim, Z. GaAs on Si epitaxy by aspect ratio trapping: Analysis and reduction of defects propagating along the trench direction. *J. Appl. Phys.* **2015**, *118*, 105307. [CrossRef]
129. Li, Q.; Ng, K.W.; Lau, K.M. Growing antiphase-domain-free GaAs thin films out of highly ordered planar nanowire arrays on exact (001) silicon. *Appl. Phys. Lett.* **2015**, *106*, 072105. [CrossRef]
130. Wei, W.Q.; Qi, F.; Wang, Z.H.; Wang, T.; Zhang, J.J. Perspective: Optically-pumped III-V quantum dot microcavity lasers via CMOS compatible patterned Si (001) substrates. *J. Semicond.* **2019**, *40*, 101303. [CrossRef]

131. Tang, G.P.; Lubnow, A.; Wehmann, H.; Zwinge, G.; Schlachetzki, A. Antiphase-Domain-Free InP on (100) Si. *Jpn. J. Appl. Phys.* **1992**, *31*, L1126–L1128. [CrossRef]
132. Loo, R.; Wang, G.; Orzali, T.; Waldron, N.; Merckling, C.; Leys, M.R.; Richard, O.; Bender, H.; Eyben, P.; Vandervoust, W. Selective Area Growth of InP on On-Axis Si (001) Substrates with Low Antiphase Boundary Formation. *J. Electrochem. Soc.* **2012**, *159*, H260–H265. [CrossRef]
133. Li, J.Z.; Bai, J.; Hydrick, J.M.; Fiorenza, J.M.; Major, C.; Carroll, M.; Shellenbarger, Z.; Lochtefeld, A. Thin Film InP Epitaxy on Si (001) using Selective Aspect Ratio Trapping. *ECS Trans.* **2009**, *18*, 887–894. [CrossRef]
134. Merckling, C.; Waldron, N.; Jiang, S.; Guo, W.; Collaert, N.; Caymax, M.; Vancoille, E.; Barla, K.; Thean, A.; Heyns, M.; et al. Heteroepitaxy of InP on Si (001) by selective-area metal organic vapor-phase epitaxy in sub-50nm width trenches: The role of the nucleation layer and the recess engineering. *J. Appl. Phys.* **2014**, *115*, 023710. [CrossRef]
135. Tommaso, O.; Vert, A.; Kim, T.W.; Hung, P.Y.; Herman, J.L.; Vivekanand, S.; Huang, G.; Kelman, M.; Karim, Z.; Hill, R.; et al. Growth and characterization of an $In_{0.53}Ga_{0.47}As$-based Metal-Oxide-Semiconductor Capacitor (MOSCAP) structure on 300 mm on-axis Si (001) wafers by MOCVD. *J. Crys. Growth* **2015**, *427*, 72–79. [CrossRef]
136. Li, S.; Zhou, X.; Li, M.; Kong, X.; Mi, J.; Wang, M.; Pan, J. Ridge InGaAs/InP multi-quantum-well selective growth in nanoscale trenches on Si(001) substrate. *Appl. Phys. Lett.* **2016**, *108*, 021902. [CrossRef]
137. Saravanan, S.; Hayashi, Y.; Soga, T.; Jimbo, T. Growth and characterization of GaAs epitaxial layers on Si/porous Si/Si substrate by chemical beam epitaxy. *J. Appl. Phys.* **2001**, *89*, 5215–5218. [CrossRef]
138. Takano, Y.; Kururi, T.; Kuwahara, K.; Fuke, S. Residual strain and threading dislocation density in InGaAs layers grown on Si substrates by metalorganic vapor-phase epitaxy. *Appl. Phys. Lett.* **2001**, *78*, 93–95. [CrossRef]
139. Zhang, J.-Y.; Wei, W.-Q.; Wang, J.-H.; Cong, H.; Feng, Q.; Wang, Z.-H.; Wang, T.; Zhang, J.-J. Epitaxial growth of InAs/GaAs quantum dots on {113}-faceted Ge/Si (001) hollow substrate. *Opt. Mater. Express* **2020**, *10*, 1045. [CrossRef]
140. Du, Y.; Wang, G.; Miao, Y.; Xu, B.; Li, B.; Kong, Z.; Yu, J.; Zhao, X.; Lin, H.; Radamson, H.H.; et al. Strain Modulation of Selectively and/or Globally Grown Ge Layers. *Nanomaterials* **2021**, *11*, 1421. [CrossRef] [PubMed]
141. Kohen, D.; Bao, S.Y.; Lee, K.H.; Tan, C.S.; Yoon, S.F.; Fitzgerald, E.A. The role of AsH_3 partial pressure on anti-phase boundary in GaAs-on-Ge grown by MOCVD—Application to a 200 mm GaAs virtual substrate. *J. Cryst. Growth* **2015**, *421*, 58–65. [CrossRef]
142. Fiorenza, J.G.; Park, J.S.; Hydrick, J.; Li, J.; Li, J.Z.; Curtin, M.; Lochtefeld, A. Aspect ratio trapping: A unique technology for integrating Ge and III-Vs with silicon CMOS. *ECS Trans.* **2010**, *33*, 963–976. [CrossRef]
143. Han, Y.; Xue, Y.; Laua, K.M. Selective lateral epitaxy of dislocation-free InP on silicon-on-insulator. *Appl. Phys. Lett.* **2019**, *114*, 192105. [CrossRef]
144. Ujiie, Y.; Nishinaga, T. Epitaxial Lateral Overgrowth of GaAs on a Si Substrate. *Jpn. J. Appl. Phys.* **1989**, *28*, L337–L339. [CrossRef]
145. Sakawa, S.; Nishinaga, T. Effect of Si Doping on Epitaxial Lateral Overgrowth of GaAs on GaAs-Coated Si Substrate. *Jpn. J. Appl. Phys.* **1992**, *31*, L359–L361. [CrossRef]
146. Tsaur, B.-Y.; McClelland, R.W.; John, C.C.; Gale, R.P.; Salerno, J.P.; Vojak, B.A.; Bozler, C.O. Low-dislocation-density GaAs epilayers grown on Ge-coated Si substrates by means of lateral epitaxial overgrowth. *Appl. Phys. Lett.* **1982**, *41*, 347–349. [CrossRef]
147. He, Y.R.; Wang, J.; Hu, H.Y.; Wang, Q.; Huang, Y.Q.; Ren, X.M. Coalescence of GaAs on (001) Si nano-trenches based on three-stage epitaxial lateral Overgrowth. *Appl. Phys. Lett.* **2015**, *106*, 20. [CrossRef]
148. Naritsuka, S.; Nishinaga, T. Epitaxial lateral overgrowth of InP by liquid phase epitaxy. *J. Cryst. Growth* **1995**, *146*, 314–318. [CrossRef]
149. Naritsuka, S.; Nishinaga, T. Spatially resolved photoluminescence of laterally overgrown InP on InP-coated Si substrates. *J. Cryst. Growth* **1997**, *174*, 622–629. [CrossRef]
150. Kochiya, T.; Oyama, Y.; Kimura, T.; Suto, K.; Nishizawa, J. Dislocation-free large area InP ELO layers by liquid phase epitaxy. *J. Cryst. Growth* **2005**, *281*, 263–274. [CrossRef]
151. Metaferia, W.; Junesand, C.; Gau, M.-H.; Lo, I.; Pozina, G.; Hultman, L.; Lourdudoss, S. Morphological evolution during epitaxial lateral overgrowth of indium phosphide on silicon. *J. Cryst. Growth.* **2011**, *332*, 27–33. [CrossRef]
152. Han, Y.; Lau, K.M. III-V lasers selectively grown on (001) silicon. *J. Appl. Phys.* **2020**, *128*, 200901. [CrossRef]
153. Parillaud, O.; Gil-Lafon, E. High quality InP on Si by conformal growth. *Appl. Phys. Lett.* **1996**, *68*, 2654. [CrossRef]
154. Metaferia, W.; Kataria, H.; Sun, Y.T.; Lourdudoss, S. Growth of InP directly on Si by corrugated epitaxial lateral overgrowth. *J. Phys. D Appl. Phys.* **2015**, *48*, 045102. [CrossRef]
155. Omanakuttan, G.; Sacristán, O.M.; Marcinkevičius, S.; Uždavinys, T.; Jiménez, J.; Ali, H.; Lourdudoss, S.; Sun, Y.T. Optical and interface properties of direct InP/Si heterojunction formed by corrugated epitaxial lateral overgrowth. *Opt. Mater. Express* **2019**, *9*, 1488–1500. [CrossRef]
156. Nakamura, S.; Senoh, M.; Nagahama, S.; Iwasa, N.; Yamada, T.; Matsushita, T.; Kiyoku, H.; Suginoto, Y.; Kozaki, T.; Umemoto, H.; et al. Present status of InGaN/GaN/AlGaN-based laser diodes. *J. Cryst. Growth* **1998**, *189*, 820–825. [CrossRef]
157. Deura, M.; Kondo, Y.; Takenaka, M.; Takagi, S.; Nakano, Y.; Sugiyama, M. Twin free InGaAs thin layer on Si by multi-step growth using micro-channel selective-area MOVPE. *J. Cryst. Growth* **2010**, *312*, 1353–1358. [CrossRef]
158. Chuang, L.C.; Moewe, M.; Chase, C.; Kobayashi, N.P.; Hasnain, C.C. Critical diameter for III-V nanowires grown on lattice-mismatched substrates. *Appl. Phys. Lett.* **2007**, *90*, 043115. [CrossRef]
159. Ghalamestani, S.G.; Johansson, S.; Borg, B.M.; Lind, E.; Dick, K.A.; Wernersson, L.-E. Uniform and position-controlled InAs nanowires on 2" Si substrates for transistor applications. *Nanotechnology* **2012**, *23*, 015302. [CrossRef] [PubMed]

160. Hertenberger, S.; Rudolph, D.; Bichler, M.; Finley, J.J.; Abstreiter, G.; Koblmuller, G. Growth kinetics in position-controlled and catalyst-free InAs nanowire arrays on Si(111) grown by selective area molecular beam epitaxy. *J. Appl. Phys.* **2010**, *108*, 114316. [CrossRef]
161. Björk, M.T.; Schmid, H.; Breslin, C.M.; Gignac, L.; Riel, H. InAs nanowire growth on oxide-masked <111> silicon. *J. Cryst. Growth* **2012**, *344*, 31–37. [CrossRef]
162. Grégoire, G.; Zeghouane, M.; Goosney, C.; Goktas, N.I.; Staudinger, P.; Schmid, H.; Moselund, K.E.; Taliercio, T.; Tournie, E.; Trassoudaine, A.; et al. Selective Area Growth by Hydride Vapor Phase Epitaxy and Optical Properties of InAs Nanowire Arrays. *Cryst. Growth Des.* **2021**, *21*, 5151–5163. [CrossRef]
163. Nishi, K.; Yokoyama, M.; Yokoyama, H.; Takenaka, M.; Takagi, S. Thin Body GaSb-OI P-MOSFETs on Si Wafers Fabricated by Direct Wafer Bonding. In Proceedings of the IEEE International Conference on Indium Phosphide and Related Materials (IPRM), Montpellier, France, 11–15 May 2014; pp. 14–15. [CrossRef]
164. Takei, K.; Madsen, M.; Fang, H.; Kapadia, R.; Chuang, S.; Kim, H.S.; Liu, C.H.; Nah, J.; Krishna, S.; Chueh, Y.L.; et al. Nanoscale InGaSb Heterostructure Membranes on Si Substrates for High Hole Mobility Transistors. *Nano Lett.* **2012**, *12*, 2060–2066. [CrossRef] [PubMed]
165. Yang, Z.; Han, N.; Fang, M.; Lin, H.; Cheung, H.Y.; Yip, S.; Wang, E.; Hung, T.; Wong, C.Y.; Ho, J. Surfactant-assisted chemical vapour deposition of high-performance small-diameter GaSb nanowires. *Nat. Commun.* **2014**, *5*, 55249. [CrossRef]
166. Schmid, H.; Borg, M.; Moselund, K.; Gignac, L.; Breslin, C.M.; Bruley, J.; Cutaia, D.; Riel, H. Template-assisted selective epitaxy of III-V nanoscale devices for co-planar heterogeneous integration with Si. *Appl. Phys. Lett.* **2015**, *106*, 233101. [CrossRef]
167. Borg, M.; Schmid, H.; Gooth, J.; Rossell, M.; Cutaia, D.; Knoedler, M.; Bologna, N.; Wirths, S.; Moselund, K.E.; Riel, H. High-Mobility GaSb Nanostructures Cointegrated with InAs on Si. *ACS Nano* **2017**, *11*, 2554–2560. [CrossRef]

Article

Core-Shell Dual-Gate Nanowire Charge-Trap Memory for Synaptic Operations for Neuromorphic Applications

Md. Hasan Raza Ansari, Udaya Mohanan Kannan and Seongjae Cho *

Graduate School of IT Convergence Engineering, Gachon University, Seongnam 13120, Korea; hasanrazaadnan@gmail.com (M.H.R.A.); kannan.um@gmail.com (U.M.K.)
* Correspondence: felixcho@gachon.ac.kr; Tel.: +82-31-750-8722

Abstract: This work showcases the physical insights of a core-shell dual-gate (CSDG) nanowire transistor as an artificial synaptic device with short/long-term potentiation and long-term depression (LTD) operation. Short-term potentiation (STP) is a temporary potentiation of a neural network, and it can be transformed into long-term potentiation (LTP) through repetitive stimulus. In this work, floating body effects and charge trapping are utilized to show the transition from STP to LTP while de-trapping the holes from the nitride layer shows the LTD operation. Furthermore, linearity and symmetry in conductance are achieved through optimal device design and biases. In a system-level simulation, with CSDG nanowire transistor a recognition accuracy of up to 92.28% is obtained in the Modified National Institute of Standards and Technology (MNIST) pattern recognition task. Complementary metal-oxide-semiconductor (CMOS) compatibility and high recognition accuracy makes the CSDG nanowire transistor a promising candidate for the implementation of neuromorphic hardware.

Keywords: short-term potentiation (STP); long-term potentiation (LTP); charge-trap synaptic transistor; band-to-band tunneling; pattern recognition; neural network; neuromorphic system

Citation: Ansari, M..H.R.; Kannan, U.M.; Cho, S. Core-Shell Dual-Gate Nanowire Charge-Trap Memory for Synaptic Operations for Neuromorphic Applications. *Nanomaterials* 2021, 11, 1773. https://doi.org/10.3390/nano11071773

Academic Editor: Francisco Javier García Ruiz

Received: 1 June 2021
Accepted: 6 July 2021
Published: 7 July 2021

Publisher's Note: MDPI stays neutral with regard to jurisdictional claims in published maps and institutional affiliations.

Copyright: © 2021 by the authors. Licensee MDPI, Basel, Switzerland. This article is an open access article distributed under the terms and conditions of the Creative Commons Attribution (CC BY) license (https://creativecommons.org/licenses/by/4.0/).

1. Introduction

Modern day computer architectures suffer from the Von Neumann bottleneck where the separation of memory and processing units impose a fundamental limit on the maximum achievable processing speeds. In addition, the high levels of energy consumption in the conventional computing architecture are a major drawback especially for data intensive applications like big data analytics, machine learning etc. The human brain on the other hand has a highly energy efficient design where the storage and processing are carried out locally using a hugely parallel network of neurons and synapses [1,2]. Neuromorphic systems are gaining research attention due to their potential to design computer chips that can mimic the human brain in merging memory and processing [1,2]. The brain functions (observation, reorganization, learning, and memorization) are performed by neurons (computing elements) and synapses (memory elements) [1,2]. In the neuromorphic system, an artificial synaptic device plays a key role in linking the artificial neurons and modulating the connection strength (synaptic weight) between neurons [3–15]. In order to realize brain-like computing, different types of artificial synaptic devices have been proposed for artificial intelligence applications [3–22]. The major applications for these artificial synaptic transistors are neuromorphic in-memory computing chip, artificial sensory perception, humanoid robotics, memorize, and recognize massive and unstructured data through parallel and power-efficient ways [3–22]. Charge tapping/de-trapping based artificial synapse are favorable for in-memory computing applications due to their stable analogue conductance state and nonvolatile characteristic [12].

Among these electronic artificial synapses, two terminal non-volatile memory devices such as resistive random access memory (RRAM) and phase change memory (PCM) are strong candidates due to their small form factor [5,6,21,22]. However, due to variability and reliability issues in these devices, the recognition rate undergoes fast degradation.

These issues can be resolved with synaptic devices based on complementary metal-oxide-semiconductor (CMOS) field-effect transistors (FET) [15–20]. These FET-based electronic synaptic devices operate with a charge trap layer, which is an attractive candidate with many advantages, (i) low synaptic current; (ii) good reliability; (iii) high integration density; (iv) a large conductance window; and (v) process compatibility with CMOS [15–20]. These FET-type devices show the feasibility of artificial synapse but may have difficulties in scaling due to short channel effect and band-to-band tunneling in nanoscale regime. These effects degrade the performance of a synaptic transistor and reduces state retention. These issues can be resolved with multigate transistors [23]. Among these transistors, gate all around (GAA) transistor shows the better performance due better controllability over the gate [23,24]. Recently, a novel GAA transistor utilizing a nanotube with a core gate has been proposed to improve the gate controllability and enhance the device performances with same effective silicon film thickness [24–28] and also deal with better scaling over the nanowires [24,25].

Therefore, in this work, we emulate biological synaptic properties such as short-term potentiation (STP), long-term potentiation (LTP), and long-term depression (LTD) in an artificial synaptic device with a novel core-shell gate all around transistor. In neuromorphic systems, STP plays a key role in learning mechanism of the human brain [1,2]. STP is observed due to floating body effect (non-volatile characteristic) and LTP and LTD is observed due to trapping and detrapping of the holes from the nitride layer (volatile characteristic). In order to evaluate the inference capabilities of the proposed synaptic device, the weights (conductance values) extracted from the LTP/LTD characteristics of the device are utilized for pattern recognition using a simulated artificial neural network. The designed neural network recorded a high degree of recognition accuracy of 92.28% for the synaptic device.

2. Device Design Strategies and Models Calibration

Figure 1a–c show the schematic representation of the core-shell dual-gate (CSDG) nanowire transistor in 3D and biological synapse, 2D and top view, respectively. The simulated device dimensions and parameters are optimized and illustrated in Table 1. The device consists of two gates (core (inner) and shell (outer) gate) in gate all around or surrounded gate manner, which helps to accumulate the carriers from the silicon film and makes the deeper potential well for charge storage [24–28]. The dual gate in the device increases the gate controllability and shows better scalability [24–28]. In this work, core gate with thinner oxide (SiO_2 of 2 nm), which is responsible for the floating body effect for STP while thicker oxide (SiO_2 of 6 nm/nitride of 4 nm/SiO_2 of 2 nm (O/N/O)) is utilized for charge trapping to show LTP and LTD operations [17]. The device is simulated with the Silvaco ATLAS (Santa Clara, CA, USA) simulation tool with calibrated models. The physical simulation models are calibrated with experimental transfer characteristics of gate all around and double gate inversion mode transistor as shown in Figure 2a,b, respectively.

Table 1. Device specification for CSDG device as synapse.

Device Parameters	Values
Gate length (L_g)	100 nm–50 nm
Silicon core channel radius (T_{Si})	20 nm
Tunneling oxide thickness (T_{TOX})	2 nm (SiO_2)
Nitride layer thickness (T_{NOX})	4 nm (Si_3N_4)
Blocking oxide thickness (T_{BOX})	6 nm (SiO_2)
Oxide thickness (T_{OX})	2 nm (SiO_2)
Core-gate workfunction ($\phi_{m,Core}$)	4.6 eV
Shell-gate workfunction ($\phi_{m,Shell}$)	4.8 eV
Channel doping (N_A)	10^{15} cm^{-3}
Source/drain doping (N_D)	10^{20} cm^{-3}

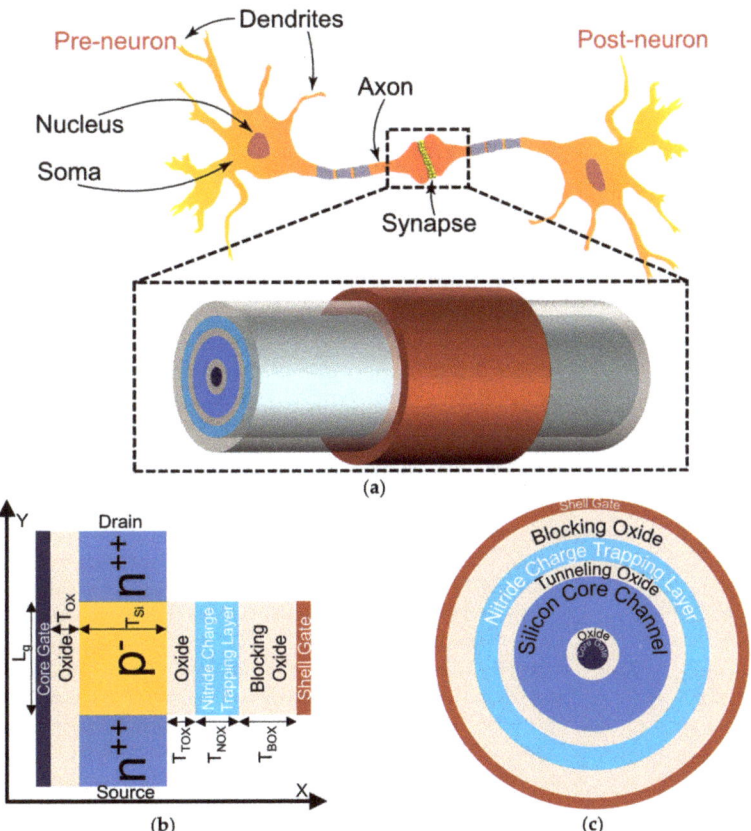

Figure 1. (**a**) Schematic representation of biological synapse and 3D illustration of core-shell dual-gate (CSDG) nanowire transistor. Schematic representation in (**b**) 2D and (**c**) top view of CSDG nanowire transistor for artificial synapse device.

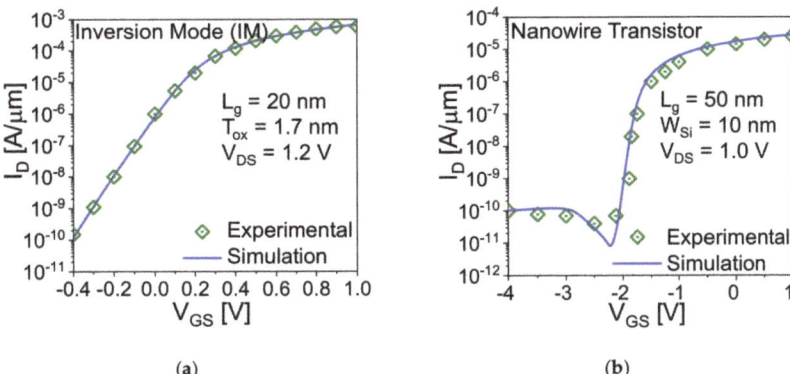

Figure 2. Comparison of simulated transfer characteristic with experimental (**a**) double-gate inversion mode [29] and (**b**) nanowire transistor [30].

Synaptic transistor operations (STP, LTP, and LTD) are based on charge generation, recombination, trapping, and detrapping of charge from the device [15–17]. In order to generate the charge in the device, non-local band-to-band-tunneling (BTBT) and impact ionization models are incorporated while for trapping and detrapping the charge from the nitride layer, a macro model (DYNASONOS) is embedded in the simulation, along with hot carrier injection, Fowler–Nordheim (F–N) tunneling, and Poole–Frenkel emission models have also been activated. Other physical models are also incorporated such as the Fermi–Dirac statistics model, concentration-dependent, Shockley–Read–Hall generation, and recombination model, Auger recombination model, density gradient quantum effect, bandgap narrowing model, concentration and temperature-dependent carrier lifetime model, Lombardi's mobility model.

3. Results and Discussion

Thanks to core-shell dual gate nanowire transistor, which creates a deeper potential well for charge storage and enhances the retention characteristic of a capacitorless dynamic random access memory (1T DRAM) cell [31]. In this work, the biasing and timing schemes are optimized to achieve STP, LTP, and LTD in the device and mimic the core-shell dual gate transistor as an electronic synapse. The working principle of the device as synapse is based on floating body effect and charge trapping and de-trapping from the nitride layer [15–17,32,33]. These operations are based on band-to-band-tunneling, impact ionization, hot carrier injection, and F–N tunneling in the device [15–17,19,32,33]. In this work, program operation is based on the BTBT mechanism due to its low power consumption and better reliability issue. In order to achieve the transition from STP to LTP at $\geq$ 10th pulse, core-gate voltage is optimized with fixed drain voltage of 1.4 V and shell gate voltage of -1.0 V. Achieving STP in the device not only shows the capability for both STP and LTP, but also consumes lower voltage for LTP operation [19]. The reason for the optimization of the core-gate voltage is that band-to-band-tunneling in the device takes place near the channel and drain due to thinner oxide for core-gate. Figure 3a,b show the transfer characteristics of the device with independent gate operation, respectively. It is clearly observed from Figure 3a (drain current-core-gate voltage) that at lower gate voltage, drain current is increasing due to tunnelling between channel and drain compared to Figure 3b (drain current-shell-gate voltage).

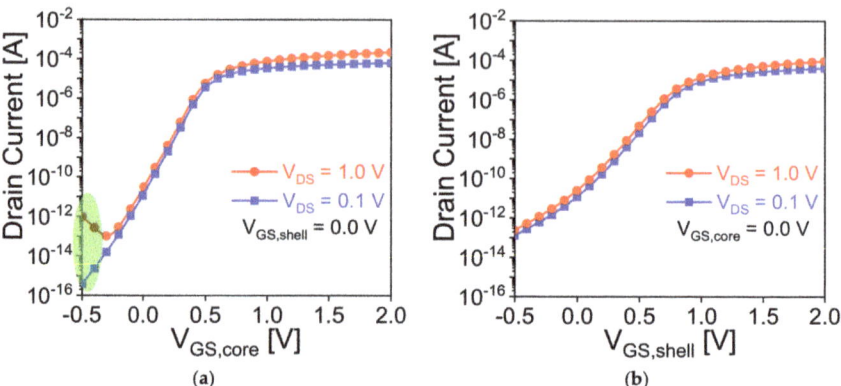

Figure 3. Dual-gate operations of the synaptic memory device. Transfer characteristics at V_{DS} = 0.1 V and 1 V. (**a**) I_D-$V_{GS,core}$ curves at $V_{GS,shell}$ = 0 V. (**b**) I_D-$V_{GS,shell}$ curves at $V_{GS,core}$ = 0 V. L_g = 100 nm and T_{Si} = 20 nm.

Figure 4a shows the voltage waveform during potentiation operation to find the optimized bias with fixed drain and shell gate voltage. A repetitive pulse with pulse and interval width of 2 µs are applied to mimic the device as synapse and shows the transformation from STP to LTP through trapping the charge in the nitride layer. To achieve efficient neuromorphic computational functions, 35 consecutive pulses are applied to stimulate potentiation and, then 35 pulses for depression. Figure 4b,c show the variation of electric field (E field) and energy band diagram, respectively, with different core-gate voltages (−0.1 V, −0.2 V, and −0.3 V). The E field and energy band diagram extracted 1 nm below core gate oxide. CB and VB indicate the conduction and valence band energies. Figure 4b,c reveal that increase in core-gate voltage (in negative magnitude) increases the E field and reduces the tunneling width between the core-gate and drain junctions, which helps to enhance the tunneling in the device and generates more electron hole pairs. The generated holes are stored in the lower potential region, and furthermore, these stored holes trigger the impact ionization in the device and start trapping the holes in the nitride layer. In Figure 4c, the barrier between source and channel is lower for $V_{GS,core}$ of −0.3 V compared to lower gate bias. This confirms that the stored holes contribute positive potential and trigger impact ionization in the device and achieves transition at lower pulse as shown in Figure 4d. Figure 4d shows the variation of trapped charge in the nitride layer during potentiation pulse. Similarly, in the case of LTD, shell gate voltage plays a crucial role to de-trap the holes from the nitride layer. Figure 5a shows the voltage waveform during the depression operation to de-trap the charge. Figure 5b shows the energy band diagram of the device during depression operation with different shell gate voltage. Increase in shell gate voltage increases the F–N tunneling probability in the device and starts de-trapping the trapped charges from the nitride layer as shown in Figure 5c.

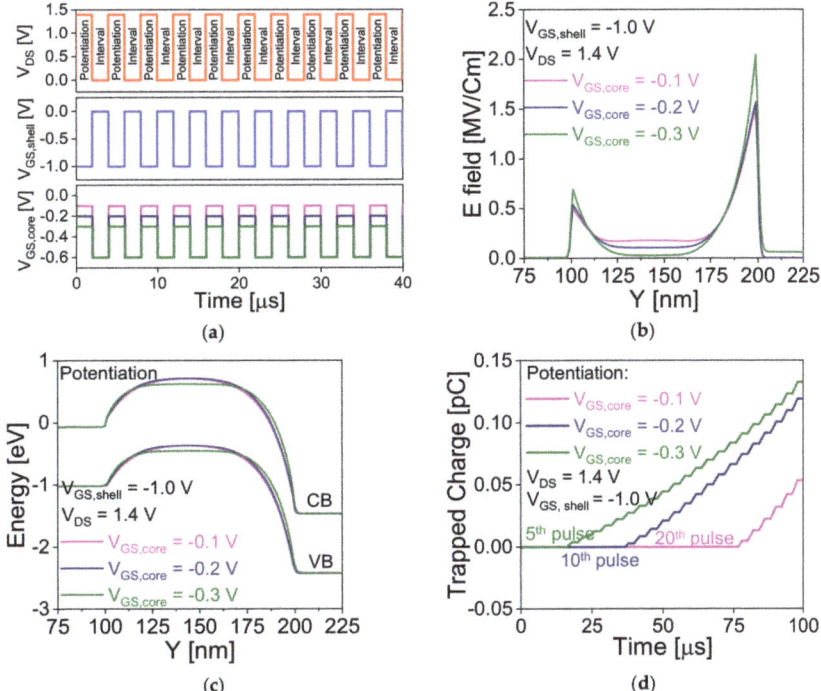

Figure 4. Potentiation operation. (**a**) voltage waveform during potentiation operation. Variation of (**b**) electric field (E filed) and (**c**) energy band diagram with core-gate voltage along the y-direction. (**d**) Variation of trapped charge for different $V_{GS,core}$. E field and energy band diagram are extracted 1 nm below of the core gate oxide.

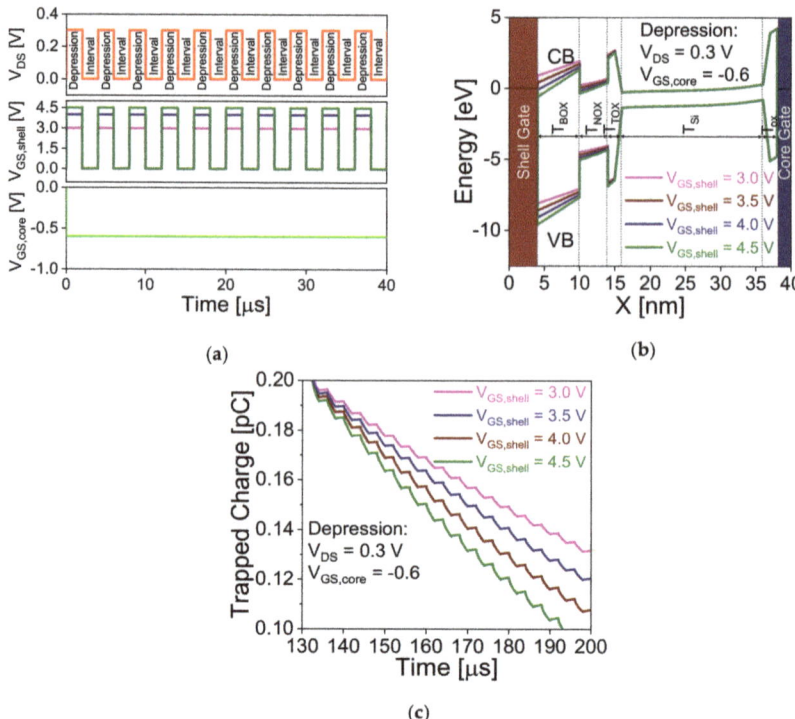

Figure 5. Depression. (**a**) voltage waveform during depression operation. Variation of (**b**) energy band diagram with core-gate voltage along the *x*-direction. (**c**) variation of trapped charge for different $V_{GS,shell}$.

Figure 6a,b show the transient analysis and trapped charge in the nitride layer of the device during potentiation and depression operation, respectively. During potentiation, a repetitive pulse start trapping the holes in the nitride layer due to hot hole injection and F-N tunnelling at lower bias, which shows the transformation from STP to LTP at the 10th pulse. At the 10th pulse, a sharp transition is observed in the drain current and nitride layer, which confirms that the device is in LTP state. Figure 7a–d show the contour plots of impact ionization rate in the device and charge trapped in the nitride layer during potentiation operation at different pulses (1st pulse, 5th pulse, 10th pulse, and 20th pulse). Initially (1st pulse) for STP, the BTBT mechanism is utilized to generate the holes in the device by applying drain voltage (V_{DS}) = 1.4 V, core gate voltage ($V_{GS,core}$) = −0.2, and shell gate voltage ($V_{GS,shell}$) = −1.0 V. The generated holes are accumulated at a lower potential region and electrons start drifting towards the drain side. Further, on increasing the number of repetitive pulses, electrons obtain sufficient energy to trigger the impact ionization in the device and generates more number of electrons-holes pairs in the device. At the 10th pulse, the generated holes get sufficient energy to get trapped in the nitride layer due to F–N tunneling even at lower bias. At the 20th pulse, it can be observed that impact ionization rate is constant while trapped charge in the nitride layer is increasing with increase in pulse. This confirms that after the 10th pulse, generated holes with the energy at the Fermi-Dirac distribution tail have higher probabilities of injection into nitride layer due to hot hole injection. De-trapping the holes from the nitride layer is performed through F–N tunnelling. LTD operation is performed by applying a lower drain bias, V_{DS} = 0.3 V, $V_{GS,core}$ = −0.6, and higher $V_{GS,shell}$ = 4.0 V.

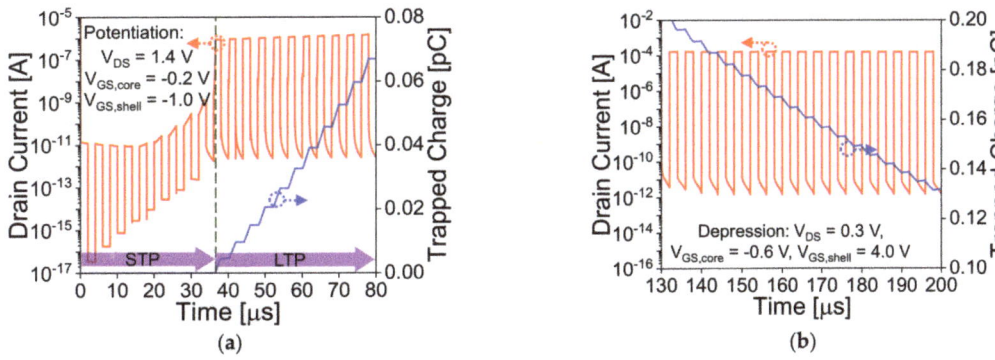

Figure 6. Transient analysis and trapped charges in the nitride layer during (**a**) potentiation and (**b**) depression. A sharp transition in current and trapped charges in the transient analysis of potentiation shows the transformation from short-term potentiation (STP) to long-term potentiation (LTP).

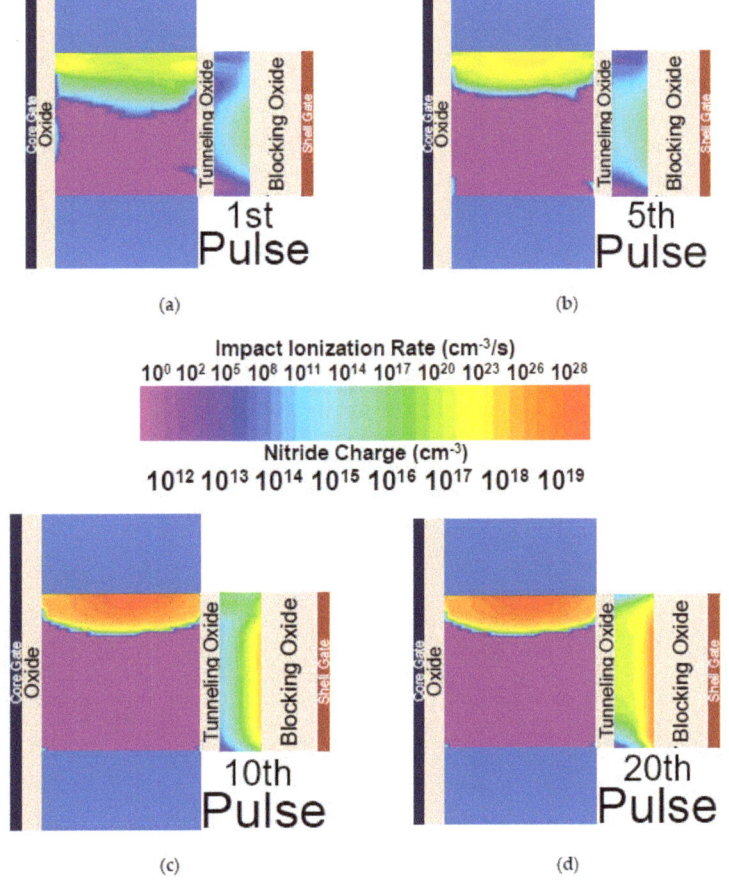

Figure 7. Contour plots of Impact Ionization and charge trapped in the nitride layer during potentiation at (**a**) 1st, (**b**) 5th, (**c**) 10th, and (**d**) 20th pulse.

The transformation from STP to LTP can also be observed from the transfer characteristics (Figure 8a,b) and transient analysis (Figure 8c,d) of the device during inference (read) operation. Inference in the biological brain is analogous to the read operation in an artificial synaptic transistor. To avoid the disturbance and for non-destructive read a lower bias is applied. Figure 8a,b show the drain current—shell gate voltage curve at $V_{DS} = V_{GS,core} = 0.1$ V for different pulses of potentiation and depression, respectively. In the case of potentiation operation, as shown in Figure 8a the transfer characteristics of the device are unchanged during STP states (from 0 to 9th pulse) while from the 10th pulse (LTP), threshold voltage (V_{TH}) is decreasing with increase in pulse due to increase in trapped charge in the nitride layer (Figures 4d and 6a).

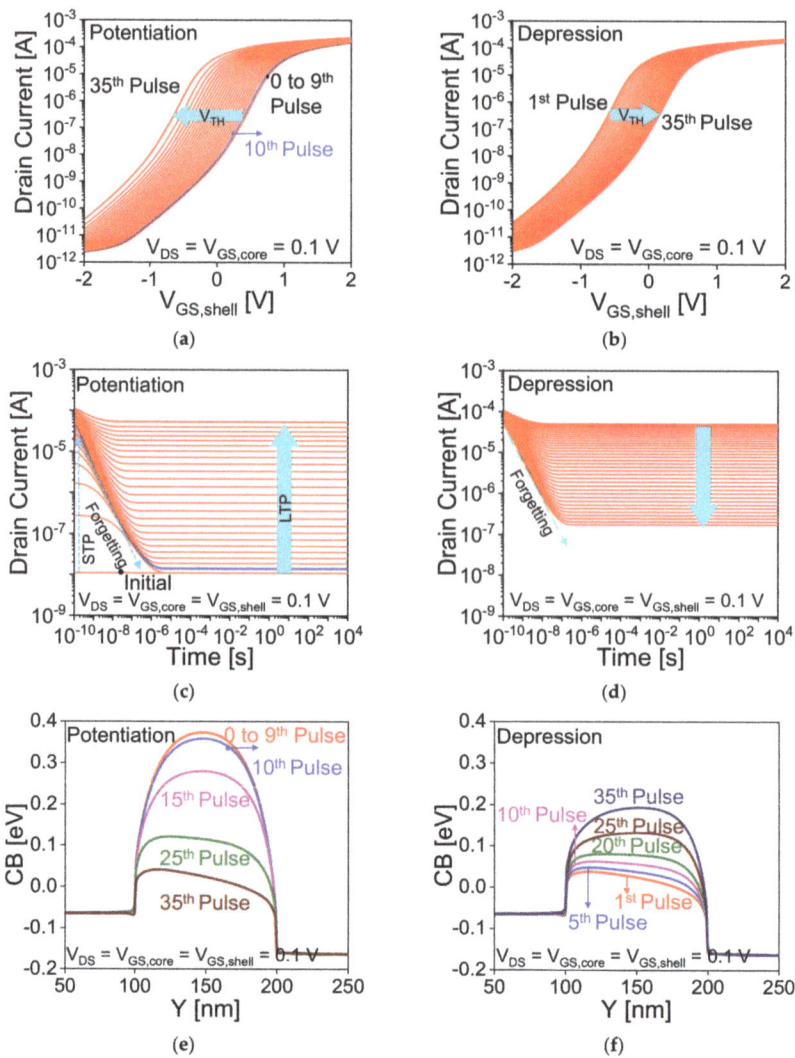

Figure 8. Transfer characteristics of the device during inference (read) operation for (**a**) potentiation (**b**) depression at different pulses. Transient analysis during read operation for (**c**) potentiation (**d**) depression at different pulses. Variation in conduction band (CB) during inference operation for different (**e**) potentiation and (**f**) depression pulses.

The trapped charges in the nitride layer lower the V_{TH} due to the increase in channel potential (lower the barrier for electron), and hence increases the drain current (higher weight) with increase in pulse. Figure 8e,f show the CB energy diagram during inference operation for different potentiation and depression pulses, respectively. Conduction band energies are extracted at below 1 nm of oxide/nitride/oxide (O/N/O) at $V_{DS} = V_{GS,core} = V_{GS,shell} = 0.1$ V. The same is sensed from Figure 8c for STP that drain current increases for a short time and start decreasing (forgetting) due to recombination of carriers during interval in the device, and thus approaches to the initial level (no pulse (device is at equilibrium condition)). For LTP (from the 10th pulse), current is higher than the STP current and retained up to 10^4 s due to trapped holes in the nitride layer, which helps to increase the channel potential during inference operation. These results (Figure 8a,c) confirm that this CSDG nanowire transistor is capable of both STP and LTP functions. In the case of LTD, the reverse process is observed, increase in repetitive pulse reduces the channel potential (increases the barrier as shown in Figure 8f) of the device due to de-trapping the holes from the nitride layer (Figures 5c and 6b), and hence increases the threshold voltage (lower weight) of the device as shown in Figure 8c. The same is illustrated in Figure 8d, increase in pulse reduces the current due to recombination of the carriers in the device, and thus current approaches to the initial state with increasing pulse.

Figure 9a shows the conduction band (CB) diagram at zero bias condition for different gate length (100 nm, 75 nm, and 50 nm). Reduction in gate length reduces the storage area for floating based memory, which reduces the retention time [34–39]. The operation of this synaptic transistor is based on the floating body effect, and charge trapping/de-trapping from the nitride layer. Thus, reduction in gate length reduces the minimum required potentiation pulses by which STP-to-LTP transit occurs as a function of gate length. Downscaling of gate length increases the electric field between the channel and source/drain junctions, which increases the tunneling in the device, and thus, minimum number of pulses are required for the STP-to-LTP transition decreases as shown in Figure 9b. As we scale down the device dimensions, the potentiation behaviour of the device remains the same except for the reduction in pulse number required for STP to LTP transition. Figure 10a shows the variation of conductance (weight) value of LTP and LTD operations with different pulses, respectively. In inset of Figure 10a shows the conductance value in logarithmic scale. From Figure 6b, it is evident that the conductance value for LTP operation is relatively linear compared to the LTD state because the charge trapped in the nitride layer is increasing logarithmically with an increase in the number of pulses (Figures 4d and 6a). In the case of LTD operation, conductance value is estimated with different shell gate voltage ($V_{GS,shell}$ of 3.0, 4.0, and 4.5 V). This shows that for shell gate voltage of 4.5 V, more number of holes are de-trapped from the nitride layer, and thus reduces the conductance value abruptly. For shell gate voltage of 4.0 V, the conductance is linear compared to 3.0 and 4.5 V because the amount holes de-trapped rate from the nitride layer is lower. Although an increase in $V_{GS,shell} > 4.5$ V (increasing the F–N tunneling probability), increases the memory window during depression but degrades the conductance value abruptly. In the same way, the LTP weight can also be modulated, and it is more prominent by core gate voltage ($V_{GS,core}$) because the tunneling is govern by ($V_{GS,core}$). Thus, to obtain the linear conductance value, voltages during operations (potentiation, depression, and inference) need to be optimized. The linearity in the conductance of the LTP and LTD curves affects the inference accuracy of the neural network because it is related to the degree of the synaptic weight change.

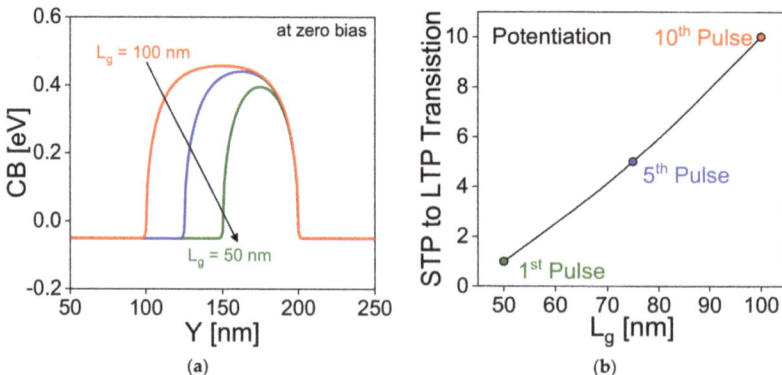

Figure 9. Downscaling of gate length. (**a**) Variation of CB energy. (**b**) number of pulses required for the transformation from STP to LTP. CB and VB indicate conduction and valence band, respectively.

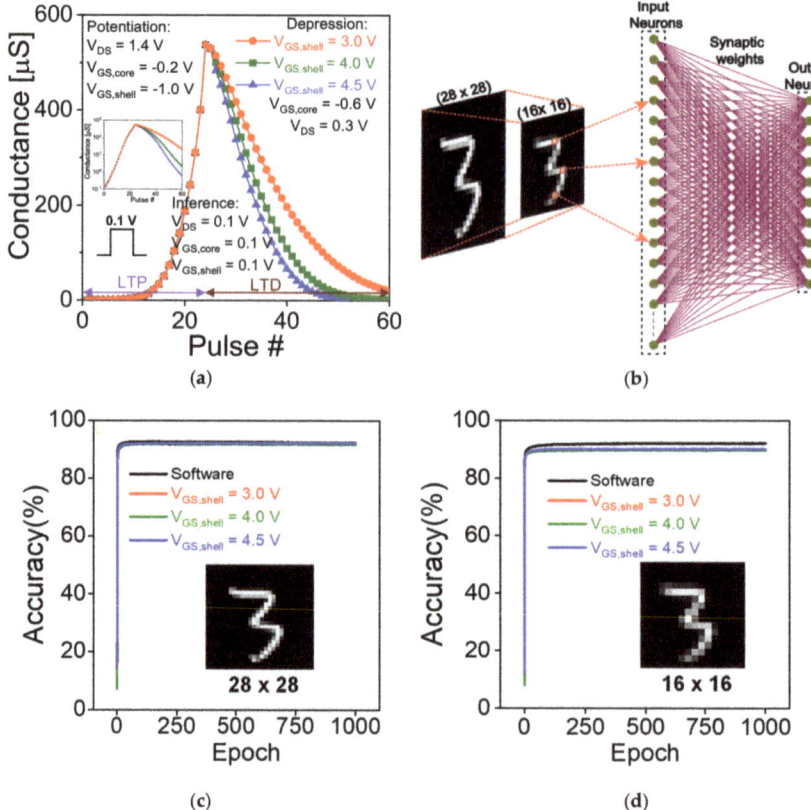

Figure 10. (**a**)Variation in conductance value for Long term potentiation (LTP) and Long-term depression (LTD) characteristics of the CSDG device (**b**) schematic of the single layer neural network with the CSDG transistor as the synapse for Modified National Institute of Standards and Technology (MNIST) digit recognition. Digit recognition accuracy (%) as a function of the number of training epochs at three distinct depression voltages of the synaptic device for (**c**) 28 × 28 and (**d**) 16 × 16 MNIST dataset. Inset of (**c**,**d**) shows the MNIST image of digit "3" in 28 × 28 and 16 × 16 resolutions, respectively.

Finally, to investigate the learning and inference capabilities of the proposed synaptic device for hardware-based neural networks (HNN), we have simulated a single layer neural network (NN) [40] consisting of one input layer and one output layer as shown in Figure 10b. The synaptic weights (conductance values) obtained from the Potentiation/Depression data in Figure 10a was used for off-chip training of the NN. The designed NN was used to classify image data from a Modified National Institute of Standards and Technology (MNIST) dataset which consists of 60,000 training images and 10,000 test images of handwritten digits from "0" to "9". All the images are in grayscale format with a resolution of 28 × 28 pixels. The normalized pixel intensities in the interval [0,1] are linearized to form a column matrix with 784 elements which is then fed to the input of the NN. The input values (x_i) undergo vector matrix multiplication [27] with the corresponding weight values (w_{ij}) and is summed up to form $\sum_{i,j} w_{ij} x_i$ at each of the output neurons. i and j denote the number of input nodes and output nodes, respectively. The output at each of these neurons is then transformed using a rectified linear unit (ReLU) activation function. The output neuron with the highest probability for a particular input image is considered as the corresponding prediction from the NN. In the present work, we have also trained the NN using a 16 × 16 downscaled version of the MNIST dataset, so that the effective number of input nodes is reduced to 256. Figure 10c,d shows the variation of accuracy of digit recognition with the number of epochs for the 28 × 28 MNIST dataset using randomly initialized weight distribution (software-based) and device weights extracted for 3 distinct values of depression voltages ($V_{GS,shell}$) i.e., 3.0 V, 4.0 V and 4.5 V. The final accuracy of digit recognition for the weight update from devices with $V_{GS,shell}$ = 3.0 V, 4.0 V and 4.5 V after 1000 epochs was 91.88%, 91.91% and 92.28% respectively which is very close to the ideal software based NN accuracy of 92.36%. This high accuracy in image recognition reveals that the proposed synaptic device is highly suited for neuromorphic inference applications. Similarly, 16 × 16 MNIST images were used for the devices with $V_{GS,shell}$ = 3.0 V, 4.0 V and 4.5 V yielding an accuracy of 89.94%, 89.65% and 90.17% respectively. In comparison to the software-based recognition accuracy of 92.25%, there is only a marginal drop in accuracy for these devices, which indicates the high reproducibility of our synaptic devices for inference applications requiring reduced input data.

4. Conclusions

In this work, we have simulated a novel core-shell dual gate nanowire transistor as an artificial synaptic transistor with calibrated simulation models. The dual gate helps to achieve a deeper potential well for charge retention of the device. The analysis highlights that the combination of floating body effect and charge trapped in the nitride achieves short-term potentiation and long-term potentiation and depression. The results of the CSDG nanowire indicate the following.

1. Transformation from STP to LTP occurs at the 10th pulse and it can be modulated through core gate voltage ($V_{GS,core}$) because the tunneling is governed through $V_{GS,core}$.
2. The trade-off between change in threshold voltage, and linearity in, conductance is observed during depression operation.
3. We can investigate the learning and inference capabilities of the proposed synaptic device for hardware based neural networks (HNN).
4. A reliable and consistent digit recognition accuracy of 92.28% is achieved by a single layer neural network on the MNIST dataset.

Furthermore, the analysis highlights the feasibility of the proposed synaptic device for inference applications pertinent to neuromorphic computing.

Author Contributions: Design and optimization of the synaptic device structure, validation of the synaptic operation schemes by device simulation, and preparation of the manuscript draft, M.H.R.A.; design of the artificial neural network based on the proposed synaptic transistor array, evaluation of the system accuracy, and preparation of the manuscript draft, U.M.K.; conception of the device structure and operation schemes, orientation of the directions of this research, organization of the workflow managing the overall research projects, and preparation of the final manuscript, S.C. All authors have read and agreed to the published version of the manuscript.

Funding: This research was supported by the National Research Foundation of Korea (NRF) funded by the Ministry of Science and ICT (MSIT) (No. 2016M3A7B4910348, Nano · Material Technology Development Program, 50%) and was partly supported by Institute of Information & Communications Technology Planning & Evaluation (IITP) grant funded by the Korean government (MSIT) (No. 2020-0-01294, Development of IoT based edge computing ultra-low power artificial intelligent processor, 50%).

Data Availability Statement: The data presented in this study are available on request from the corresponding author.

Conflicts of Interest: The authors declare no conflict of interest.

References

1. Kuzum, D.; Yu, S.; Philip Wong, H.-S. Synaptic electronics: Materials, devices and applications. *Nanotechnology* **2013**, *24*, 38200. [CrossRef] [PubMed]
2. Indiveri, G.; Liu, S.-C. Memory and Information Processing in Neuromorphic Systems. *Proc. IEEE* **2015**, *103*, 1379–1397. [CrossRef]
3. Kim, S.; Yoon, J.; Kim, H.-D.; Choi, S.-J. Carbon Nanotube Synaptic Transistor Network for Pattern Recognition. *ACS Appl. Mater. Interfaces* **2015**, *7*, 25479–25486. [CrossRef] [PubMed]
4. Tang, B.; Hussain, S.; Xu, R.; Cheng, Z.; Liao, J.; Chen, Q. Novel Type of Synaptic Transistors Based on a Ferroelectric Semiconductor Channel. *ACS Appl. Mater. Interfaces* **2020**, *12*, 24920–24928. [CrossRef]
5. Ambrogio, S.; Ciocchini, N.; Laudato, M.; Milo, V.; Pirovano, A.; Fantini, P.; Ielmini, D. Unsupervised Learning by Spike Timing Dependent Plasticity in Phase Change Memory (PCM) Synapses. *Front. Neurosci.* **2016**, *10*, 1–12. [CrossRef] [PubMed]
6. Burr, G.W.; Shelby, R.M.; Sebastian, A.; Kim, S.; Kim, S.; Sidler, S.; Virwani, K.; Ishii, M.; Narayanan, P.; Fumarola, A.; et al. Neuromorphic computing using non-volatile memory. *Adv. Phys. X* **2017**, *2*, 89–124. [CrossRef]
7. Kang, D.; Jang, J.T.; Park, S.; Ansari, M.H.R.; Bae, J.-H.; Choi, S.-J.; Kim, D.M.; Kim, C.; Cho, S.; Kim, D.H. Threshold-Variation-Tolerant Coupling-Gate α-IGZO Synaptic Transistor for More Reliably Controllable Hardware Neuromorphic System. *IEEE Access* **2021**, *9*, 59345–59352. [CrossRef]
8. Sun, B.; Guo, T.; Zhou, G.; Ranjan, S.; Jiao, Y.; Wei, L.; Zhou, Y.N.; Wu, Y.A. Synaptic devices based neuromorphic computing applications in artificial intelligence. *Mater. Today Phys.* **2021**, *18*, 100393. [CrossRef]
9. Ryu, J.-H.; Kim, B.; Hussain, F.; Ismail, M.; Mahata, C.; Oh, T.; Imran, M.; Min, K.K.; Kim, T.-H.; Yang, B.-D.; et al. Zinc Tin Oxide Synaptic Device for Neuromorphic Engineering. *IEEE Access* **2020**, *8*, 130678–130686. [CrossRef]
10. Lee, D.K.; Kim, M.-H.; Bang, S.; Kim, T.-H.; Choi, Y.-J.; Kim, S.; Cho, S.; Park, B.-G. HfOx-based nano-wedge structured resistive switching memory device operating at sub- μ A current for neuromorphic computing application. *Semicond. Sci. Technol.* **2020**, *35*, 055002. [CrossRef]
11. Kim, M.-H.; Cho, S.; Park, B.-G. Nanoscale wedge resistive-switching synaptic device and experimental verification of vector-matrix multiplication for hardware neuromorphic application. *Jpn. J. Appl. Phys.* **2021**, *60*, 050905. [CrossRef]
12. Cho, S.W.; Kwon, S.M.; Kim, Y.-H.; Park, S.K. Recent Progress in Transistor-Based Optoelectronic Synapses: From Neuromorphic Computing to Artificial Sensory System. *Adv. Intell. Syst.* **2021**, *2000162*, 2000162. [CrossRef]
13. Sun, B.; Ranjan, S.; Zhou, G.; Guo, T.; Du, C.; Wei, L.; Zhou, Y.N.; Wu, Y.A. A True Random Number Generator Based on Ionic Liquid Modulated Memristors. *ACS Appl. Electron. Mater.* **2021**, *3*, 2380–2388. [CrossRef]
14. Kim, D.; Jang, J.T.; Yu, E.; Park, J.; Min, J.; Kim, D.M.; Choi, S.J.; Mo, H.S.; Cho, S.; Roy, K.; et al. Pd/IGZO/p+-Si Synaptic Device with Self-Graded Oxygen Concentrations for Highly Linear Weight Adjustability and Improved Energy Efficiency. *ACS Appl. Electron. Mater.* **2020**, *2*, 2390–2397. [CrossRef]
15. Kim, H.; Hwang, S.; Park, J.; Park, B.G. Silicon synaptic transistor for hardware-based spiking neural network and neuromorphic system. *Nanotechnology* **2017**, *28*. [CrossRef] [PubMed]
16. Yu, E.; Cho, S.; Park, B.-G. A Silicon-Compatible Synaptic Transistor Capable of Multiple Synaptic Weights toward Energy-Efficient Neuromorphic Systems. *Electronics* **2019**, *8*, 1102. [CrossRef]
17. Kim, H.; Park, J.; Kwon, M.-W.; Lee, J.-H.; Park, B.-G. Silicon-Based Floating-Body Synaptic Transistor With Frequency-Dependent Short- and Long-Term Memories. *IEEE Electron Device Lett.* **2016**, *37*, 249–252. [CrossRef]
18. Yu, R.; Li, E.; Wu, X.; Yan, Y.; He, W.; He, L.; Chen, J.; Chen, H.; Guo, T. Electret-Based Organic Synaptic Transistor for Neuromorphic Computing. *ACS Appl. Mater. Interfaces* **2020**, *12*, 15446–15455. [CrossRef]
19. Yu, E.; Cho, S.; Roy, K.; Park, B.G. A Quantum-Well Charge-Trap Synaptic Transistor with Highly Linear Weight Tunability. *IEEE J. Electron Devices Soc.* **2020**, *8*, 834–840. [CrossRef]
20. Seo, Y.-T.; Lee, M.-S.; Kim, C.-H.; Woo, S.Y.; Bae, J.-H.; Park, B.-G.; Lee, J.-H. Si-Based FET-Type Synaptic Device with Short-Term and Long-Term Plasticity Using High-κ Gate-Stack. *IEEE Trans. Electron Devices* **2019**, *66*, 917–923. [CrossRef]

21. Moon, K.; Lim, S.; Park, J.; Sung, C.; Oh, S.; Woo, J.; Lee, J.; Hwang, H. RRAM-based synapse devices for neuromorphic systems. *Faraday Discuss.* **2019**, *213*, 421–451. [CrossRef]
22. Ielmini, D. Brain-inspired computing with resistive switching memory (RRAM): Devices, synapses and neural networks. *Microelectron. Eng.* **2018**, *190*, 44–53. [CrossRef]
23. Ferain, I.; Colinge, C.A.; Colinge, J.P. Multigate transistors as the future of classical metal-oxide-semiconductor field-effect transistors. *Nature* **2011**, *479*, 310–316. [CrossRef] [PubMed]
24. Fahad, H.M.; Smith, C.E.; Rojas, J.P.; Hussain, M.M. Silicon Nanotube Field Effect Transistor with Core–Shell Gate Stacks for Enhanced High-Performance Operation and Area Scaling Benefits. *Nano Lett.* **2011**, *11*, 4393–4399. [CrossRef] [PubMed]
25. Fahad, H.M.; Hussain, M.M. Are Nanotube Architectures More Advantageous Than Nanowire Architectures For Field Effect Transistors? *Sci. Rep.* **2012**, *2*, 475. [CrossRef] [PubMed]
26. Sahay, S.; Kumar, M.J. Nanotube Junctionless FET: Proposal, Design, and Investigation. *IEEE Trans. Electron Devices* **2017**, *64*, 1851–1856. [CrossRef]
27. Tekleab, D. Device Performance of Silicon Nanotube Field Effect Transistor. *IEEE Electron Device Lett.* **2014**, *35*, 506–508. [CrossRef]
28. Musalgaonkar, G.; Sahay, S.; Saxena, R.S.; Kumar, M.J. Nanotube Tunneling FET With a Core Source for Ultrasteep Subthreshold Swing: A Simulation Study. *IEEE Trans. Electron Devices* **2019**, *66*, 4425–4432. [CrossRef]
29. Vinet, M.; Poiroux, T.; Widiez, J.; Lolivier, J.; Previtali, B.; Vizioz, C.; Guillaumot, B.; Le Tiec, Y.; Besson, P.; Biasse, B.; et al. Bonded planar double-metal-gate NMOS transistors down to 10 nm. *IEEE Electron Device Lett.* **2005**, *26*, 317–319. [CrossRef]
30. Choi, S.-J.; Moon, D.-I.; Kim, S.; Duarte, J.P.; Choi, Y.-K. Sensitivity of Threshold Voltage to Nanowire Width Variation in Junctionless Transistors. *IEEE Electron Device Lett.* **2011**, *32*, 125–127. [CrossRef]
31. Ansari, M.H.R.; Kim, D.; Cho, S.; Lee, J.-H.; Park, B.-G. Core-Shell Dual-Gate Nanowire Synaptic Transistor with Short/Long-Term Plasticity. In Proceedings of the 5th IEEE Electron Devices Technology & Manufacturing Conference (EDTM), Chengdu, China, 8–11 April 2021; pp. 1–3.
32. Navlakha, N.; Lin, J.-T.; Kranti, A. Retention and Scalability Perspective of Sub-100-nm Double Gate Tunnel FET DRAM. *IEEE Trans. Electron Devices* **2017**, *64*, 1561–1567. [CrossRef]
33. Ansari, M.H.R.; Cho, S. Performance Improvement of 1T DRAM by Raised Source and Drain Engineering. *IEEE Trans. Electron Devices* **2021**, *68*, 1577–1584. [CrossRef]
34. Yoshida, E.; Tanaka, T. A capacitorless 1T-DRAM technology using gate-induced drain-leakage (GIDL) current for low-power and high-speed embedded memory. *IEEE Trans. Electron Devices* **2006**, *53*, 692–697. [CrossRef]
35. Ansari, M.H.R.; Navlakha, N.; Lin, J.T.; Kranti, A. Doping Dependent Assessment of Accumulation Mode and Junctionless FET for 1T DRAM. *IEEE Trans. Electron Devices* **2018**, *65*, 1205–1210. [CrossRef]
36. Yu, E.; Cho, S.; Shin, H.; Park, B.-G. A Band-Engineered One-Transistor DRAM With Improved Data Retention and Power Efficiency. *IEEE Electron Device Lett.* **2019**, *40*, 562–565. [CrossRef]
37. Han, D.C.; Jang, D.J.; Lee, J.Y.; Cho, S.; Cho, I.H. Investigation of Modified 1T DRAM with Twin Gate Tunneling Field Effect Transistor for Improved Retention Characteristics. *J. Semicond. Technol. Sci.* **2020**, *20*, 145–150. [CrossRef]
38. Yu, E.; Kim, Y.; Lee, J.; Cho, Y.; Lee, W.J.; Cho, S. Processing and Characterization of Ultra-thin Poly-crystalline Silicon for Memory and Logic Application. *J. Semicond. Technol. Sci.* **2018**, *18*, 172–179. [CrossRef]
39. Ha, J.; Lee, J.Y.; Kim, M.; Cho, S.; Cho, I.H. Investigation and Optimization of Double-gate MPI 1T DRAM with Gate-induced Drain Leakage Operation. *J. Semicond. Technol. Sci.* **2019**, *19*, 165–171. [CrossRef]
40. Martí, D.; Rigotti, M.; Seok, M.; Fusi, S. Energy-Efficient Neuromorphic Classifiers. *Neural Comput.* **2016**, *28*, 2011–2044. [CrossRef]

Organotrialkoxysilane-Functionalized Prussian Blue Nanoparticles-Mediated Fluorescence Sensing of Arsenic(III)

Prem. C. Pandey [1], Shubhangi Shukla [1] and Roger J. Narayan [2,*]

[1] Department of Chemistry, Indian Institute of Technology (BHU), Varanasi 221005, India; pcpandey.apc@iitbhu.ac.in (P.C.P.); shubhangi.rs.chy14@itbhu.ac.in (S.S.)
[2] Joint Department of Biomedical Engineering, University of North Carolina, Chapel Hill, NC 27599, USA
* Correspondence: rjnaraya@ncsu.edu

Abstract: Prussian blue nanoparticles (PBN) exhibit selective fluorescence quenching behavior with heavy metal ions; in addition, they possess characteristic oxidant properties both for liquid–liquid and liquid–solid interface catalysis. Here, we propose to study the detection and efficient removal of toxic arsenic(III) species by materializing these dual functions of PBN. A sophisticated PBN-sensitized fluorometric switching system for dosage-dependent detection of As^{3+} along with PBN-integrated SiO_2 platforms as a column adsorbent for biphasic oxidation and elimination of As^{3+} have been developed. Colloidal PBN were obtained by a facile two-step process involving chemical reduction in the presence of 2-(3,4-epoxycyclohexyl)ethyl trimethoxysilane (EETMSi) and cyclohexanone as reducing agents, while heterogeneous systems were formulated via EETMSi, which triggered in situ growth of PBN inside the three-dimensional framework of silica gel and silica nanoparticles (SiO_2). PBN-induced quenching of the emission signal was recorded with an As^{3+} concentration (0.05–1.6 ppm)-dependent fluorometric titration system, owing to the potential excitation window of PBN (at 480–500 nm), which ultimately restricts the radiative energy transfer. The detection limit for this arrangement is estimated around 0.025 ppm. Furthermore, the mesoporous and macroporous PBN-integrated SiO_2 arrangements might act as stationary phase in chromatographic studies to significantly remove As^{3+}. Besides physisorption, significant electron exchange between Fe^{3+}/Fe^{2+} lattice points and As^{3+} ions enable complete conversion to less toxic As^{5+} ions with the repeated influx of mobile phase. PBN-integrated SiO_2 matrices were successfully restored after segregating the target ions. This study indicates that PBN and PBN-integrated SiO_2 platforms may enable straightforward and low-cost removal of arsenic from contaminated water.

Keywords: prussian blue nanoparticles; organotrialkoxysilane; silica beads; arsenite; arsenate; water decontamination

1. Introduction

Large numbers of people in Bangladesh and India are exposed to arsenic contamination in potable water. Metallurgical, agricultural, and industrial processes result in the discharge of arsenic into soil and water [1,2]. Long-term exposure to arsenic, even at low concentrations, can lead to oncological, immunological, neurological, and endocrine effects [3]. The World Health Organization recently set an arsenic limit of 10 μg/L for drinking water (Holm, 2002) [4]. Natural water predominantly contains the inorganic species arsenate [$HAsO_4^{2-}$, As(V)] and arsenite [AsO_2^-, As(III)]. Inorganic As(III) was noted to be more toxic (10 times), mobile, and water-soluble (4–10 times) than As(V) [5]. The conversion rate of As(III) (arsenite) to As(V) (arsenate) in oxygenated water is a slow process, which depends on certain specific conditions [6]. Consequently, there is an alarming need to develop novel methods for sensing and removal of arsenic from drinking water [7].

Prussian blue nanoparticles (PBN) contain metal in two different oxidation states, Fe^{+3} and Fe^{+2}; these materials are known for their advanced peroxidase mimetic activity [8–10].

The charge transfer between the two iron species is responsible for the deep blue color of the complex [11]. Bi-metallic coordination compound PBN are a well-known inorganic material for electrocatalytic applications [12–15]. Several reports demonstrated the formation of mixed metal analogues, which involve straightforward replacement of the ferric/ferrous ion with another metal having a similar chemical state [16–18]. The properties of Prussian blue complex can be readily modified depending upon the nature of the constituent metal pair. Iron hexacyanoferrate synthesized via traditional synthetic routes (e.g., co-precipitation and electrosynthesis) do not exhibit appropriate processability for technical applications. We processed PBN from a single precursor involving the active role of the organotrialkoxysilane, which not only controlled the nucleation and solubility but also provided stability to the contents of reaction medium [19]. In addition, the PBN made from a single precursor were found to act as a light quenching material [20]; the photoactivity of the materials was examined using fluorescence imaging. Earlier studies show that organotrialkoxysilanes such as 3-aminopropyltrimethoxysilane (APTMS) allow the conversion of a single precursor, potassium hexacyanoferrate, to Prussian blue; this material was used for electrocatalytic detection of dopamine [21]. We further examined the use of another organotrialkoxysilane, 2-(3,4-epoxycyclohexyl) ethyltrimethoxysilane (EETMSi), in the presence of cyclohexanone for the controlled synthesis of PBN as a light quenching material.

Several methods, including iron oxide-coated sand, manganese greens, and iron ores, were previously described for arsenic removal [22]. Spectrophotometric and fluorometric methods have been previously studied to estimate the trace amounts of arsenite in water [23–25]. PBN, which include iron of two different oxidation states in a metal framework, may undergo specific interactions with As(III). Thus, we examined the fluorescence quenching ability of the PBN in the presence of arsenic(III). A novel result based on fluorescent sensing of arsenic was recorded, indicating the interaction between PBN and arsenic(III). PBN within a matrix were subsequently studied for use in arsenic removal.

Silica (SiO_2) beads are a non-toxic and inexpensive matrix, which may be used as a template to synthesize PBN using organotrialkoxysilane. PBN were inserted into mesoporous SiO_2; the PBN became embedded in the accessible SiO_2 pores. The PBN@SiO_2 was used for As(III) removal and its subsequent oxidation into arsenate through an interaction with the iron species in the material. This adsorption–oxidation process was demonstrated with PBN@SiO_2 under different pH conditions to analyze the efficacy of the oxidant system. The high uptake efficiency of PBN@SiO_2 (95%) indicated that this material is attractive for use in As(III) removal. XPS, ICP, and HPLC techniques were used to detect and quantify As(III) species. The PBN@SiO_2 was separated easily through centrifugation; this recycled material also showed As(III) removal activity. The proposed As(III) removal process is more cost effective over those reported to date. The ability to recycle PBN@SiO_2 adds to the economic viability of this process.

2. Experimental Section

2.1. Materials

Potassium ferricyanide was purchased from Merck India (Bengaluru, Karnataka, India). Silica beads (50 μm) and silica nanoparticles (200 nm) were purchased from Sigma-Aldrich (Bengaluru, Karnataka, India). Sodium arsenite was purchased from S D Fine-chem Limited (Mumbai, Maharashtra, India), and Azure-B was obtained from Sisco Research Laboratories Pvt. Ltd. (Mumbai, Maharashtra, India). 2-(3,4-epoxycyclohexyl)ethyltrimethoxysilane (EETMSi) and cyclohexanone were obtained from Sigma-Aldrich (Bengaluru, Karnataka, India). In addition, the remaining chemicals were of analytical grade and procured from commercial sources. The working solution of As(III) was freshly prepared with Milli-Q water using sodium arsenite ($NaAsO_2$) and stored in a dark freezer. Milli-Q water was used throughout the experiment to avoid interference from contaminants.

2.2. Synthesis of PBN, PBN@SiO$_2$ and PBN@MSNP Mediated through EETMSi

2.2.1. EETMSi-Mediated Formation of PBN

The synthesis of PBN was accomplished using [2-(3,4-Epoxycyclohexyl)ethyl]trimethoxysilane(EETMSi) and cyclohexanone from the single precursor potassium ferricyanide via chemical reduction. The homogeneous colloidal sol of Prussian blue nanoparticles (PBN) was prepared by adding 20 µL of EETMSi (0.1 M) to 100 µL of potassium ferricyanide (0.03 M) under stirring conditions. Subsequently, 20 µL of cyclohexanone was added to the reaction mixture; this mixture was kept in an oven at 343 K for 8 h. The blue-colored colloidal suspension of PBN was characterized byX-ray diffraction (XRD), Transmission electron micrpscopy (TEM), etc.

2.2.2. EETMSi-Mediated Formation of Prussian Blue Nanoparticles Modified Silica (PBN@SiO$_2$)

Mesoporous silica was used to obtain PBN-confined mesoporous silica (PBN@SiO$_2$). Typical synthesis involved a multistep procedure as follows: At first, 10 mg of mesoporous silica beads were suspended in 100 mL of EETMSi (1.2 M) aqueous solution under constant stirring conditions. After 3 h, un-adsorbed EETMSi was extracted with methanol, followed by centrifugation. 200 mL of potassium ferricyanide [$K_3Fe(CN)_6$] aqueous solution (0.03 M) was added to the alkoxysilane-modified SiO$_2$ suspension under vigorous stirring conditions (800 rpm). Cyclohexanone was added to the alkoxysilane-modified $K_3Fe(CN)_6$@SiO$_2$ suspension under vigorous stirring and left to stand in oven at 338 K overnight. The unreacted $K_3Fe(CN)_6$ and unabsorbed PBN were removed via washing (five times) with methanol/water (2:1) solvent. The residual material was collected after centrifugation; a drying step was subsequently performed.

2.2.3. EETMSi-Mediated Formation of PBN@MSNPs

Mesoporous silica nanoparticles (MSNPs) were used to prepare Prussian blue nanoparticle-embedded mesoporous silica nanoparticles (PBN@MSNP). Ten milligrams of mesoporous silica nanoparticles (average particle size 200 nm and pore size 6 nm) were suspended in 100 mL of EETMSi (1.2 M) aqueous solution under stirring conditions. After 3 h, un-adsorbed EETMSi was removed with methanol, followed by centrifugation. Two hundred milliliters of potassium ferricyanide aqueous solution (0.03 M) were added to the alkoxysilane-modified MSNP suspension under vigorous stirring conditions (800 rpm). Cyclohexanone was added to the alkoxysilane-modified $K_3Fe(CN)_6$@MSNPs suspension under continuous stirring and left to stand in oven at 338 K overnight. The unreacted $K_3Fe(CN)_6$ and unabsorbed PBN were removed via washing (five times) with methanol/water (2:1) solvent. The residual material (PBN@MSNPs) was collected after centrifugation; a drying step was subsequently performed.

2.3. Materials Characterization

The particle size and morphology of as-synthesized PBN/PBN@SiO$_2$ and PBN@MSNP were analyzed using high-resolution transmission electron microscopy (HRTEM) with 800 and 8100 instruments (Hitachi, Tokyo, Japan) at an acceleration voltage of 200 kV. The topographical properties of as-synthesized PBN over SiO$_2$ were analyzed using a field emission scanning electron microscopy instrument (FEI (S.E.A.) Pte Ltd., Singapore). The elemental confirmation and mapping analyses were accomplished with an EDX attachment (Oxford Instruments plc, Abingdon, UK). A Rigaku X-ray diffractometer (Rikagu, Tokyo, Japan) with Cu Kα radiation (λ = 1.5406 A^0) was used to evaluate diffraction data. The XRD analysis was performed over the scan range of 10–90° for PBN. FTIR spectra were recorded on an ALFA-ATR Fourier transform infrared spectrometer (Bruker, Ettington, Germany). XPS analysis was performed using an ESCA/AES System (Surface Nano Analysis, GmbH, Berlin, Germany), which was equipped with an Al-Kα (1486.6 eV) X-ray source operating at a power of 385 W and a PHOBIOS 150 3D energy hemispherical analyzer with a delayline detector (SPECS Surface Nano Analysis GmbH, Berlin, Germany). The C-1s peak (284.5 eV) was used as an internal reference to calibrate the absolute binding energy. The quantitative detection of elements was performed through ICP techniques. Fluorescence analysis was performed using a 7100 spectrophotometer (Hitachi, Tokyo, Japan). Arsenic speciation was performed using high-performance liquid chromatography (HPLC) with a Shim-packed GIST C18 chromatography column encompassing a hydrophobic (non-polar) stationary phase (column length = 75 mm, inner diameter = 7.6 mm) for the determination of all species. Ammonium phosphate solution was used as an eluent for the entire HPLC experiment. The HPLC mobile phases of ammonium phosphate solution with pH 6.9 were prepared by mixing monobasic ($NH_4H_2PO_4$) and dibasic (($NH_4)_2HPO_4$) salt solutions with an appropriate ratio.

2.4. Fluorometric Method

A fluorometric method was used for the determination of As(III) species. Fluorescein (Flo) was used as a probe molecule (λex = 480 nm, λem = 510 nm) for the estimation of As(III) species. The fluorescence experiment was performed under neutral pH (6.8) conditions using Milli-Q water. Different concentrations of As(III) standard solution (10 ppm to 320 ppm) were prepared by adding appropriate amounts of sodium arsenite to Milli-Q water. The result was obtained using the effective concentration of Flo, PBN, and As(III).

3. Results and Discussion

3.1. Organotrialkoxysilane-Mediated Synthesis of PBN Analogs

Organotrialkoxysilane with an amine functional group, APTMS, in the presence of cyclohexanone was previously used for the controlled conversion of a single precursor, $K_3[Fe(CN)_6]$, into Prussian blue nanoparticles under ambient conditions [20]. Subsequently, 2-(3,4-epoxycyclohexyl) ethyltrimethoxysilane (EETMSi) was used to make PBN in the absence of cyclohexanone [9]. Although the process enabled efficient conversion of single precursor, $K_3[Fe(CN)_6]$, into Prussian blue nanoparticles, the duration was substantially longer. Accordingly, we attempted to use cyclohexanone along with EETMSi to obtain PBN from a single precursor pathway. Indeed, the process enabled the rapid formation of PBN, as shown in Figure 1; additional details on this process are provided below.

3.1.1. PBN as a Homogeneous Suspension

Slow decomposition under hydrothermal conditions via single precursor synthesis readily produced a blue-colored solution of PBN. The TEM micrographs in Figure 1A,B revealed well-dispersed nanocubes of PBN with an average diameter of 30 nm. The histogram (inset of Figure 1A) shows broad size distribution of crystalline nanoparticles, ranging between 27 and 53 nm. The average width of nanoparticles may be altered by modifying the EETMSi/Fe^{3+}/cyclohexanone feed ratio and thermal conditions. Accordingly, we investigated the role of EETMSi in combination with a ketonic reducing agent. The EDX and TEM data provided information on the chemical composition and nanoparticle structure, respectively. Figure 1D shows the contributions to the EDX spectrum from the Fe Kα peak at 6.4–7.0 keV and 0.9 keV, the Cu peak at 7.8–9.0 keV, and the Si peak at 7.057 keV; Fe peak and the peaks for N, O, and C are also noted. The XRD spectrum shown in Figure 1E reveals nearly all the planes assigned to 2θ values as per JCPDS # 73-0687, 17.4° (200), 24.7° (220), 35.3° (400), 39.6° (420), and 43.7° (422), 50.0° (440), 53.9° (600), 57.2° (620), 66.1° (640), and 68.9° (642).

3.1.2. PBN Confined in Mesoporous Silica (PBN@SiO_2)

We also attempted to insert PBN into mesoporous silica through the synergistic action of EETMSi and cyclohexanone; the product is represented as PBN@SiO_2. Mesoporous silica with a particle size of 50 micrometers and a pore diameter of 6 nm was used for the synthetic insertion of PBN; these materials have use in column chromatography. SEM micrographs in Figure 2A–C show the topographical features of PBN@SiO_2 and the narrow size distribution of the PBN in the SiO_2 matrix. Figure 2D shows the EDX data for

PBN@SiO$_2$, which shows a silicon content of 31.9% elemental weight; the inset of Figure 2A shows photographic images of mesoporous SiO$_2$ (I) and PBN@SiO$_2$ (II). The XRD spectra of as-made PBN@SiO$_2$ and mesoporous SiO$_2$ are shown in Figure 2E(a–b). The results for as-made PBN@SiO$_2$ and mesoporous SiO$_2$ demonstrate a broad peak, which is assigned to the 101 plane of amorphous SiO$_2$ (Figure 2E(a)); additional peaks for as-made PBN@SiO$_2$ are assigned to 220, 220 and 400 lattice planes of crystalline PBN. After exposure to the EETMSi-mediated PBN-laden formulation, an alteration in the SiO$_2$ pore size was detected via BET analysis (Table 1).

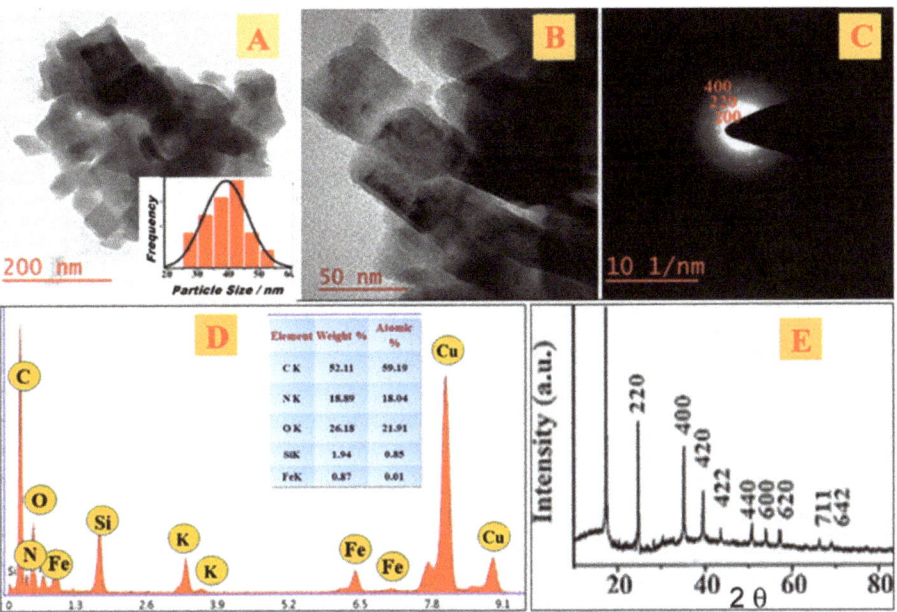

Figure 1. TEM images of PBN at different magnifications (**A**,**B**). Bar histogram displaying the particle size distribution curve of the nanoparticles (inset of Figure 1A). SAED pattern of as-synthesized particles (**C**), EDX profile with all the anticipated elements (**D**), XRD of EETMSi-functionalized PBN (**E**).

Table 1. Parameters calculated from BET nitrogen gas adsorption isotherm.

Sample Name	Surface Area ×10^4 cm^2/g	Pore Size (nm)
SiO$_2$ Bead	474.8	6.1
PB@SiO$_2$ Bead	426.8	4.3

3.1.3. PBN-Doped Mesoporous Silica Nanoparticles (PBN@MSNP)

We also undertook the synthetic incorporation of PBN within mesoporous silica nanoparticles. Silica nanoparticles (MSNP) with an average particle size of 200 nm and a pore size of 6 nm were used for this purpose. The porous nanocomposite was obtained primarily in two steps: (a) surface functionalization of the matrix by EETMSi, followed by (b) the uniform distribution of metal precursor throughout the network and subsequent reduction to form nanoscale particles. The in situ growth of PBN was pH controlled. The soluble Fe^{3+} species easily adhered to the pore channels in the presence of capping agent EETMSi. The HRTEM micrograph of bare MSNPs (Figure 3a(A)) shows a porous skeleton of spherical morphology. Figure 3a(B) shows PBN inside the mesoporous silica nanoparticles (encircled in red) [26]. The selected area electron diffraction (SAED) pattern of the corresponding hybrid nanoparticle assembly (PBN@MSNP) is shown in Figure 3a(C).

The zeta potential value was obtained from dynamic light scattering (DLS) data to understand the solution stability of particles. As shown in Figure 3a(D), the value of zeta potential is nearly −23 mV (i.e., towards the negative side); hence, the PBNPs are also negatively charged.

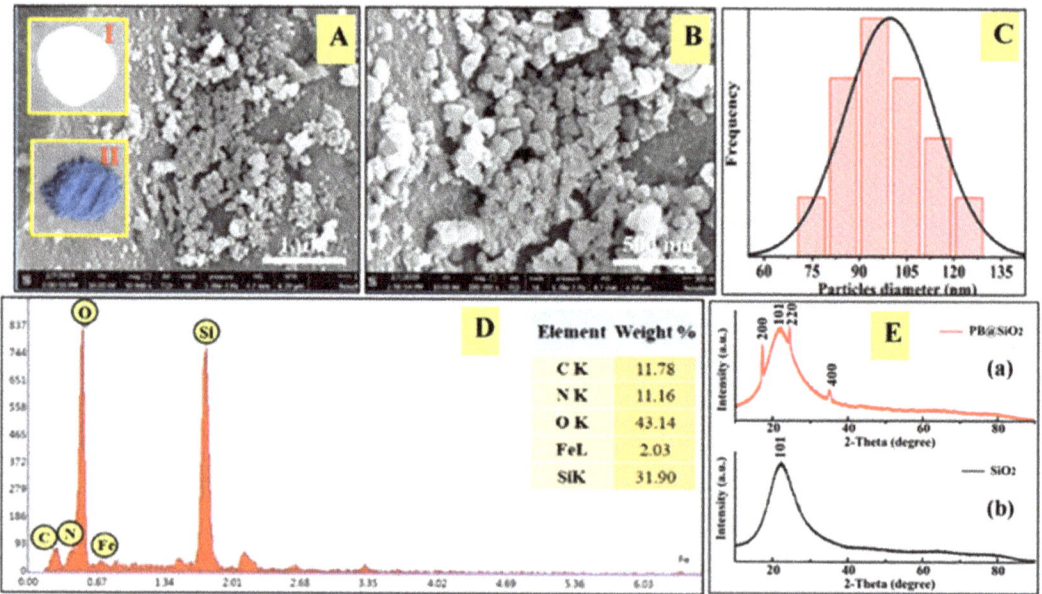

Figure 2. (**A**,**B**) HRSEM images of PBN@SiO$_2$ at two different magnifications. Inset of (**A**) shows photographic images of mesoporous silica (I) and PBN-inserted mesoporous silica (II). (**C**) The particle size distribution of PBN within mesoporous silica. (**D**) The EDX spectrum of PBN-inserted mesoporous silica. (**E**) XRD spectra of SiO$_2$ (a) and PBN@SiO$_2$ (b).

The EDX spectrum of PBN@MSNPs is shown in Figure 3a(E). The EDX mapping of the organotrialkoxysilane-functionalized PBN@MSNPs with the elemental composition of (B) carbon, (C) nitrogen, (D) oxygen (E) iron, and (F) silicon is shown in Figure 3b. The crystallographic data for as-prepared PBN@MSNPs and blank MSNPs are shown in Figure 3a(F). The peaks indexed at 2θ values of 17.6° (200), 24.3° (220), and 37.8° (400) indicated the successful insertion of crystalline PBN within the SiO$_2$ matrix; per JCPDS # 73-0687, 17.4° (200), 24.7° (220), 35.3° (400), 39.6° (420), and 43.7° (422), 50.0° (440), 53.9° (600), 57.2° (620), 66.1° (640), 68.9° (642), and 77.2° (820) can be indexed as the PB cubic space group Fm3m.

3.2. FTIR Analysis of PBN@SiO$_2$

The peak at 2086 cm^{-1} in the FTIR spectrum (Figure 4A of PBN may be attributed to the CN stretching mode of the Fe(II)-C-N-(III)Fe moiety in PBN. The broad bands at 3402 cm^{-1} and 1642 cm^{-1} in the spectrum correspond to OH-stretching and H$_2$O bending mode of the interstitial water molecule, respectively, within the PBN lattice. The strong band near 2885–2990 cm^{-1} corresponds to the C-H stretching vibration of sp^2-hybridized carbon in cyclohexanone.

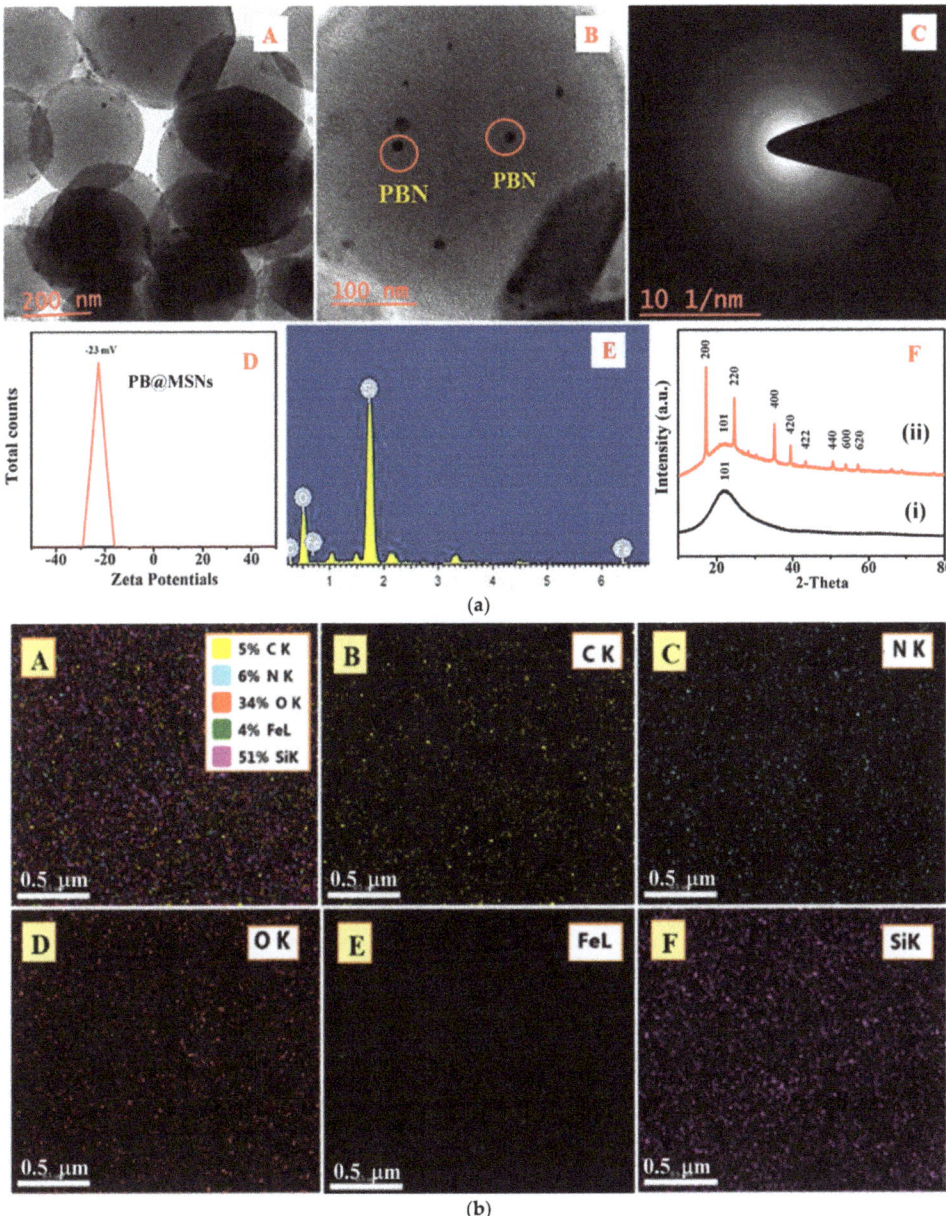

Figure 3. (**a**) (**A**) HRTEM image of organotrialkoxysilane-functionalized Prussian blue nanoparticles (PBN@MSN), (**B**) micrograph showing the magnified view of bulk-confined PBN (encircled in red) in mesoporous silica, (**C**) SAED pattern of the corresponding hybrid nanoparticle assembly (PBN@MSN), (**D**) stability profile of PBN@MSN in terms of zeta potential measurement, and (**E**) EDX spectrum of PBN@MSNP. (**F**) XRD profile for MSNPs (i) and as-synthesized PBN@MSNPs (ii). (**b**) (**A**) Mapping analysis of organotrialkoxysilane-functionalized Prussian blue nanoparticles with elemental composition (**B**) carbon, (**C**) nitrogen, (**D**) oxygen (**E**) iron, and (**F**) silicon.

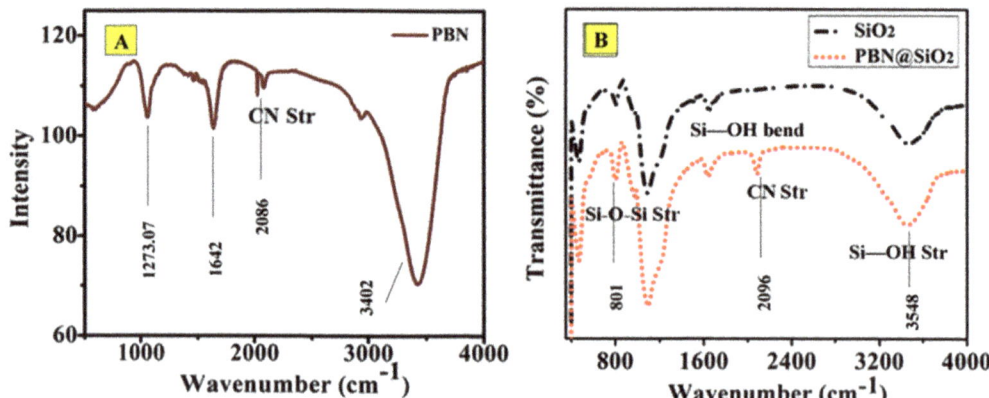

Figure 4. (**A**) FTIR spectrum of as-synthesized PBN; (**B**) FTIR spectra of (black line) SiO$_2$ and (red line) as-synthesized PBN@SiO$_2$.

The broad bands centered at 3548 cm^{-1} and at 1632 cm^{-1} are assigned to the stretching and bending vibrations of silanol groups (Si–OH), respectively, in the silica beads [27]. The bands at 1093 cm^{-1} and 801 cm^{-1} in the spectrum are associated with the anti-symmetric and symmetric stretching modes (Si–O–Si) of SiO$_4$ units. The prominent peak at 2096 cm^{-1} (Figure 4B is attributed to the stretching mode of Fe(II)-CN-(III)-Fe moiety in PBN [28] and indicates the successful formation of nanoparticles over SiO$_2$.

3.3. Fluorometric Study

3.3.1. Effect of the Addition of As(III) on the Fluorescent Intensity of Fluorescein

Since PBN have already been established as light quenching material, the PBN-mediated fluorescence quenching of fluorescein was evaluated. The impact of As(III) on fluorophore activity was studied via adding a different concentration of As(III) solution to a fixed Flo concentration. Subsequently, 0.01 mL of As(III) (10–320 ppm) and 10 μL of Flo solution (0.2 mM) were transferred into 2 mL of Milli-Q water and allowed to stand for 2 min at room temperature prior to fluorescence analysis. The fluorescence intensity of Flo was found to be enhanced as the function of As(III) (Figure 5A,B). At a lower As(III) concentration, a less pronounced enhancement phenomenon was observed. This result revealed that the extent of the interaction between Flo molecule and As(III) occurred to a higher extent at a higher As(III) concentration (up to 1.5 fold). The emission intensity was found to enhanced three-fold when the concentration of As(III) was elevated from 0.05 ppm to 2 ppm (effective concentration).

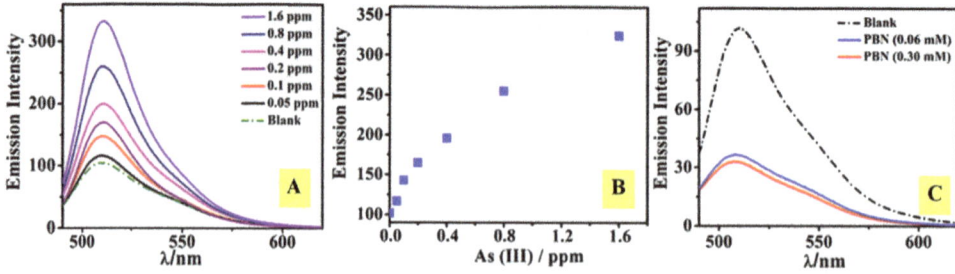

Figure 5. Fluorescence emission spectra of Flo (blank) in the presence of different As(III) concentrations from 0.05 ppm to 1.6 ppm (**A**). Plot of fluorescence intensity of Flo (0.2 mM) after exposure to a variable concentration (0.05–1.6 ppm) of As(III) (**B**). Effect of PBN addition (0.06 mM, 0.3 mM) over Flo (0.2 mM) fluorescence (**C**).

3.3.2. Interaction of Fluorophore with PBN

PBN were employed to observe the effect of the nano-sized particles over Flo. A constant amount of PBN nanosol (0.3 mM) was added to a known concentration of Flo (0.2 mM); the mixture was allowed to stand at room temperature for 2 min. The results revealed that PBN quenched the fluorescence property of Flo as shown in Figure 5C. Furthermore, the effect of PBN concentration over the emission intensity of Flo was investigated. Mixtures containing various concentrations of PBN with Flo were used to understand the interaction of nanoparticles with the fluorophore. The mixtures (i) PBN (0.06 mM) with Flo (0.2 mM) and (ii) PBN (0.3 mM) with Flo (0.2 mM) were evaluated. It was shown that EETMSi functionalized PBN acted as a quencher for Flo since the intensity of the fluorophore was found to diminish in the presence of PBN (Figure 5C). On increasing the concentration of PBN from 0.06 mM to 0.3 mM, only a small reduction in the emission intensity was observed (Figure 5C).

3.3.3. Effect of As(III)/PBN System over Flo Intensity

To understand the active role of PBN over As(III) interaction, we performed two experiments. In the first experiment, different concentrations of As(III) varying from 0.05 ppm to 1.6 ppm (effective concentration) were added to the fixed content of Flo (0.3 mM); the Flo-As(III) system was then exposed to a constant amount of PBN (0.3 mM) (as shown in Figure 6). In the second experiment, PBN (0.3 mM) were initially added to the Flo solution (0.2 mM); a variable concentration of As(III) between 0.05 ppm and 1.6 ppm (effective concentration) was then added to the PBN-Flo system (Figure 7). The substantial fluorescence quenching of the Flo-As(III) system in the presence of PBN was calculated using the relation F_0/F, where F_0 and F denote the fluorescent intensity of the Flo-As(III) system in the absence and in the presence of PBN, respectively (Figure 8A). Similarly, the substantial fluorescence quenching of the Flo system in the presence of PBN was calculated using the relation F_0/F, where F_0 denotes the fluorescence intensity of the Flo system in the absence of PBN and As(III), and F denotes the fluorescence intensity of Flo in the presence of PBN and As(III) (Figure 8B).

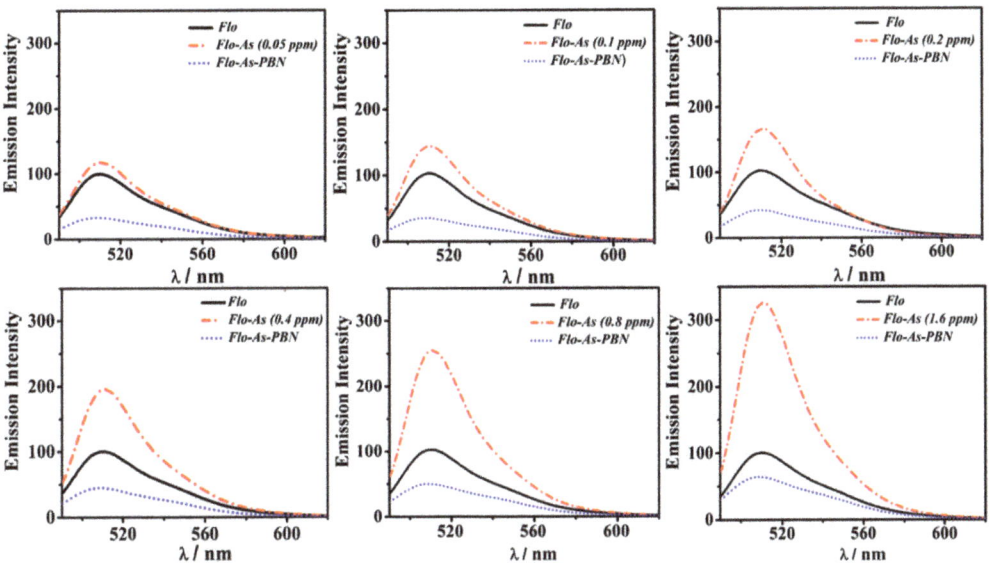

Figure 6. Study of the impact of PBN addition (10^{-1}) over the emission intensity of Flo-As(III) system by varying the As(III) solution (0.05–1.6 ppm) and keeping a constant concentration of Flo (0.2 mM) and PBN (0.3 mM).

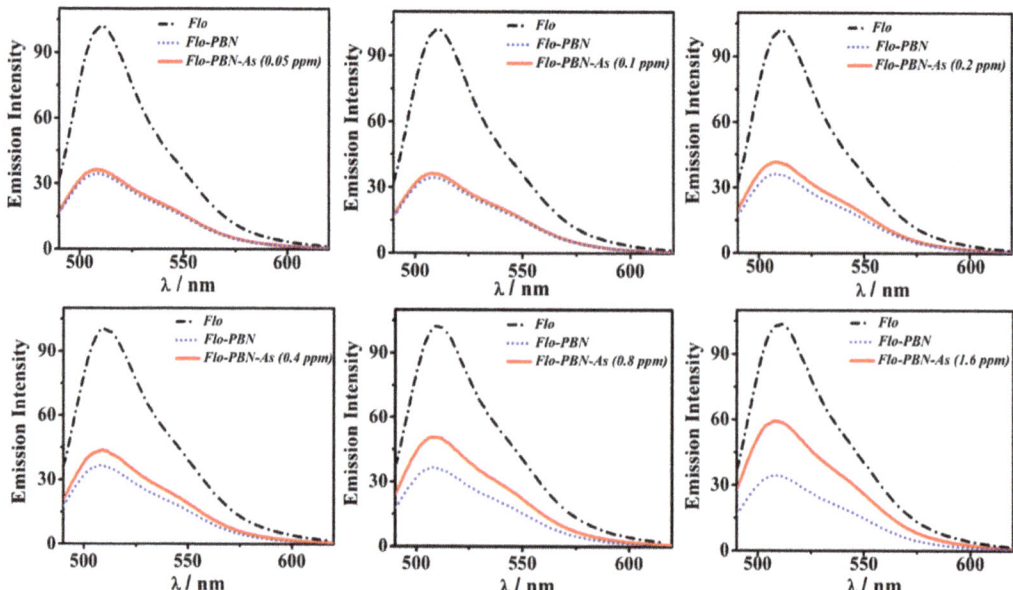

Figure 7. Study of the influence of As(III) addition over the emission spectra of the quenched Flo-PBN system by adding different concentrations of As(III) solution (0.05–1.6 ppm) and keeping a constant concentration of Flo (0.2 mM) and PBN (0.3 mM).

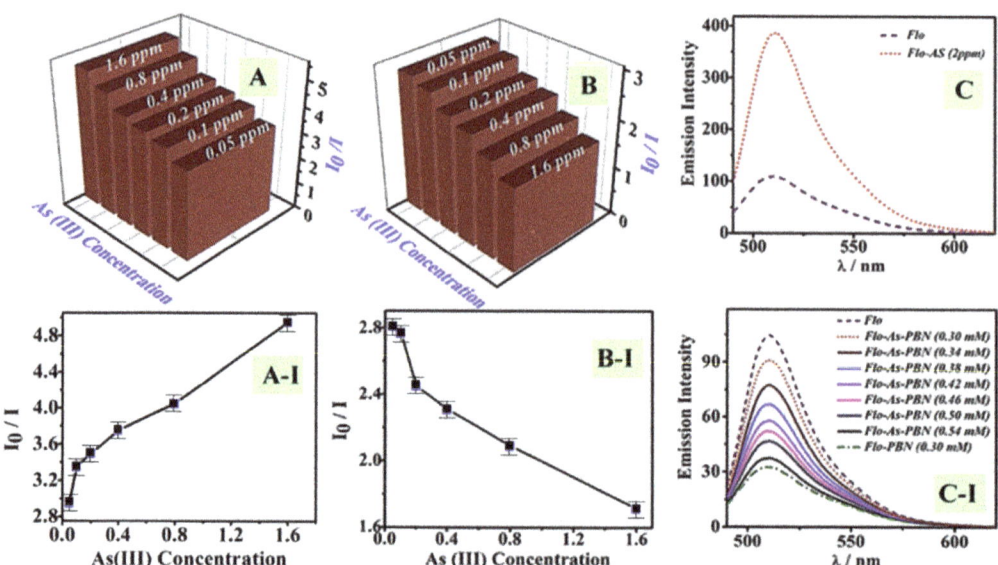

Figure 8. Bar diagram displaying fluorescence quenching (FQ) with respect to the variable concentration (effective concentration) of As(III) for the PBN@Flo-As(III) system (**A**) and the As(III)@PBN-Flo system (**B**). Plots of fluorescence intensity of the Flo-PBN system with respect to a variable content of As(III), showing concentration-dependent fluorescence quenching in both circumstances (**A-I,B-I**), with error bars representing the standard deviation. Emission spectra displaying the effect of adding 2 ppm of As(III) to the emission intensity of Flo (**C**). The required amount of PBN for complete quenching by adding various amounts of PBN to the Flo-As(III) system (**C-I**) and Flo.

It was observed that As(III) interacted predominately with the available quantity of the PBN moiety; the residual available As(III) was associated with the rise in emission intensity after interacting with Flo (Figure 8C-I). The subsequent addition of PBN achieved maximum quenching after interacting with available As(III), as displayed in Figure 8C-I. This result indicates that 0.54 mM PBN was sufficient to obtain complete interaction with 2 ppm arsenite. A similar concentration of 2 ppm of As(III) was added to the Flo-PBN system, which contained both PBN and Flo. The results (Figure 8A-I and Figure 8B-I) showed a decrease in fluorescent intensity of Flo-As(III) and Flo as a function of PBN; PBN altered the fluorescence influencing properties of As(III). It is surmised that Prussian blue interacted with As(III) more efficiently than the Flo-As system throughout the fluorescence process. A separate experiment was performed to discover the PBN loading for complete removal of As(III) from a concentration of 2 ppm (effective concentration). For this study, primary emission spectra of Flo-As(III) were recorded while adding As(III) aqueous solution (2 ppm) to the blank solution containing Flo (0.2 Mm) only (Figure 8C). A similar concentration of 2 ppm of As(III) was added to the Flo-PBN system, which contained both PBN.

3.4. As(III) Decontamination from Aqueous Solution Using PBN@SiO$_2$

The heterogeneous PBN@SiO$_2$ system was studied in order to understand the dynamic interaction occurring between As(III) and PBN. Accordingly, inexpensive and non-reactive silica beads were used for the modulation of active PBN in the formulation of the heterogeneous matrix. Heterogeneous methods are considered to play an influential role in catalysis due to their straightforward separation and large-scale applicability. For As(III) decontamination, the as-synthesized PBN@SiO$_2$ (0.05 g) was successfully packed in a column of 10 mm diameter. The standard As(III) solution (10 ppm) was prepared via adding an appropriate amount of sodium arsenite salt in Milli-Q water; 10 mL of the solution was passed through the PBN@SiO$_2$ enclosed column. The fluorescence analysis of separated supernatant (PBN@SiO$_2$ processed) was performed using Flo (0.2 mM) under similar conditions. In this study, 10 μL of as-eluted supernatants (PBN@SiO$_2$ processed and unprocessed As(III) solution) were added separately with Flo and left to stand at room temperature for 2 min. Their emission spectra were recorded to understand PBN@SiO$_2$ interactions with As(III). Unprocessed As(III) solution was observed to enhance the emission intensity of Flo many-fold (Figure 9A) as compared to the PBN@SiO$_2$ processed As(III) solution (Figure 9B).

PBN@SiO$_2$ was shown to significantly remove the As(III) from the contaminated solution. The ICP analysis of PBN@SiO$_2$ processed As(III) aqueous solution was performed to quantify the arsenic concentration in the solution. The result showed 0.0018 ppm arsenic (As) content for the PBN@SiO$_2$ processed As(III) solution. In addition, 0.13 ppm Fe content was also detected in the processed As(III) solution. The ICP analysis indicated that some of the iron species of PBN ([Fe^{III}[Fe^{II}(CN)$_6$]]) leached out with the eluent during interaction with the As(III) species. To investigate the presence of active iron species in the eluent, we studied the addition of ferrous sulfate with active ferrous species (Fe^{+2}) to the colorless supernatant eluent. During this process, we added the ferric chloride-containing active ferric species to the colorless supernatant eluent. We observed that colorless supernatant changed immediately to an intense blue color (resembling the Prussian blue color) when ferric chloride was added. However, no such changes were observed when ferrous sulfate (containing Fe^{+2}) was added.

We performed a fluorometric experiment in which supernatant (SN) was employed to observe its modulation of the Flo fluorescence properties. Fluorescent emission spectra were recorded after adding Flo (10 μL) to 10 μL of PBN@SiO$_2$ processed supernatant (SN). A small change in intensity (I_o = 102.23) was observed with respect to the Flo (I_o = 98.74) as seen in Figure 9C (1 and 2). A study that involved adding ferric chloride to the Flo-supernatant (Flo-SN) mixture showed that the Flo fluorescence property was quenched (I = 31.73) when compared to Flo (I = 98.74), as shown in Figure 9C (3). Supernatant-containing ferrocyanide species had an instant interaction with the added ferric chloride, which instantly converted into PBN.

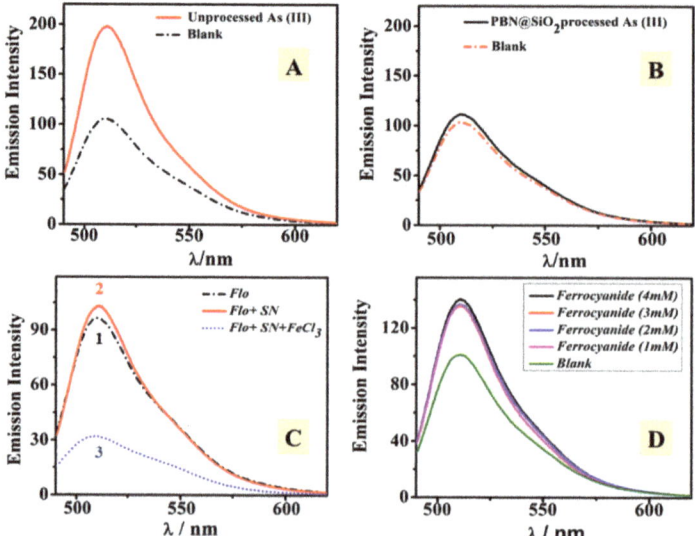

Figure 9. Fluorescence emission spectra of Flo with unprocessed (**A**) and PBN@SiO$_2$ processed (**B**) As(III) solution. Identification of ferrocyanide species in supernatant on the addition of ferric chloride via fluorescence quenching (**C**). Study of the effect of ferrocyanide species (1–4 mm) over Flo emission spectra (**D**).

To analyze the role of the residual ferrocyanide species in the supernatant over the emission spectra, a fluorescence experiment centered on the ferrocyanide concentration was conducted. We prepared and added different amounts (1 mM to 4 mM) of ferrocyanide solution to a constant amount of Flo (0.2 mM) to analyze the influence of the solution over the Flo emission intensity. Ferrocyanide acted as a weak enhancer (Figure 9D). These results indicate that the As(III) was supposed to undergo oxidation into arsenate in the presence of PBN. The iron species in the PBN undergo reduction into Fe^{+2} throughout the As(III) removal process. On the addition of active ferric species to the supernatant, an immediate reaction leads to the formation of PBN after the interaction with residual ferrous species. The collected PBN@SiO$_2$ was characterized with XPS to observe the significance of arsenic treatment over the PBN@SiO$_2$ phase (as discussed in a subsequent section). Moreover, the resultant eluent was collected into separate vials and underwent HPLC analysis for the detection of arsenic species.

3.5. HPLC Results on PBN@SiO$_2$ Treated Arsenic(III)

All the separated species were noted in the ion-chromatogram at their respective retention time such as arsenobetaine (AsB) at 2.17/2.42/2.55 min, dimethylarsinic acid (DMA) at 3.57 min, As(III) at 3.8/3.9 min, and As(V) at 7.7 min. The chromatogram shown in Figure 10A–D was obtained as the result of HPLC separation of the arsenic species after treatment with PBN@SiO$_2$ at different pH values (2.2–8.5). HPLC analysis illustrates that the removal efficiency of As(III) ((Figure 10A) by PBN@SiO$_2$ increased from 33.52% (Figure 10B) to 59.90% (Figure 10C) with a pH increase from 2 to 6.5; this improved to 95.13% (Figure 10D) under a mild alkaline condition (pH-8.5). The ion chromatogram results also showed an insignificant peak at a retention time of 7.35 min ((Area% = 1.4) at pH = 6.5 and (Area% = 9.09) at pH = 8.5), which was associated with leaching of As(V) in an aqueous solution during the oxidation–adsorption process. All of the arsenic species (As(III), DMA, AsB) identified at various retention times along with their relative concentration in a HPLC environment are shown in Table 2.

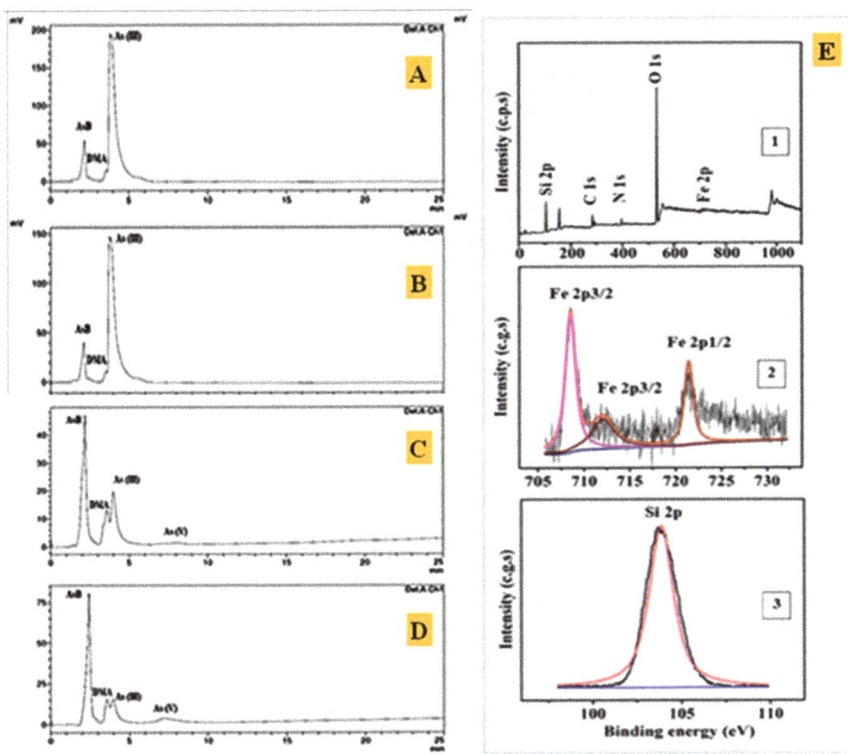

Figure 10. Ion chromatogram obtained during HPLC speciation of species present in the sample. (**A**) Standard As(III) (5 ppm) solution at pH = 6.5). (**B**) Standard As(III) sample (5 ppm) after PBN@SiO$_2$ treatment in acidic medium (pH = 2). (**C**) As(III) standard solution (5 ppm) after PBN@SiO$_2$ treatment in neutral medium (pH = 6.5). (**D**) As(III) standard sample (5 ppm) after PBN@SiO$_2$ treatment in alkaline medium (pH = 8.5). (**E**) XPS analysis of PBN@SiO$_2$. (**1**) A complete survey scan with all recognized species. (**2**) Fe^{2+} and Fe^{3+} species XPS peak in PBN@SiO$_2$. (**3**) Identified Si(IV) chemical states in SiO$_2$.

Table 2. All arsenic species (As(III), DMA, AsB) identified at different retention times along with their relative concentration during HPLC.

Species	Description	Molecular Formula	Height (%)	Area (%)	System pH	Retention Time	Figure
As^{III}	Sodium arsenite (NaAsO$_2$)		47.1	57.03	2.2	3.86 min	Figure 10B
			23.86	30.65	6.8	3.97 min	Figure 10C
			6.91	5.23	8.5	3.95 min	Figure 10D
AsB	Arsenobetaine (AsB)		45.56	34.39	2.2	2.55 min	Figure 10B
			58.29	49.74	6.8	2.17 min	Figure 10C
			70.02	56.95	8.5	2.42 min	Figure 10D
DMA	Dimethylarsinic acid (DMA)		6.85	4.21	2.2	3.58 min	Figure 10B
			15.81	16.21	6.8	3.57 min	Figure 10C
			13.45	11.79	8.5	3.57 min	Figure 10D

AsB, which frequently existed in the zwitterionic form due to the interaction between the positively charged arsenic and the negatively charged carboxylic group, starts to migrate immediately after interacting with the hydrophobic C18 Shim-pack column. However, As(III) is a neutral species (pKa = 9.2) up to a pH of 8, which eluents slowly with the solvent front. Consequently, negatively charged DMA and As(V) species feasibly eluent by a variety of interactions (e.g., H bonding and ion-exchange) along with hydrophobic effects. The obtained result was acquired after a total run time of 25 min and repeated twice to minimize the experimental error.

3.6. XPS Analysis of PBN@SiO$_2$

XPS survey scans indicated the presence of Si, O, Fe, and C in blank PBN@SiO$_2$ and As, Si, O, Fe, and C in As(III)-PBN@SiO$_2$. The peaks were assigned as follows: Fe $2p_{3/2}$—708 eV; Fe $2p_{3/2}$—713 eV; Fe $2p_{1/2}$—722 eV; As(III) 3d—44.2 eV and As(IV) 3d—47 eV, respectively. The peak position of the Si 2p spectrum corresponds to a binding energy of 103.63 eV and shows the characteristics of Si(IV) in a SiO$_2$-type compound [29].

3.6.1. Fe(II) and Fe(III) Identification in PBN@SiO$_2$

After peak fitting, the spectrum can be de-convolved into three peaks. Figure 10E shows the XPS peaks centered on binding energies of 721.27 and 708.34 eV for Fe $2p_{1/2}$ and Fe $2p_{3/2}$, respectively; these features are characteristic of the Fe^{+2} moiety in Prussian blue. In addition, a spectrum shows a peak at a binding energy of 712.12 eV, which corresponds to Fe^{+3} species. The position of these peaks is in good agreement with the results in the literature for the characteristic Fe^{+3} and Fe^{+2} components of Prussian blue compounds [30].

3.6.2. As(III)-PBN@SiO$_2$ and As(V)-PBN@SiO$_2$

The cation As(III) and the oxidized species As(V) detected on the PBN@SiO$_2$ substrate with XPS after a decontamination process are shown in Figure 11. The binding energy values (in eV) for O (1s), Si (2p), Fe (2p), and N (1s) in PBN@SiO$_2$ and As-PBN@SiO$_2$ are listed in Table 3. The XPS survey scan as shown in Figure 11D shows peaks at a binding energy of 49.03 eV, which are associated with the presence of As(V) and indicate the successful sorption of As(V) by PBN@SiO$_2$ [31]. The other peak located at 43.23 eV is associated with the adsorption of As(III) over SiO$_2$ prior to the oxidation process [32]. However, the peak positions observed for the Fe^{+3} and Fe^{+2} core level (2p) spectra of PBN@SiO$_2$ are shifted slightly to a lower binding energy relative to the unreacted and unabsorbed PBN@SiO$_2$ species. This shift in the peaks for Fe^{2+} $2p_{3/2}$ (binding energy of 708.19 eV) and Fe^{2+} $2p_{1/2}$ (binding energy of 721.42 eV) may be attributed to arsenic adsorption [33]. The shift in the peak position (with a reduction in intensity) of Fe^{+3} $2p_{3/2}$ (binding energy of 712.86 eV) relative to pure PBN@SiO$_2$ suggests the reduction of the material during arsenic oxidation [34]. The position of the characteristic peak of Fe^{+2} (binding energy of 55.04 eV, 3p) remained unchanged throughout the As(III) oxidation and adsorption process [31]. A peak emerged at a binding energy of 398.99 eV, which was attributed to the presence of nitrogen in the environment. Alterations in the peak position of PBN@SiO$_2$ relative to that of As-PBN@SiO$_2$ were associated with chemical adsorption by PBN@SiO$_2$ of arsenic species.

Table 3. XPS data of PBN@SiO$_2$ and Arsenic-PBN@SiO$_2$.

Sample	Si(2p)	O	Fe^{+3}	Fe^{+2}	Fe^{+2}	N	As(III)	As(V)
		1s	$2p_{3/2}$	$2p_{1/2}$	$2p_{3/2}$	1s	3d	3d
PBN@SiO$_2$	103.63	532.62	712.12	721.27	708.34	397.07	-	-
As-PBN@SiO$_2$	103.49	531.99	712.86	721.42	708.19	397.05 and 398.99	43.23	49.03

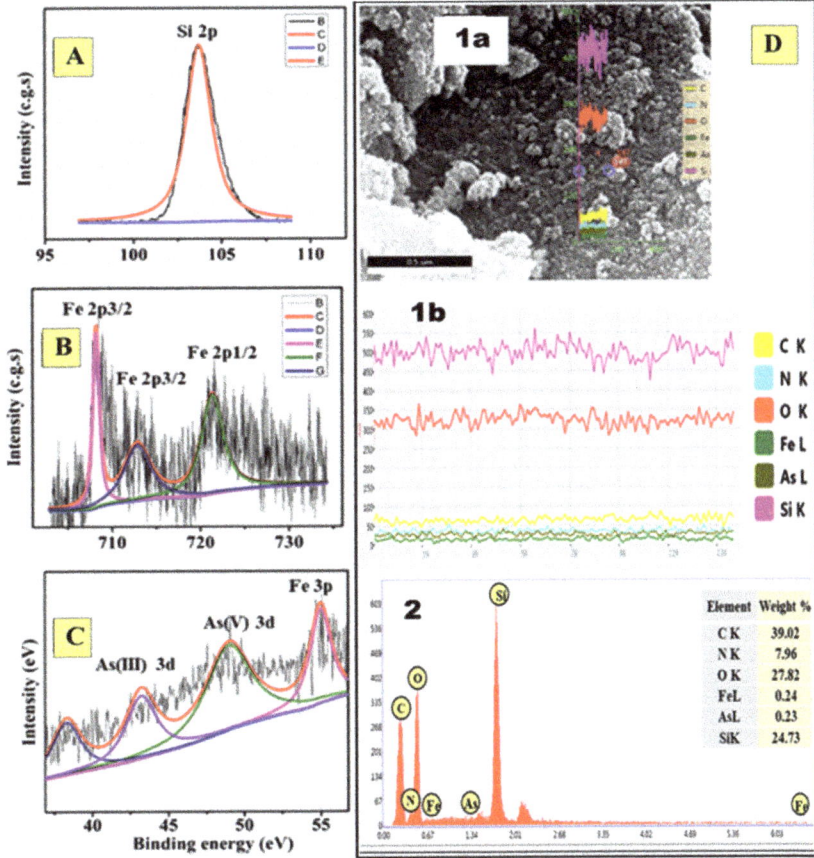

Figure 11. X-ray photoelectron spectrum of As-PBN@SiO$_2$ after arsenic exposure. (**A**) Identified Si(IV) chemical states in SiO$_2$. (**B**) Fe^{2+} and Fe^{3+} species XPS peak in PBN@SiO$_2$. (**C**) Arsenic species (As(V) and As(III)) at PBN@SiO$_2$. (**D**) (**1a,1b**) EDX line scan measurement comprised of an inner SiO$_2$ and an outer ferric hexacyanoferrate with As(III) enrichment over the PBN@SiO$_2$ surface. (**2**) EDX analysis shows all of the anticipated elements.

3.7. Effect of pH on Arsenic Removal

The results in Figure 10 illustrate the effects of pH on the removal of As(III) using PBN@SiO$_2$. As can be observed, As(III) removal was dependent on pH; the greatest removal efficiency occurred under moderate pH (pH = 6–9) and was found to diminish at highly acidic pH (pH < 3). As reported earlier, surfaces of silica beads were positively charged in highly acidic conditions and acquired a negative charge in the pH range of 3–10 [35]. Subsequently, moderate pH was found to be favorable for a sorbent surface since decreased protonation is supposed to enlarge the attraction force between the negatively charged PBN@SiO$_2$ surface and the positively charged As(III) cationic species. This result is similar to earlier findings by Gupta et al. who reported a significant increase in As(III) adsorption onto iron oxide-coated quartz sand with an increase in pH from 4.5 to 7.5 [36]. At a highly acidic pH (<3), repulsion occurred between the positively charged adsorbent sites and the adsorbate species (As^{+3}), which prevented the adsorption and arsenic oxidation processes. No substantial rise in As(III) removal efficiency was observed with an elevation in pH.

3.8. Analysis of PBN@SiO$_2$ Surface through SEM-EDX after As(III) Remediation

After the As(III) removal process, the PBN@SiO$_2$ surface was analyzed using SEM. The result showed the change in morphology of PBN (cubic to spherical) after arsenic interaction (Figure 11E). The EDX results suggest that the material is comprised of an inner SiO$_2$ chemistry and an outer ferric hexacyanoferrate (Fe^{+3}[Fe^{+2}(CN)$_6$]) chemistry, with some As(III) enrichment over the PBN@SiO$_2$ surface. The presence of the anticipated elements was confirmed through EDX analysis.

3.9. Recyclability and Proposed Mechanism

It has been well established that Prussian blue is comprised of Fe metal in Fe^{+2} (low spin) and Fe^{+3} (high spin) states, which are linked via CN bridges. Prussian blue can undergo reduction to what is known as Prussian white (Fe^{II}–C≡N–Fe^{II}) or oxidation to what is known as Prussian yellow (Fe^{III}–C≡N–Fe^{III}) [37,38]. Reduction of Prussian blue to Prussian white on the surface of silica gel was found to facilitate the As(III) oxidation to As(V) and their subsequent removal. Conversion of Fe^{3+} to Fe^{+2} was shown during the decontamination in PBN and validate the altered chemical environment due to arsenic interactions. The addition of ferric chloride to the white-blue colored arsenite-treated PBN@SiO$_2$ residue instantly generated a blue color; this phenomenon is attributed to the conversion of hexacyanoferrate $\left[\text{Fe}^{II}(\text{CN})_6\right]$ species of K$_2$[Fe^{II}Fe^{II}(CN)$_6$] into Prussian blue $\left(\text{K}[\text{Fe}^{III}\text{Fe}^{II}(\text{CN})_6]\right)$ through the interaction with ferric species (ferric chloride). The reaction during Prussian blue synthesis has been shown as:

$$4\text{K}_2[\text{Fe}^{II}\text{Fe}^{II}(\text{CN})_6] + 4\text{FeCl}_3 4\left(\text{K}[\text{Fe}^{III}\text{Fe}^{II}(\text{CN})_6]\right) + 4\text{KCl}$$

The variation of surface charge of SiO$_2$ with a change in pH was found to be the fundamental framework for PBN activity over the course of arsenic removal.

3.10. Mechanism of PBN Based Fluorescence Sensing of As(III)

The findings as shown in Figures 4–6 revealed an analyte-dependent intervalence transition in iron hexacyanoferrate [Fe^{III}$_4$[Fe^{II}(CN)$_6$]$_3$] between Fe^{2+} and Fe^{3+} as shown below:

$$\text{Fe}^{2+} + \text{Fe}^{3+} + \text{light energy} \rightarrow \text{Fe}^{3+} + \text{Fe}^{2+}$$

The intervalence transition may be evaluated based on changes to the absorption spectrum. The fluorescein–PBN interaction is associated with fluorescence resonance energy transfer as recently described [39,40]; this material is capable of quenching the emitted fluorescence of fluorescein. When PBN undergo interaction with As(III), there is a conversion of PBN into Prussian white nanoparticles, followed by a conversion of As(III) to As(V); thus, the quenching ability is lost. The Prussian white nanoparticles can further be converted into PBN after treating the same with acid as discussed above. This scheme provides an effective and inexpensive method for PBN-mediated removal of As(III) udermvisible light.

3.11. Characterization of Recyclable PBN@SiO$_2$

After arsenic elimination, the recycled PBN@SiO$_2$ was investigated with XRD, SEM, and FTIR to understand the effect of the recycling process on PBN@SiO$_2$ morphology, size, and crystallinity. An SEM image of the recycled PBN@SiO$_2$ is shown in Figure 12. More spherical-shaped than cube-shaped particles were observed; the aggregation of PBN with no precise shape was also observed. XRD analysis demonstrated a shifting inward of the peak position (θ) as compared with the unused PBN@SiO$_2$. The presence of anticipated elements was identified via EDX analysis. In addition, characteristic CN stretching in PBN was noted; this feature was noted at a considerably lower wavenumber (2054 cm^{-1}), which is attributed to particle aggregation [39]. The PBN characteristics before and after the recycling process as obtained from XRD, SEM, and FTIR analysis are listed in Table 4.

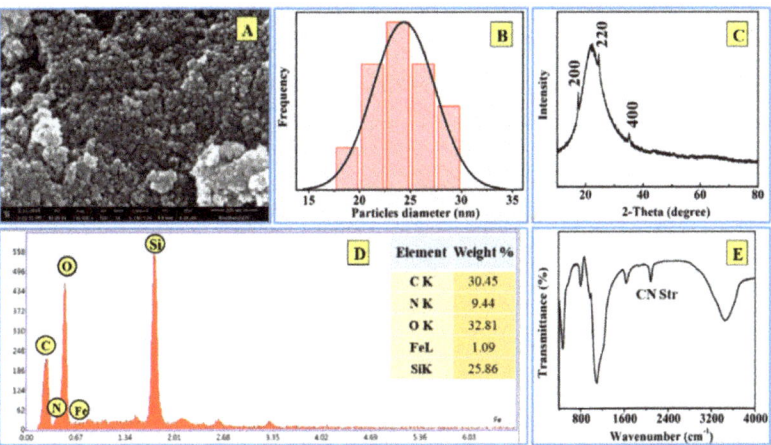

Figure 12. Analysis of recycled PBN@SiO$_2$ following acid treatment. (**A**) An alteration in nanoparticle morphology as identified through SEM imaging, (**B**) A decrease in particle size (26 nm) as calculated from the displayed histogram, (**C**) Minor shift in peak position with similar planes as identified through XRD analysis, (**D**) Spectrum with the entire anticipated element as detected by EDX, (**E**) Shift in CN stretching peak observed using FTIR.

Table 4. Table with HRSEM, XRD and FTIR data displaying the variations between unused and recycled PBN@SiO$_2$.

Analysis	Property	Unused PBN@SiO$_2$	Recycled PBN@SiO$_2$
HRSEM	Shape	Nanocubic (82%) and Spherical (18%)	Nanocubic (19%) and Spherical (81%)
HRSEM	Size	70–20 nm	17–26 nm
XRD	2-Theta (Planes)	17.6 (200), 24.3 (220), 37.83 (400)	17.4 (200), 24.6 (220), 35.12 (400)
FTIR	CN Str.	2096 cm^{-1}	2054 cm^{-1}

4. Conclusions

PBN are a light-sensitive material that is processed through functional alkoxysilane- and cyclohexanone-mediated conversion of a single precursor, potassium hexacyanoferrate. The synthetic incorporation of PBN within mesoporous silica (PBN@SiO$_2$) was also studied; the morphology of these particles was characterized using TEM, SEM, XRD, and XPS. The as-made PBN were studied as a fluorescent quencher. The quenching ability of the materials is found to be a function of arsenic(III) concentration; this result suggested a novel application of PBN for fluorescence sensing of arsenic. In addition, XPS studies confirmed that arsenic is adsorbed on PBN@SiO$_2$ as arsenite (As(III)) and arsenate (As(V)) irrespective of the initial oxidation state of the material; this result indicated a novel application of PBN for the removal of arsenic(III) from a given sample.

Author Contributions: Conceptualization, P.C.P., S.S. and R.J.N.; methodology, P.C.P., S.S. and and R.J.N.; software, P.C.P., S.S. and R.J.N.; validation, P.C.P., S.S. and R.J.N.; formal analysis, P.C.P., S.S. and R.J.N.; investigation, P.C.P., S.S. and R.J.N.; resources, P.C.P., S.S. and R.J.N.; data curation, P.C.P., S.S. and R.J.N.; writing—original draft preparation, P.C.P., S.S. and R.J.N.; writing—review and editing, P.C.P., S.S. and R.J.N.; visualization, P.C.P., S.S. and R.J.N.; supervision, P.C.P., S.S. and R.J.N.; project administration, P.C.P., S.S. and R.J.N.; funding acquisition, P.C.P., S.S. and R.J.N. All authors have read and agreed to the published version of the manuscript.

Funding: This research was funded by [SERB, DST and DRDO] grant number [VJR/2017/000034 and LSRB-316 respectively].

Institutional Review Board Statement: The study was conducted according to the guidelines of the Declaration of Indian Institute of Technology (BHU), and approved by the Institutional Review Board.

Informed Consent Statement: Informed consent was obtained from all subjects involved in the study.

Data Availability Statement: Data supporting reported results can be found in the laboratory of Prof. Prem C Pandey of IIT(BHU). https://iitbhu.ac.in/dept/apc/people/faculty.

Acknowledgments: The authors acknowledge IIT (BHU), Varanasi for financial assistance. Thanks are given to SERB for the VAJRA Fellowship and DRDO for granting LSRB-316.

Conflicts of Interest: The authors declare that they have no known competing financial interests or personal relationships that influence the work reported in this paper.

References

1. NRC. *Subcommittee to Update the 1999 Arsenic in Drinking Water Report, Committee on Toxicology*; NRC: Washington, DC, USA, 2002.
2. Nriagu, J.O.; Azcue, J. *Arsenic in the Environment. Part I: Cycling and Characterization*; John Wiley & Sons: New York, NY, USA, 1994; Volume 26.
3. Roberts, L.C.; Hug, S.J.; Ruettimann, T.; Billah, M.; Khan, A.W.; Rahman, M.T. Arsenic Removal with Iron(II) and Iron(III) in Waters with High Silicate and Phosphate Concentrations. *Environ. Sci. Technol.* **2004**, *38*, 307–315. [CrossRef] [PubMed]
4. Holm, T.R. Effects of CO_3^{2-}/bicarbonate, Si, and PO_4^{3-} on Arsenic sorption to HFO. *J. Am. Water Work. Assoc.* **2002**, *94*, 174–181. [CrossRef]
5. Jain, C.; Ali, I. Arsenic: Occurrence, toxicity and speciation techniques. *Water Res.* **2000**, *34*, 4304–4312. [CrossRef]
6. Edwards, M. Chemistry of arsenic removal during coagulation and Fe-Mn oxidation. *J. Am. Water Work. Assoc.* **1994**, *86*, 64–78. [CrossRef]
7. Shih, M.-C. An overview of arsenic removal by pressure-driven membrane processes. *Desalination* **2005**, *172*, 85–97. [CrossRef]
8. Pandey, P.C.; Pandey, A.K. Tetrahydrofuran hydroperoxide mediated synthesis of Prussian blue nanoparticles: A study of their electrocatalytic activity and intrinsic peroxidase-like behavior. *Electrochim. Acta* **2014**, *125*, 465–472. [CrossRef]
9. Pandey, P.C.; Singh, S.; Sawant, S.N. Functional alkoxysilane mediated controlled synthesis of Prussian blue nanoparticles, enabling silica alginate bead development; nanomaterial for selective electrochemical sensing. *Electrochim. Acta* **2018**, *287*, 37–48. [CrossRef]
10. Karyakin, A.A.; Karyakina, E.E.; Gorton, L. Amperometric biosensor for glutamate using Prussian blue-based "artificial peroxidase" as a transducer for hydrogen peroxide. *Anal. Chem.* **2000**, *72*, 1720–1723. [CrossRef] [PubMed]
11. Robin, M.B. The Color and Electronic Configurations of Prussian Blue. *Inorg. Chem.* **1962**, *1*, 337–342. [CrossRef]
12. Pintado, S.; Goberna-Ferron, S.; Escudero-Adan, E.C.; Galan-Mascaros, J.R. Fast and persistent electrocatalytic water oxidation by Co–Fe Prussian blue coordination polymers. *J. Am. Chem. Soc.* **2013**, *135*, 13270–13273. [CrossRef]
13. Qiu, J.-D.; Peng, H.-Z.; Liang, R.-P.; Li, J.; Xia, X.-H. Synthesis, Characterization, and Immobilization of Prussian Blue-Modified Au Nanoparticles: Application to Electrocatalytic Reduction of H_2O_2. *Langmuir* **2007**, *23*, 2133–2137. [CrossRef] [PubMed]
14. Karyakin, A.A.; Puganova, E.A.; Budashov, I.A.; Kurochkin, I.N.; Karyakina, E.E.; Levchenko, V.A.; Matveyenko, V.N.; Varfolomeyev, S.D. Prussian blue based nanoelectrode arrays for H_2O_2 detection. *Anal. Chem.* **2004**, *76*, 474–478. [CrossRef]
15. Pandey, P.C.; Singh, S.; Walcarius, A. Palladium-Prussian blue nanoparticles; as homogeneous and heterogeneous electrocat-alysts. *J. Electroanal. Chem.* **2018**, *823*, 747–754. [CrossRef]
16. Itaya, K.; Uchida, I.; Neff, V.D. Electrochemistry of polynuclear transition metal cyanides: Prussian blue and its analogues. *Acc. Chem. Res.* **1986**, *19*, 162–168. [CrossRef]
17. Han, L.; Yu, X.-Y.; Lou, X.W. Formation of Prussian-Blue-Analog Nanocages via a Direct Etching Method and their Conversion into Ni-Co-Mixed Oxide for Enhanced Oxygen Evolution. *Adv. Mater.* **2016**, *28*, 4601–4605. [CrossRef] [PubMed]
18. Pandey, P.C.; Pandey, A.K. Electrochemical sensing of dopamine and pyrogallol on mixed analogue of Prussian blue nano-particles modified electrodes—Role of transition metal on the electrocatalysis and peroxidasemimetic activity. *Electrochim. Acta* **2013**, *109*, 536–545. [CrossRef]
19. Pandey, P.C.; Pandey, A.K. Novel synthesis of super peroxidase mimetic polycrystalline mixed metal hexacyanoferrates nanoparticles dispersion. *Analyst* **2013**, *138*, 2295–2301. [CrossRef] [PubMed]
20. Pandey, P.C.; Pandey, A.K. Novel synthesis of Prussian blue nanoparticles and nanocomposite sol: Electro-analytical ap-plication in hydrogen peroxide sensing. *Electrochim. Acta* **2013**, *87*, 1–8. [CrossRef]
21. Pandey, P.C.; Upadhyay, B.C. Studies on differential sensing of dopamine at the surface of chemically sensitized Ormosil-modified electrodes. *Talanta* **2005**, *67*, 997–1006. [CrossRef]
22. Zhang, W.; Singh, P.; Paling, E.; Delides, S. Arsenic removal from contaminated water by natural iron ores. *Miner. Eng.* **2004**, *17*, 517–524. [CrossRef]
23. Morita, K.; Kaneko, E. Spectrophotometric Determination of Trace Arsenic in Water Samples Using a Nanoparticle of Ethyl Violet with a Molybdate–Iodine Tetrachloride Complex as a Probe for Molybdoarsenate. *Anal. Chem.* **2006**, *78*, 7682–7688. [CrossRef] [PubMed]

24. Yuan, B.; Qu, J.; Lin, Q. Fluorimetric Determination of Arsenite and Arsenate in Water Using Fluorescein and Iodine. *Int. J. Environ. Anal. Chem.* **2002**, *82*, 31–36. [CrossRef]
25. Zhu, R.H.; Chen, L. Spectrofluorimetric determination of arsenic(III) in water samples. *Asian J. Chem.* **2011**, *23*, 5271–5274.
26. Dutta, D.; Chatterjee, S.; Pillai, K.; Pujari, P.; Ganguly, B. Pore structure of silica gel: A comparative study through BET and PALS. *Chem. Phys.* **2005**, *312*, 319–324. [CrossRef]
27. Antony, R.; Manickam, S.T.D.; Kollu, P.; Chandrasekar, P.V.; Karuppasamy, K.; Balakumar, S. Highly dispersed Cu(II), Co(II) and Ni(II) catalysts covalently immobilized on imine-modified silica for cyclohexane oxidation with hydrogen peroxide. *RSC Adv.* **2014**, *4*, 24820–24830. [CrossRef]
28. Ayers, J.B.; Waggoner, W.H. Synthesis and properties of two series of heavy metal hexacyanoferrates. *J. Inorg. Nucl. Chem.* **1971**, *33*, 721–733. [CrossRef]
29. Cros, A.; Saoudi, R.; Hollinger, G.; Hewett, C.; Lau, S. An X-ray photoemission spectroscopy investigation of oxides grown on Au_xSi_{1-x} layers. *J. Appl. Phys.* **1990**, *67*, 1826–1830. [CrossRef]
30. Datta, M.; Datta, A. In situ FTIR and XPS studies of the hexacyanoferrate redox system. *J. Phys. Chem.* **1990**, *94*, 8203–8207. [CrossRef]
31. Stec, W.J.; Morgan, W.E.; Albridge, R.G.; van Wazer, J.R. Measured binding energy shifts of "3p" and "3d" electrons in arsenic compounds. *Inorg. Chem.* **1972**, *11*, 219–225. [CrossRef]
32. Fantauzzi, M.; Atzei, D.; Elsener, B.; Lattanzi, P.; Rossi, A. XPS and XAES analysis of copper, arsenic and sulfur chemical state in enargites. *Surf. Interface Anal.* **2006**, *38*, 922–930. [CrossRef]
33. Ding, M.B.; de Jong, B.H.; Roosendaal, S.J.; Vredenberg, A.X. XPS studies on the155 electronic structure of bonding between solid and solutes: Adsorption of arsenate, chromate, phosphate, Pb2+, and Zn2+ ions on amorphous blackferric oxyhydroxide. *Geochim. Cosmochim. Acta* **2000**, *64*, 1209–1219. [CrossRef]
34. Brundle, C.; Chuang, T.; Wandelt, K. Core and valence level photoemission studies of iron oxide surfaces and the oxidation of iron. *Surf. Sci.* **1977**, *68*, 459–468. [CrossRef]
35. Júnior, J.A.A.; Baldo, J.B. The Behavior of Zeta Potential of Silica Suspensions. *New J. Glas. Ceram.* **2014**, *4*, 29–37. [CrossRef]
36. Gupta, V.; Saini, V.; Jain, N. Adsorption of As(III) from aqueous solutions by iron oxide-coated sand. *J. Colloid Interface Sci.* **2005**, *288*, 55–60. [CrossRef]
37. Zen, J.-M.; Chen, P.-Y.; Kumar, A.S. Flow Injection Analysis of an Ultratrace Amount of Arsenite Using a Prussian Blue-Modified Screen-Printed Electrode. *Anal. Chem.* **2003**, *75*, 6017–6022. [CrossRef] [PubMed]
38. Zen, J.-M.; Kumar, A.S.; Chen, H.-W. Electrochemical behavior of stable cinder/Prussian blue analogue and its mediated nitrite oxidation. *Electro-Anal. Int. J. Devoted Fundam. Pract. Asp. Electroanal.* **2001**, *13*, 1171–1178.
39. Farah, A.M.; Shooto, N.D.; Thema, F.T.; Modise, J.S.; Dikio, E.D. Fabrication of Prussian blue/multi-walled carbon nanotubes modified glassy carbon electrode for electrochemical detection of hydrogen peroxide. *Int. J. Electrochem. Sci.* **2012**, *7*, e13.
40. Pandey, P.C.; Shukla, S.; Pandey, G.; Narayan, R.J. Organotrialkoxysilane-mediated controlled synthesis of noble metal nanoparticles and their impact on selective fluorescence enhancement and quenching. *J. Vac. Sci. GY B Nanotechnol. Microelectron. Mater. Process. Meas. Phenom.* **2020**, *38*, 052801.

Article

Flexible Carbon Nanotubes Confined Yolk-Shelled Silicon-Based Anode with Superior Conductivity for Lithium Storage

Na Han [1,†], Jianjiang Li [1,†], Xuechen Wang [1], Chuanlong Zhang [1], Gang Liu [1], Xiaohua Li [1], Jing Qu [1], Zhi Peng [1], Xiaoyi Zhu [1,*] and Lei Zhang [2,*]

[1] School of Material Science and Engineering, School of Environmental Science and Engineering, Chemical Experimental Teaching Center, School of Automation, Qingdao University, No. 308, Ningxia Road, Qingdao 266071, China; 2018020395@qdu.edu.cn (N.H.); jjli@qdu.edu.cn (J.L.); 2018020384@qdu.edu.cn (X.W.); 2018205858@qdu.edu.cn (C.Z.); 2019025785@qdu.edu.cn (G.L.); 2019020442@qdu.edu.cn (X.L.); 2017201339@qdu.edu.cn (J.Q.); pengzhi@qdu.edu.cn (Z.P.)

[2] Key Laboratory of Materials Physics, and Anhui Key Laboratory of Nanomaterials and Nanotechnology, Institute of Solid State Physics, Chinese Academy of Sciences, Hefei 230031, China

* Correspondence: xyzhu@qdu.edu.cn (X.Z.); lei.zhang@issp.ac.cn (L.Z.)
† These authors equally contributed to this work.

Abstract: The further deployment of silicon-based anode materials is hindered by their poor rate and cycling abilities due to the inferior electrical conductivity and large volumetric changes. Herein, we report a silicon/carbon nanotube (Si/CNT) composite made of an externally grown flexible carbon nanotube (CNT) network to confine inner multiple Silicon (Si) nanoparticles (Si NPs). The in situ generated outer CNTs networks, not only accommodate the large volume changes of inside Si NPs but also to provide fast electronic/ionic diffusion pathways, resulting in a significantly improved cycling stability and rate performance. This Si/CNT composite demonstrated outstanding cycling performance, with 912.8 mAh g^{-1} maintained after 100 cycles at 100 mA g^{-1}, and excellent rate ability of 650 mAh g^{-1} at 1 A g^{-1} after 1000 cycles. Furthermore, the facial and scalable preparation method created in this work will make this new Si-based anode material promising for practical application in the next generation Li-ion batteries.

Keywords: silicon; yolk−shell structure; anode; lithium-ion batteries

1. Introduction

Silicon (Si) is the most promising anode candidate in lithium-ion batteries (LIBs) due to its high theoretical specific capacity (~4200 mAh g^{-1}) and cut-price [1–4]. However, the large volume changes (over 400% expansion after full lithiation) induced poor structural stability and continuous breaking and regenerating of the solid-electrolyte interphase (SEI) cause's short working life for Si-based anodes [5–7]. Moreover, the low electrical conductivity of the Si limits its rate performance under high current densities [8–10]. Up till now, introducing a reserved void space and conductive framework into silicon-based materials has been regarded as the most effective strategy to fundamentally improve the electrochemical behavior of Si-based anodes [11–13]. The introduced reserved space can buffer the huge changes in volume of Si during cycling, leading to the enhanced structural integrity and cycling stability [14–16]. Additionally, the conductive framework within the composites increases the overall conductivity of the electrodes, resulting in the high-rate capacities under high current densities [17–20].

Among various Si-based composites, the yolk-shelled Si/carbon (Si/C) composites are the most promising candidate because of their distinctive advantages over the existing Si-based composites in terms of cycling stability and rate behavior [21–24]. Many previous reports confirmed the effective structure [25–27]. For these yolk-shelled Si/C composites, the Si-yolk was encapsulated within a hollow C-shell with reserved space between the Si-yolk and the C-shell. Therefore, the volume changes of inner Si-yolk can be accommodated by the void space and confined within the hollow C-shell, leading to increased structural stability and limited formation of the outer generated SEI film [25–27]. However, the introduced void space limits the conductive contact between Si-yolk and C-shell and further decreases the tap density of the composite [28–30]. Carbon nanotubes (CNTs) with excellent mechanical properties and high electrical conductivity are regarded as another hopeful carbon matrix to increase the overall behavior of Si-based materials [31–36]. Currently, most of the reported Si/CNT anodes are synthesized by directly using expensive commercialized CNTs to mix with Si nanoparticles (Si NPs), causing increased production cost [37–39]. Moreover, it is difficult to achieve the uniform distribution between CNTs and Si NPs due to their large surface area [40,41]. Currently, new Si/CNT anodes composites have been developed via a chemical vapor deposition (CVD) process, which provides distinguished structural stability and electrochemical performance, enhances the overall conductivity of the electrode, and increases the safety of the battery [32,42]. Moreover, it is remaining a great challenge to prepare promising Si/CNT composites with low-cost methods while preserving the unique volume change containment functionality of Si/C yolk–shell structures.

Herein, we overcome these obstacles by developing new Si/CNTs anodes (Scheme 1). Si NPs were successively double-coated with rigid carbon and silica layers (Si@C@SiO$_2$) to better encapsulate the incorporated multiple Si NPs to realize good safety levels. Furthermore, the SiO$_2$ coating layer on the outer surface of Si@C@SiO$_2$ further provided active position for in situ CNTs grown via a CVD method, resulting in a new Si/CNT composite. For this new Si/CNT, the flexible CNT networks were grown on the surface of Si@C@SiO$_2$ particles. Therefore, the aggregation for both the CNTs and Si NPs can be significantly suppressed due to the external in-situ grown CNTs networks. Additionally, compared with the traditional yolk-shell structure, the CNT networks and the carbon coating shell effectively increase the conductive contact, not only between the inner Si-yolks and CNT networks but also among different Si/CNT microparticles, leading to increased electronic conductivity and rate capacities. Moreover, the overall structural stability and integrity of this new Si/CNT can also be enhanced by flexible porous CNT networks and rigid carbon coating [42].

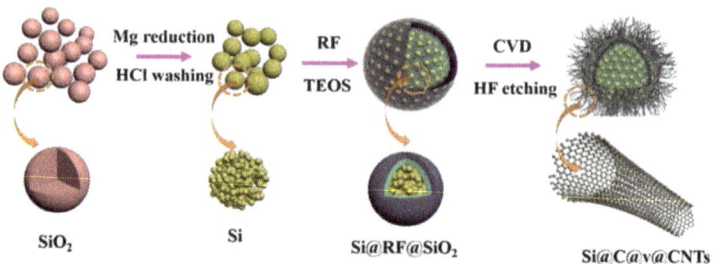

Scheme 1. Schematic illustration of the preparation of the Si@C@v@CNTs.

2. Materials and Methods

2.1. Synthesis of Si NPs

All reagents in this paper were purchased from Sinopharm Co (Shanghai, China). The nano-sized silica (SiO$_2$) spheres were firstly synthesized by the well-established Stöber method. In the following magnesiothermic reduction (MR) process, Mg powders (99%) and the obtained SiO$_2$ spheres were uniformly mixed and placed in one side of a crucible boat. After that, a certain amount of NaCl (AR) (SiO$_2$:NaCl = 1:10) was placed in the other side of the crucible boat. The crucible boat was then placed in the center of the tube furnace (OTF-1200X, Shenzhen kejing-zhida Co, Shenzhen, China) and increased to 700 °C under an Ar/H$_2$ (95:5 vol. %) flow and retained for 6 h. After cooling down to normal condition, the obtained sample was dispersed in 1 M HCl for several hours to remove NaCl and byproduct MgO. The final porous Si NPs powders were obtained after a wash and vacuum dry.

2.2. Synthesis of Si@RF@SiO$_2$

The above prepared Si NPs were modified with 3-aminopro-pyltrimethoxysilane (APTES) (AR) to positively charge the surfaces. In total, 0.4 g Si NPs were uniformly dispersed in 300 mL ethanol (AR), containing 4 mL of APTES, and stirred for 5 h to obtain APTES-Si NPs. The above APTES-Si NPs was re-dispersed in an alkaline mixture of 150 mL deionized water and 30 mL ethanol, containing 1 mL of aqueous ammonia (AR) under magnetic stirring for 30 min. After that, 0.6 g resorcinol (AR) and 0.8 mL formaldehyde (AR) were separately added to the reaction system and continued stirring for 10 h to form a homogeneous phenolic resin (RF) coating layer under room temperature. The Si@RF powders were obtained via centrifugation treatment of the reaction solution. For the SiO$_2$-coated Si@RF composite (named Si@RF@SiO$_2$), 400 mg Si@RF was mixed with 1.5 mL tetraethyl orthosilicate (TEOS) (AR) hydrolysis under alkaline condition to form SiO$_2$ coating layer on the outer surface of Si@RF.

2.3. Synthesis of Si@C@v@CNTs

Carbon nanotubes were grown in situ via a CVD method using Iron(III) nitrate nonahydrate (AR) as the catalyst (Fe) and acetylene (5%) as carbon precursors at 900 °C for 2 h under Ar/H$_2$ in a tube furnace. The catalyst was loaded on the precursor microspheres of Si@RF@SiO$_2$ prior to the deposition procedure to ensure that the CNTs could be grown in situ on the active positions during the CVD process. Finally, the SiO$_2$ coating layer was removed with dilute hydrofluoric acid (AR) solution and followed by centrifugation treatment and ethanol washing. After removing the SiO$_2$ sacrificial coating layers, the final composite was named Si@C@v@CNTs ("v" stands for "void"). For comparison purposes, the Si@C@v@C ("v" stands for "void") without in-situ grown CNTs was also synthesized via the same CVD method but in the absence of a Fe(NO$_3$)$_3$·9H$_2$O catalyst, and the Si@v@CNT without an inner carbon layer was also synthesized via the same CVD method in the absence of the RF-layer.

2.4. Characterizations

The XRD patterns of samples were obtained via a DX-2007 X-ray diffraction (XRD) experiment apparatus (λ = 1.5418 Å) (Dandong Haoyuan Instrument Co, Dandong, China) to confirm the phases and crystallinity. The Si content of composites was characterized by a thermogravimetric analyzer (TGA4000) (NSK LTD, Tokyo, Japan) at a heating rate of 10 °C min^{-1} in air atmosphere. The nitrogen adsorption/desorption isotherm curves were obtained via a Micromeritics ASAP-2020M nitrogen adsorption/desorption apparatus (Best Instrument Technology Co, Beijing, China) to confirm the porosity character. The micro-structures and morphologies of the materials were collected via a JSM-6700F scanning electron microscope (SEM) (JEOL, Tokyo, Japan) with an IE300X energy-dispersive X-ray spectrometer and a JEM-2100F transmission electron microscope (TEM) (JEOL, Tokyo, Japan). X-ray photoelectron spectroscopy (XPS) curves were obtained by applying an electron spectrometer (ESCALab250) (Thermo Fay, Boston, MA, USA) to analyze the surface of composites. Raman spectra was collected with a Raman spectrometer (JobinYvon HR800) (Renishaw, London, UK).

2.5. Electrochemical Measurements

The CR2016 coin-type cells were assembled in the glove box under inert atmosphere conditions without water and oxygen to test the electrode performance by using polypropylene films as separators (the thickness of the separator was 25 μm) to separate the working and counter electrodes (lithium wafer) in an electrolyte. The electrolyte was the 1 M LiPF$_6$

dissolved in the solvent of ethylene carbonate and dimethyl carbonate (the amount of the electrolyte used in assembling the coin-type cell was 70 µL). The working electrode was made by coating the slurry of the above active materials containing a proportional conductive carbon black and polyvinylidene fluoride (PVDF) binder on the copper foil (the mass loading of the electrode was about 1.2 mg cm^{-2}), which was then dried under a 120 °C vacuum oven for 14 h (the thickness of the active electrode layer was about 20 µm). Finally, the copper foil was cut into wafers with a uniform size of 1 cm. The galvanostatic cycling measurements were conducted by a CT 2001A battery tester at determinate voltage windows. Cyclic voltammogram (CV) tests were performed by using an electrochemical workstation within a fixed voltage range and scan rate.

3. Results and Discussions

Figure S1 shows the XRD patterns of reduced Si NPs, Si@RF, and Si@RF@SiO$_2$. Three sharp diffraction peaks, which are located at 2θ values of 28.4, 47.2, and 56.1°, were attributed to the planes of (111), (220), and (311) for the crystal Si phase, respectively (JCPDS NO. 27-1402), indicating that the amorphous SiO$_2$ synthesized by the well-established Stöber method were fully reduced to crystalline Si in the MR process [43]. Another broad peak at ~25° corresponded with the amorphous carbon and silica coming from the double coat with RF and silica layers. As shown in Figure 1a, for Si@C@v@CNTs and Si@v@CNTs, after the CVD process, CNT grown in situ across Si-CNPs and the diffraction peak at ~25° was observed, corresponding to the (002) plane of the crystalline carbon [40]. Figure 1b displays the Raman spectra of Si@C@v@CNTs, Si@v@CNTs, and Si@C@v@C. The sharp peak at about 500 cm^{-1} could be appointed to the Si peaks. Additionally, the weak peaks at about 1345 cm^{-1} (D-band) and 1595 cm^{-1} (G-band) could be associated with the vibration modes of sp^3-bonded carbon atoms in amorphous carbon and sp^2-bonded carbon atoms in typical graphite, respectively [44]. As calculated, the I_D/I_G was 0.98, 0.96, 0.95 for Si@C@v@C, Si@v@CNTs, and Si@C@v@CNTs, respectively. Si@C@v@CNTs had a relatively higher graphitization degree due to the microcrystalline structure of CNTs. Thermogravimetric analysis (TGA) was tested to confirm the proportion of carbon and silicon for the samples (Figure 1c). The weight losses occurred from 500 to 800 °C were ascribed to the carbon combustion and calculated to be 54.3%, 70.6%, and 85.9% for Si@C@v@C, Si@v@CNTs, and Si@C@v@CNTs, respectively. As the temperature continued to rise, Si NPs were further oxidized, leading to a weight increase in the TGA curves. The Si content in the three samples were 45.7%, 29.4%, and 14.1% for Si@C@v@C, Si@v@CNTs, and Si@C@v@CNTs, respectively. Compared with the other two samples, high C and CNT content in Si@C@v@CNTs provided more electric contact between particles, thus enhanced overall electronic conductivity and superior performance could be expected. The elemental compositions and valence states in the composites were determined by X-ray photoelectron spectroscopy (XPS) spectra. Figure 1d reveals the whole spectrum of Si@C@v@CNTs, confirming the coexistence of Si, C, and O. The high-resolution spectrum of Si 2p is shown in Figure 1e. Three peaks of Si-Si (98.6 eV), Si-C (101.6 eV), and Si-O (102.98 eV) originated from monatomic Si, residual silica, or slight oxidation of the Si NPs [45]. In addition, the high-resolution C 1s and O 1s in Figure S2 shows that the peak at 283.75 eV was related to the graphite-like sp^2 hybridized carbon and another peak at 283.3 eV was assigned to C-O. The O 1s peak at 537 eV may be attributed to adsorbed oxygen and the residual silica layer [46].

The Brunauer−Emmett−Teller (BET) specific surface area and pore volume for the samples are illustrated in Table S1. Compared with Si@C@v@C, the Si@C@v@CNTs had a relatively higher BET specific surface area of 106.98 m^2 g^{-1}, with a pore volume of 0.3212 cm^3 g^{-1}. We believe that the higher surface area was mainly due to the existence of reserved void space and in-situ grown CNT networks. The nitrogen adsorption–desorption isotherm curve of the samples is displayed in Figure 1f. All the curves had a distinct hysteresis loop, suggesting the existence of a mesoporous structure [47]. The pore size distribution curve (Figure S3) shows a diverse pore structure between 2 and 10 nm, which incorporate a series of micro- and mesopores derived from CNT networks [48]. Furthermore, these multiple pore structures not only afford fast and shortened electronic/ionic diffusion pathways but also absorb the huge expansion in volume inside Si NPs during cycling, resulting in a significantly enhanced overall structural integrity and electrochemical performance.

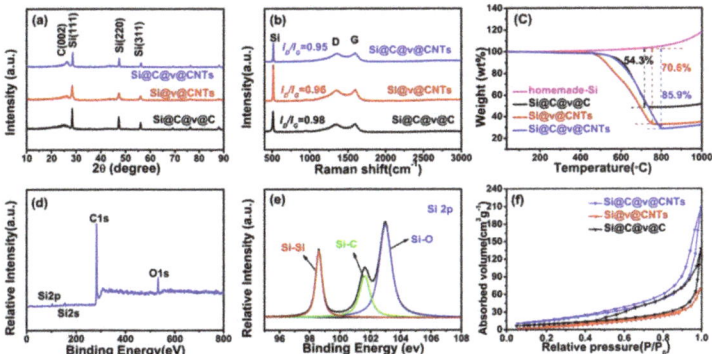

Figure 1. XRD patterns (**a**), Raman spectra (**b**), and thermogravimetric analysis (TGA) curves (**c**) of Si@C@v@C, Si@v@CNTs, and Si@C@v@CNTs; X-ray photoelectron spectroscopy (XPS) of Si@C@v@CNTs (**d**), High-resolution XPS spectrum of Si 2p (**e**), and N_2 adsorption-desorption isotherms curves of Si@C@v@C, Si@v@CNTs, and Si@C@v@CNTs (**f**).

Figure 2 shows the scanning electron microscopy (SEM) images of the structural evolution of Si@C@v@CNTs at different synthesis stages. The SEM image of SiO_2 spheres firstly synthesized by the well-established Stöber method is shown in Figure 2a. The pristine SiO_2 spheres present a monodispersed spherical shape with a uniform diameter of approximately 300 nm. After the Mg-reduction process, porous Si NPs with well-preserved monodispersed spherical morphology were successfully prepared (Figure 2b). Figure 2c,d show the SEM images of Si@RF and Si@RF@SiO_2. The diameters of as-prepared Si@RF and Si@RF@SiO_2 precursors were increased to 400 nm and 500 nm, respectively, indicating the successful coating of RF and SiO_2 layers on the Si cores, resulting in Si@C@SiO_2 particles. Figure 2e,f reveals a large amount of tangled CNTs with various diameters in the range of 50–100 nm externally grown in situ across the Si@C@SiO_2 particles. The multiple Si NPs were well supported by the in-situ generated flexible porous CNTs networks, contributing to better electric contact, not only between the inner Si-yolks but also among the Si-CNT microparticles. Figure 2g reveals the uniform elemental distribution of the Si, C, and Fe in Si@C@v@CNTs. It is clearly observed the Si-yolks were well coated by the outer C-shell and distributed across the flexible CNT networks. Figure S4 shows the SEM images of Si@C@v@C. The observed wrinkles over the entire surface confirm the full encapsulation of Si@C@SiO_2 by the carbon layer. The SEM images of Si@v@CNTs in Figure S5 reveal the similar tangled CNTs with Si@C@v@CNTs.

Figure 3a–c shows the transmission electron microscopy (TEM) images of homemade SiO_2 and reduced Si NPs. The high-resolution TEM image (HRTEM), taken from one of the Si NPs, reveals that the obtained Si NPs had high crystallinity [49,50]. The TEM images of Si@RF and Si@RF@SiO_2, as shown in Figure 3d–e, confirm that the Si NPs were well wrapped by the RF carbon-SiO_2 double coating layers. The TEM and HRTEM images of Si@C@v@CNTs presented in Figure 3f–i confirm CNTs grown in situ on the outer layer of the Si@RF NPs with a void between the two. The carbon layer derived from RF was about 5 nm and could accelerate electron transfer between Si NPs and CNTs, leading to enhanced structural stability. The void generated due to the etching of SiO_2 could effectively alleviate the expansion of the inner Si NPs. According to the TEM images of a single CNT shown in Figure S6, the outer diameter of this single CNT was approximately 35 nm and the tube wall was 12 nm. The TEM image of Si@C@v@C in Figure S7 shows that the Si NPs were well wrapped with double carbon layers. The TEM images of Si@v@CNTs in Figure S8 confirms that existence of similar CNTs grown to Si@C@v@CNTs but without of a carbon layer on the Si NPs.

Figure 2. SEM images of SiO$_2$ nanoparticles (NPs), synthesized by the well-established Stöber method (**a**), reduced Si NPs in the magnesiothermic reduction (MR) process (**b**), SEM images of the in-process samples: Si@RF (**c**), Si@RF@SiO$_2$ (**d**), Si@C@v@CNTs (**e**,**f**). (**g**) the elemental mapping results of Si@C@v@CNTs.

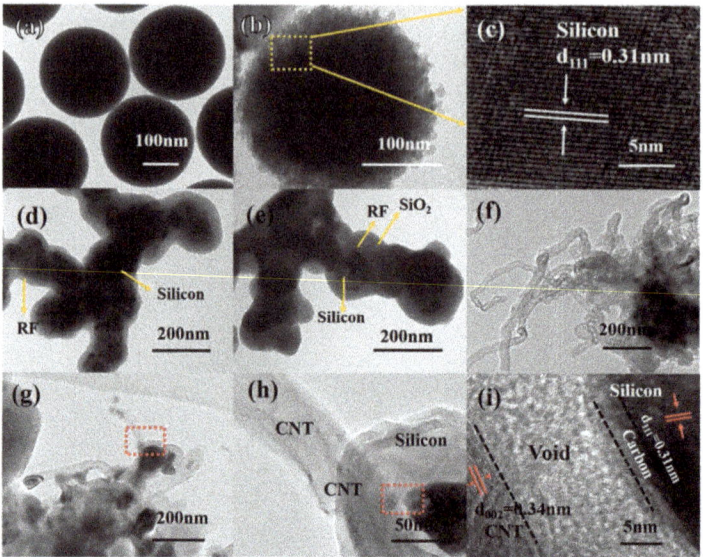

Figure 3. Transmission electron microscope (TEM) images of SiO$_2$ NPs synthesized by the well-established Stöber method (**a**), reduced Si NPs in the MR process (**b**), high resolution TEM (HRTEM) images of Si NPs (**c**), TEM images of Si@RF (**d**), Si@RF@SiO$_2$ (**e**), and Si@C@v@CNTs (**f**–**h**), and HRTEM images of Si@C@v@CNTs (**i**).

The cyclic voltammetry (CV) curve of Si@C@v@CNTs exhibited the typical electrochemical properties of Si-based anode materials (as shown in Figure 4a). In the first cathodic branch (lithiation), a distinct broad peak between 0.5 and 0.8 V was ascribed to the generated solid-electrolyte interphase (SEI) film. However, this peak disappeared after the first lithiation, suggesting the generation of the firm and stable films during the first cycle [51]. The lithiation peak at 0.18 V can be appointed to the lithiation process of Si. In the following anodic branch (delithiation), the peaks located at 0.34 and 0.5 V were ascribed to the dealloying process from Li_xSi to amorphous Si [52]. Figure 4b shows the initial charge–discharge profiles of the materials at 100 mA g^{-1}. The disappearance of the voltage plateaus between 0.5 and 0.8 V after the first cycle also confirms the generation of stable SEI films, which is in accordance with the CV results in Figure 4a [44]. The initial discharge and charge capacities were 2698.4 and 1684.2 mAh g^{-1} for Si@C@v@C, 2546.5 and 1760.6 mAh g^{-1} for Si@v@CNTs, and 2350.1 and 1787.0 mAh g^{-1} for Si@C@v@CNTs at 100 mA g^{-1}, corresponding to the initial coulombic efficiencies (ICE) of 62.41, 69.14, and 76.04%, respectively. Figure S9 shows the galvanostatic charge–discharge curves of Si@C@v@CNTs during the first five cycles at 100 mA g^{-1}. After the first cycle, the CE increased to 92.73% in the following cycling test and reached 96.04% after the fifth cycle. Most importantly, the voltage plateaus during the cycling test was well-maintained, suggesting improved electrochemical utilization of the active electrode materials. The cycling behavior in Figure 4c shows the charge capacities at 100 mA g^{-1} for Si@C@v@C.

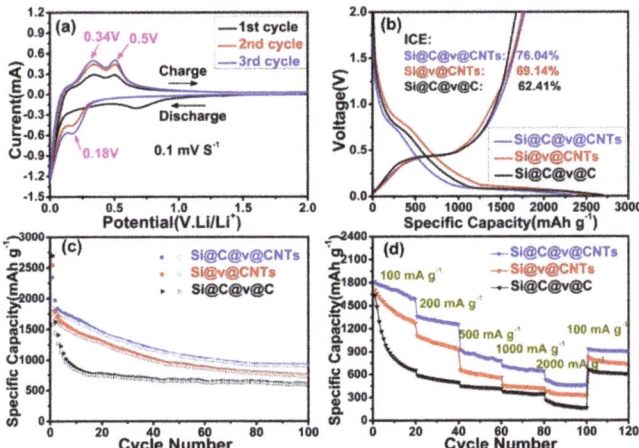

Figure 4. Electrochemical properties: cyclic voltammetry for the first three cycles of Si@C@v@CNTs at a scan rate of 0.1 mV s^{-1} between 0–2.0 V (**a**), the initial discharge–charge curves at 100 mA g^{-1} (**b**), cycling performances at 100 mA g^{-1} (**c**), the rate performances (**d**) of Si@C@v@C, Si@v@CNTs, and Si@C@v@CNTs.

Si@v@CNTs and Si@C@v@CNTs were 598.7, 736.1, and 912.8 mAh g^{-1} after 100 cycles, respectively, indicating that the Si@C@v@CNTs were endowed with the best cycling behavior. Therefore, the outer in-situ grown CNT networks can significantly improve the cycling behavior and structural integrity of the inside-coated Si NP anodes. Figure 4d shows the rate performances of Si@C@v@C, Si@v@CNTs, and Si@C@v@CNTs performed at a series of different current densities. As expected, Si@C@v@CNTs present the best rate ability, even at high current densities. Very high reversible capacity of 907.7 mAh g^{-1} was maintained when the current density was back to 100 mA g^{-1}, suggesting an excellent rate ability of Si@C@v@CNTs. Furthermore, a high reversible capacity of 650 mAh g^{-1} was retained for Si@C@v@CNTs at high 1 A g^{-1} after 1000 cycles (Figure 5a). Therefore, it can be concluded that the introduced void space and porous CNT networks can absorb the huge volume expansion inside Si NPs, leading to enhanced overall structural stability and integrity [52–54]. In addition, the CNT networks and inner rigid carbon coating provided more sufficient conductive contact to fast electronic/ionic diffusion pathways, resulting in significantly improved cycling stability and rate performance [55–57]. Figure S10 shows the SEM images of Si@C@v@CNTs after 1000 cycles at 1 A g^{-1}. No cracking can be detected for Si@C@v@CNTs, and all the active materials were still well adhered to the current collector without exfoliation (Figure S10a,b). In addition, the morphology was well maintained in Si@C@v@CNTs after the long cycling test. (Figure S10c). Figure 5b shows the schematic illustration of lithiation and delithiation processes.

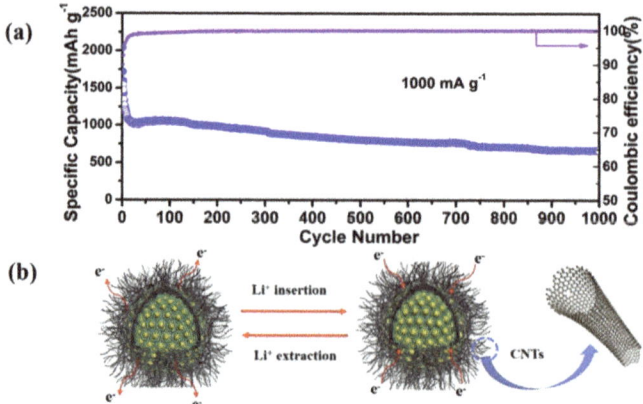

Figure 5. Cycle performance of Si@C@v@CNTs at a current density of 1000 mA g^{-1} (**a**) and a schematic illustration of lithiation and delithiation processes (**b**).

Moreover, we fabricated a full cell with our Si@C@v@CNTs as the anode and LiNi$_{0.6}$Co$_{0.2}$Mn$_{0.2}$O$_2$ (NCM622) as the cathode (Si@C@v@CNTs//NCM622). The voltage profiles of NCM622 are shown in Figure S11. The NCM622 cathode exhibited a stable reversible capacity of 160 mAh g^{-1} with a flat charging/discharging plateau at about 3.5 V. Referring to the voltage profiles of the Si@C@v@CNTs anode and the NCM622 cathode from half cells, the working potential range for the full cell was set between 2 and 4 V. The cycling performance of Si@C@v@CNTs//NCM622 is shown in Figure S12. The full cell displays a reversible capacity of 92 mAh g^{-1} at 100 mA g^{-1} after 100 cycles, indicating potential cycling stability for commercial viability.

4. Conclusions

In summary, we synthesized a yolk-shelled structured silicon/carbon nanotube composite for high performance lithium storage application. This novel Si-based anode was made of an external grown flexible CNT network to confine the inner multiple Si NPs. The in-situ generated outer CNT networks not only accommodated the huge changes in volume space inside Si nanoparticles but also provided fast electronic/ionic diffusion pathways, resulting in markedly improved cycling stability and rate ability. Furthermore, the facial and scalable preparation method created in this work could make this new Si-based anode material promising for practical application in next generation Li-ion batteries.

Supplementary Materials: The following are available online at https://www.mdpi.com/2079-4991/11/3/699/s1. Figure S1: XRD patterns of reduced Si NPs in the MR process, Si@RF, and Si@RF@SiO$_2$. Figure S2: XPS spectrum of C1s (a) and O1s (b) of Si@C@v@CNTs. Figure S3: The pore size distribution curve of Si@C@v@C, Si@v@CNTs, and Si@C@v@CNTs. Figure S4: SEM images of Si@C@v@C. Figure S5: SEM images of Si@v@CNTs. Figure S6: TEM images of a single CNT. Figure S7: TEM images of Si@C@v@C. Figure S8: TEM images of Si@v@CNTs. Figure S9: The first five discharge-charge curves of Si@C@v@CNTs at current density of 100 mA g^{-1}. Figure S10: Digital photograph (a), SEM image of Si@C@v@CNTs after the cycling test at 1.0 A g^{-1} (b,c). Figure S11: Charge/discharge profiles of NCM626 between 2.0–4.3 V. Figure S12: The electrochemical performance of the full cell using Si@C@v@CNTs as anode and LiNi$_{0.6}$Co$_{0.2}$Mn$_{0.2}$O$_2$ (NCM622) as cathode at the current density of 100 mA g^{-1}. Table S1: The Brunauer-Emmett-Teller (BET) surface area, pore volume and average pore size of the samples.

Author Contributions: Conceptualization, J.L. and Z.P.; methodology, N.H., and Z.P.; software, X.W.; validation, C.Z., J.Q., and G.L.; formal analysis, X.L.; investigation, N.H.; resources, X.Z.; data curation, N.H.; writing—original draft preparation, N.H.; writing—review and editing, N.H., X.Z., and L.Z.; supervision, X.Z.; project administration, X.Z. and Z.P. All authors have read and agreed to the published version of the manuscript.

Funding: This work was supported by the National Natural Science Foundation of China (No. 51503109) and the China Postdoctoral Science Foundation Funded Project (No. 2016M600522).

Data Availability Statement: The data presented in this study are available on request from the corresponding author.

Conflicts of Interest: The authors declare that they have no competing interests.

References

1. Li, J.; Xu, Q.; Li, G.; Yin, Y.; Wan, L.; Guo, Y. Research progress regarding Si-based anode materials towards practical application in high energy density Li-ion batteries. *Mater. Chem. Front.* **2017**, *1*, 1691–1708. [CrossRef]
2. Zuo, X.; Zhu, J.; Müller-Buschbaum, P.; Cheng, Y. Silicon based lithium-ion battery anodes: A chronicle perspective review. *Nano Energy* **2017**, *31*, 113–143. [CrossRef]
3. Luo, W.; Chen, X.; Xia, Y.; Chen, M.; Wang, L.; Wang, Q.; Li, W.; Yang, J. Surface and interface engineering of silicon-based anode materials for lithium-ion batteries. *Adv. Energy Mater.* **2017**, *7*, 1701083. [CrossRef]
4. Zhu, X.; Yang, D.; Li, J.; Su, F. Nanostructured Si-based anodes for lithium-ion batteries. *J. Nanosci. Nanotechnol.* **2015**, *15*, 15–30. [CrossRef]
5. Dou, F.; Shi, L.; Chen, G.; Zhang, D. Silicon/carbon composite anode materials for lithium-ion batteries. *Electrochem. Energy Rev.* **2019**, *2*, 149–198. [CrossRef]
6. Shen, X.; Tian, Z.; Fan, R.; Shao, L.; Zhang, D.; Cao, G.; Kou, L.; Bai, Y. Research progress on silicon/carbon composite anode materials for lithium-ion battery. *J. Energy Chem.* **2018**, *27*, 1067–1090. [CrossRef]
7. Goriparti, S.; Miele, E.; De Angelis, F.; Di Fabrizio, E.; Proietti Zaccaria, R.; Capiglia, C. Review on recent progress of nanostructured anode materials for Li-ion batteries. *J. Power Sources* **2014**, *257*, 421–443. [CrossRef]
8. Liu, Z.; Yu, Q.; Zhao, Y.; He, R.; Xu, M.; Feng, S.; Li, S.; Zhou, L.; Mai, L. Silicon oxides: A promising family of anode materials for lithium-ion batteries. *Chem. Soc. Rev.* **2019**, *48*, 285–309. [CrossRef] [PubMed]
9. Chen, Y.; Du, N.; Zhang, H.; Yang, D. Porous Si@C coaxial nanotubes: Layer-by-layer assembly on ZnO nanorod templates and application to lithium-ion batteries. *CrystEngComm* **2017**, *19*, 1220–1229. [CrossRef]
10. Jia, H.; Zheng, J.; Song, J.; Luo, L.; Yi, R.; Estevez, L.; Zhao, W.; Patel, R.; Li, X.; Zhang, J. A novel approach to synthesize micrometer-sized porous silicon as a high-performance anode for lithium-ion batteries. *Nano Energy* **2018**, *50*, 589–597. [CrossRef]
11. Cui, M.; Wang, L.; Guo, X.; Wang, E.; Yang, Y.; Wu, T.; He, D.; Liu, S.; Yu, H. Designing of hierarchical mesoporous/macroporous silicon-based composite anode material for low-cost high-performance lithium-ion batteries. *J. Mater. Chem. A* **2019**, *7*, 3874–3881. [CrossRef]
12. Chen, Y.; Hu, Y.; Shen, Z.; Chen, R.; He, X.; Zhang, X.; Li, Y.; Wu, K. Hollow core–shell structured silicon@carbon nanoparticles embed in carbon nanofibers as binder-free anodes for lithium-ion batteries. *J. Power Sources* **2017**, *342*, 467–475. [CrossRef]
13. Liang, G.; Qin, X.; Zou, J.; Luo, L.; Wang, Y.; Wu, M.; Zhu, H.; Chen, G.; Kang, F.; Li, B. Electrosprayed silicon-embedded porous carbon microspheres as lithium-ion battery anodes with exceptional rate capacities. *Carbon* **2018**, *127*, 424–431. [CrossRef]
14. Nzabahimana, J.; Guo, S.; Hu, X. Facile synthesis of Si@void@C nanocomposites from low-cost microsized Si as anode materials for lithium-ion batteries. *Appl. Surf. Sci.* **2019**, *479*, 287–295. [CrossRef]
15. Xu, Q.; Li, J.; Sun, J.; Yin, Y.; Wan, L.; Guo, Y. Watermelon-inspired Si/C microspheres with hierarchical buffer structures for densely compacted lithium-ion battery anodes. *Adv. Energy Mater.* **2017**, *7*, 1601481. [CrossRef]

16. Liu, Z.; Zhao, Y.; He, R.; Luo, W.; Meng, J.; Yu, Q.; Zhao, D.; Zhou, L.; Mai, L. Yolk@Shell SiO/C microspheres with semi-graphitic carbon coating on the exterior and interior surfaces for durable lithium storage. *Energy Storage Mater.* **2019**, *19*, 299–305. [CrossRef]
17. Wu, P.; Wang, H.; Tang, Y.; Zhou, Y.; Lu, T. Three-dimensional interconnected network of graphene-wrapped porous silicon spheres: In situ magnesiothermic-reduction synthesis and enhanced lithium-storage capabilities. *ACS Appl. Mater. Interfaces* **2014**, *6*, 3546–3552. [CrossRef] [PubMed]
18. Su, M.; Wan, H.; Liu, Y.; Xiao, W.; Dou, A.; Wang, Z.; Guo, H. Multi-layered carbon coated Si-based composite as anode for lithium-ion batteries. *Powder Technol.* **2018**, *323*, 294–300. [CrossRef]
19. Liu, H.; Shan, Z.; Huang, W.; Wang, D.; Lin, Z.; Cao, Z.; Chen, P.; Meng, S.; Chen, L. Self-assembly of silicon@oxidized mesocarbon microbeads encapsulated in carbon as anode material for lithium-ion batteries. *ACS Appl. Mater. Interfaces* **2018**, *10*, 4715–4725. [CrossRef] [PubMed]
20. Zhang, Y.; Jiang, Y.; Li, Y.; Li, B.; Li, Z.; Niu, C. Preparation of nanographite sheets supported Si nanoparticles by in situ reduction of fumed SiO_2 with magnesium for lithium-ion battery. *J. Power Sources* **2015**, *281*, 425–431. [CrossRef]
21. Chen, S.; Shen, L.; van Aken, P.A.; Maier, J.; Yu, Y. Dual-functionalized double carbon shells coated silicon nanoparticles for high performance lithium-ion batteries. *Adv. Mater.* **2017**, *29*, 1605650. [CrossRef]
22. Guan, P.; Li, J.; Lu, T.; Guan, T.; Ma, Z.; Peng, Z.; Zhu, X.; Zhang, L. Facile and scalable approach to fabricate granadilla-like porous-structured silicon-based anode for lithium-ion batteries. *ACS Appl. Mater. Interfaces* **2018**, *10*, 34283–34290. [CrossRef]
23. Guo, S.; Hu, X.; Hou, Y.; Wen, Z. Tunable synthesis of yolk–shell porous silicon@carbon for optimizing Si/C-based anode of lithium-ion batteries. *ACS Appl. Mater. Interfaces* **2017**, *9*, 42084–42092. [CrossRef] [PubMed]
24. Hu, L.; Luo, B.; Wu, C.; Hu, P.; Wang, L.; Zhang, H. Yolk-shell Si/C composites with multiple Si nanoparticles encapsulated into double carbon shells as lithium-ion battery anodes. *J. Energy Chem.* **2019**, *32*, 124–130. [CrossRef]
25. Huang, X.; Sui, X.; Yang, H.; Ren, R.; Wu, Y.; Guo, X.; Chen, J. HF-free synthesis of Si/C yolk/shell anodes for lithium-ion batteries. *J. Mater. Chem. A* **2018**, *6*, 2593–2599. [CrossRef]
26. Jiang, B.; Zeng, S.; Wang, H.; Liu, D.; Qian, J.; Cao, Y.; Yang, H.; Ai, X. Dual Core–Shell Structured Si@SiO_x@C nanocomposite synthesized via a one-step pyrolysis method as a highly stable anode material for lithium-ion batteries. *ACS Appl. Mater. Interfaces* **2016**, *8*, 31611–31616. [CrossRef]
27. Liu, N.; Wu, H.; McDowell, M.T.; Yao, Y.; Wang, C.; Cui, Y. A yolk-shell design for stabilized and scalable li-ion battery alloy anodes. *Nano Lett.* **2012**, *12*, 3315–3321. [CrossRef] [PubMed]
28. Lin, D.; Lu, Z.; Hsu, P.; Lee, H.R.; Liu, N.; Zhao, J.; Wang, H.; Liu, C.; Cui, Y. A high tap density secondary silicon particle anode fabricated by scalable mechanical pressing for lithium-ion batteries. *Energy Environ. Sci.* **2015**, *8*, 2371–2376. [CrossRef]
29. Luo, W.; Wang, Y.; Wang, L.; Jiang, W.; Chou, S.; Dou, S.; Liu, H.K.; Yang, J. Silicon/mesoporous carbon/crystalline TiO_2 nanoparticles for highly stable lithium storage. *ACS Nano* **2016**, *10*, 10524–10532. [CrossRef]
30. Zhou, X.; Tang, J.; Yang, J.; Xie, J.; Ma, L. Silicon@carbon hollow core–shell heterostructures novel anode materials for lithium-ion batteries. *Electrochim. Acta* **2013**, *87*, 663–668. [CrossRef]
31. Liu, R.; Shen, C.; Dong, Y.; Qin, J.; Wang, Q.; Iocozzia, J.; Zhao, S.; Yuan, K.; Han, C.; Li, B.; et al. Sandwich-like CNTs/Si/C nanotubes as high-performance anode materials for lithium-ion batteries. *J. Mater. Chem. A* **2018**, *6*, 14797–14804. [CrossRef]
32. Guan, P.; Zhang, W.; Li, C.; Han, N.; Wang, X.; Li, Q.; Song, G.; Peng, Z.; Li, J.; Zhang, L.; et al. Low-cost urchin-like silicon-based anode with superior conductivity for lithium storage applications. *J. Colloid Interface Sci.* **2020**, *575*, 150–157. [CrossRef]
33. Liu, Y.; Xu, Y.; Fan, B.; Yang, M.; Hamon, A.; Haghi-Ashtiani, P.; He, D.; Bai, J. Constructing 3D CNTs-SiO_2@RGO structures by using GO sheets as template. *Chem. Phys. Lett.* **2018**, *713*, 189–193. [CrossRef]
34. Su, J.; Zhao, J.; Li, L.; Zhang, C.; Chen, C.; Huang, T.; Yu, A. Three-dimensional porous Si and SiO_2 with in situ decorated carbon nanotubes as anode materials for Li-ion batteries. *ACS Appl. Mater. Interfaces* **2017**, *9*, 17807–17813. [CrossRef]
35. de las Casas, C.; Li, W. A review of application of carbon nanotubes for lithium-ion battery anode material. *J. Power Sources* **2012**, *208*, 74–85. [CrossRef]
36. Guo, H.; Ruan, B.; Liu, L.; Zhang, L.; Tao, Z.; Chou, S.; Wang, J.; Liu, H. Capillary-induced Ge uniformly distributed in N-doped carbon nanotubes with enhanced Li-storage performance. *Small* **2017**, *13*, 1700920. [CrossRef]
37. An, W.; Xiang, B.; Fu, J.; Mei, S.; Guo, S.; Huo, K.; Zhang, X.; Gao, B.; Chu, P.K. Three-dimensional carbon-coating silicon nanoparticles welded on carbon nanotubes composites for high-stability lithium-ion battery anodes. *Appl. Surf. Sci.* **2019**, *479*, 896–902. [CrossRef]
38. Ma, T.; Xu, H.; Yu, X.; Li, H.; Zhang, W.; Cheng, X.; Zhu, W.; Qiu, X. Lithiation behavior of coaxial hollow nanocables of carbon-silicon composite. *ACS Nano* **2019**, *13*, 2274–2280. [CrossRef]
39. Kim, S.K.; Chang, H.; Kim, C.M.; Yoo, H.; Kim, H.; Jang, H.D. Fabrication of ternary silicon-carbon nanotubes-graphene composites by Co-assembly in evaporating droplets for enhanced electrochemical energy storage. *J. Alloys Compd.* **2018**, *751*, 43–48. [CrossRef]
40. Wang, X.; Yushin, G. Chemical vapor deposition and atomic layer deposition for advanced lithium-ion batteries and supercapacitors. *Energy Environ. Sci.* **2015**, *8*, 1889–1904. [CrossRef]
41. Zhu, X.; Chen, H.; Wang, Y.; Xia, L.; Tan, Q.; Li, H.; Zhong, Z.; Su, F.; Zhao, X.S. Growth of silicon/carbon microrods on graphite microspheres as improved anodes for lithium-ion batteries. *J. Mater. Chem. A* **2013**, *1*, 4483–4489. [CrossRef]

42. Zhang, L.; Wang, C.; Dou, Y.; Cheng, N.; Cui, D.; Du, Y.; Liu, P.; Al-Mamun, M.; Zhang, S.; Zhao, H. A yolk-shell structured silicon anode with superior conductivity and high tap density for full lithium-ion batteries. *Angew. Chem. Int. Ed. Engl.* **2019**, *58*, 8824–8828. [CrossRef]
43. Du, F.; Ni, Y.; Wang, Y.; Wang, D.; Ge, Q.; Chen, S.; Yang, H.Y. Green fabrication of silkworm cocoon-like silicon-based composite for high-performance Li-ion batteries. *ACS Nano* **2017**, *11*, 8628–8635. [CrossRef]
44. Zhang, L.; Rajagopalan, R.; Guo, H.; Hu, X.; Dou, S.; Liu, H. A green and facile way to prepare granadilla-like silicon-based anode materials for Li-ion batteries. *Adv. Funct. Mater.* **2016**, *26*, 440–446. [CrossRef]
45. Nie, P.; Le, Z.; Chen, G.; Liu, D.; Liu, X.; Wu, H.B.; Xu, P.; Li, X.; Liu, F.; Chang, L.; et al. Graphene caging silicon particles for high-performance lithium-ion batteries. *Small* **2018**, *14*, 1800635. [CrossRef]
46. Zuo, X.; Wang, X.; Xia, Y.; Yin, S.; Ji, Q.; Yang, Z.; Wang, M.; Zheng, X.; Qiu, B.; Liu, Z.; et al. Silicon/carbon lithium-ion battery anode with 3D hierarchical macro-/mesoporous silicon network: Self-templating synthesis via magnesiothermic reduction of silica/carbon composite. *J. Power Sources* **2019**, *412*, 93–104. [CrossRef]
47. Zhang, L.; Dou, Y.; Guo, H.; Zhang, B.; Liu, X.; Wan, M.; Li, W.; Hu, X.; Dou, S.; Huang, Y.; et al. A facile way to fabricate double-shell pomegranate-like porous carbon microspheres for high-performance Li-ion batteries. *J. Mater. Chem. A* **2017**, *5*, 12073–12079. [CrossRef]
48. Zhou, J.; Lan, Y.; Zhang, K.; Xia, G.; Du, J.; Zhu, Y.; Qian, Y. In-situ growth of carbon nanotube wrapped Si composites as anodes for high performance lithium-ion batteries. *Nanoscale* **2016**, *8*, 4903–4907. [CrossRef]
49. Xu, Z.; Gang, Y.; Garakani, M.A.; Abouali, S.; Huang, J.; Kim, J. Carbon-coated mesoporous silicon microsphere anodes with greatly reduced volume expansion. *J. Mater. Chem. A* **2016**, *4*, 6098–6106. [CrossRef]
50. Zhang, J.; Tang, J.; Zhou, X.; Jia, M.; Ren, Y.; Jiang, M.; Hu, T.; Yang, J. Optimized porous Si/SiC composite spheres as high-performance anode material for lithium-ion batteries. *Chem.ElectroChem* **2019**, *6*, 450–455. [CrossRef]
51. Yu, Q.; Ge, P.; Liu, Z.; Xu, M.; Yang, W.; Zhou, L.; Zhao, D.; Mai, L. Ultrafine SiOx/C nanospheres and their pomegranate-like assemblies for high-performance lithium storage. *J. Mater. Chem. A* **2018**, *6*, 14903–14909. [CrossRef]
52. Liu, Y.; Guo, X.; Li, J.; Lv, Q.; Ma, T.; Zhu, W.; Qiu, X. Improving coulombic efficiency by confinement of solid electrolyte interphase film in pores of silicon/carbon composite. *J. Mater. Chem. A* **2013**, *1*. [CrossRef]
53. Ma, X.; Liu, M.; Gan, L.; Tripathi, P.K.; Zhao, Y.; Zhu, D.; Xu, Z.; Chen, L. Novel mesoporous Si@C microspheres as anodes for lithium-ion batteries. *Phys. Chem. Chem. Phys.* **2014**, *16*, 4135–4142. [CrossRef] [PubMed]
54. Zhang, W.; Li, J.; Guan, P.; Lv, C.; Yang, C.; Han, N.; Wang, X.; Song, G.; Peng, Z. One-pot sol-gel synthesis of Si/C yolk-shell anodes for high performance lithium-ion batteries. *J. Alloys Compd.* **2020**, *835*. [CrossRef]
55. Wang, S.; Huang, C.; Wang, L.; Sun, W.; Yang, D. Rapid fabrication of porous silicon/carbon microtube composites as anode materials for lithium-ion batteries. *RSC Adv.* **2018**, *8*, 41101–41108. [CrossRef]
56. Su, M.; Liu, Y.; Wan, H.; Dou, A.; Wang, Z.; Guo, H. High cycling performance Si/CNTs@C composite material prepared by spray–drying method. *Ionics* **2016**, *23*, 405–410. [CrossRef]
57. Zhao, T.; She, S.; Ji, X.; Jin, W.; Dang, A.; Li, H.; Li, T.; Shang, S.; Zhou, Z. In-situ growth amorphous carbon nanotube on silicon particles as lithium-ion battery anode materials. *J. Alloys Compd.* **2017**, *708*, 500–507. [CrossRef]

MDPI
St. Alban-Anlage 66
4052 Basel
Switzerland
Tel. +41 61 683 77 34
Fax +41 61 302 89 18
www.mdpi.com

Nanomaterials Editorial Office
E-mail: nanomaterials@mdpi.com
www.mdpi.com/journal/nanomaterials

www.ingramcontent.com/pod-product-compliance
Lightning Source LLC
LaVergne TN
LVHW070428100526
838202LV00014B/1552